AF552387

STP 1118

Current Practices in Ground Water and Vadose Zone Investigations

David M. Nielsen and Martin N. Sara, editors

ASTM Publication Code Number (PCN)
04-011180-38

ASTM
1916 Race Street
Philadelphia, PA 19103

Library of Congress Cataloging-in-Publication Data

Current practices in ground water and vadose zone investigations / David M. Nielsen and Martin N. Sara, editors.
(STP ; 1118)
Papers presented at a symposium held in San Diego, CA Jan. 30-Feb. 1, 1991.
Includes bibliographical references and indexes.
ISBN 0-8031-1462-1
1. Water, Underground—Quality—Measurement—Congresses. 2. Zone of aeration—Congresses. 3. Water, Underground—Sampling—Congresses. 4. Water table—Congresses. I. Nielsen, David. II. Sara, Martin N., 1946- . III. Series: ASTM special technical publication ; 1118.
TD426.C87 1992
628.1'61—dc20 92-24664
CIP

Photocopy Rights

Peer Review Policy

Each paper published in this volume was evaluated by three peer reviewers. The authors addressed all of the reviewers' comments to the satisfaction of both the technical editor(s) and the ASTM Committee on Publications.

The quality of the papers in this publication reflects not only the obvious efforts of the authors and the technical editor(s), but also the work of these peer reviewers. The ASTM Committee on Publications acknowledges with appreciation their dedication and contribution to time and effort on behalf of ASTM.

Printed in Chelsea, MI
August 1992

Foreword

This publication, *Current Practices in Ground Water and Vadose Zone Investigations,* contains papers presented at the symposium of the same name, held in San Diego, CA on 30 Jan. to 1 Feb., 1991. The symposium was sponsored by ASTM Committee D-18 on Soil and Rock, its Subcommittee D18.21 on Ground Water and Vadose Zone Investigations, the U.S. Environmental Protection Agency, and the U.S. Geological Survey. David M. Nielsen of Nielsen Ground Water Sciences, Inc. (NGWS) in Galena, OH and Martin N. Sara of Waste Management, Inc. in Oak Brook, IL presided as symposium chairmen and are the editors of the resulting publication.

Contents

Determination of Hydraulic Parameters

Statistical Evaluation of Ground-Water Data

Overview

The three-day symposium on Ground Water and Vadose Zone Monitoring was held to evaluate the experiences gained in the 1980s in ground-water and vadose zone monitoring technology and to examine the expectations for the 1990s. The symposium was broken down into nine sessions, which were as follows:

- **Session I**—Improving the Quality of Ground-Water Investigations
- **Session II**—Surface and Borehold Geophysics in Ground-Water Investigations
- **Session III**—Vadose Zone Investigation Methods
- **Session IV**—Soil Vapor Migration, Monitoring & Remediation
- **Session V**—Statistical Evaluations of Ground-Water Data
- **Session VI**—Direct Push Technology in Ground-Water Investigations
- **Session VII**—Ground-Water Monitoring Well Design and Installations
- **Session VIII**—Ground-Water Sampling & Sample Analysis
- **Session IX**—Miscellaneous Field Methods Used in Ground-Water Investigations

The papers contained in this Special Technical Publication (STP) represent a collection of some of the information being used to develop standards for the rapidly growing and evolving field of ground-water and vadose-zone monitoring. The intent of the symposium was to foster interdisciplinary communication and to make available state-of-the-art technology to those scientists and engineers engaged in ground-water and vadose-zone monitoring. A side benefit, but an important one, is that some of the papers may be useful in developing acceptable standards.

The papers published herein received at least three peer reviews and were reviewed by the editors following revisions of papers by the authors. The editors express appreciation to the reviewers who assisted so much in assuring the quality of papers in this STP. Appreciation also is expressed to the speakers at the symposium, the authors who prepared, revised, and provided final papers for publication, and the ASTM staff and officers of Committee D-18.

Improving the Quality of Ground-Water Investigations

Michael J. Barcelona, and John A. Helfrich

REALISTIC EXPECTATIONS FOR GROUND-WATER INVESTIGATIONS IN THE 1990'S

REFERENCE: Barcelona, M. J. and Helfrich, J. A., "Realistic Expectations for Ground-Water Investigations in the 1990's," Current Practices in Ground-Water and Vadose Zone Investigations, ASTM STP 1118, David M. Nielsen and Martin N. Sara, Eds., American Society for Testing and Materials, Philadelphia, 1992.

ABSTRACT: Significant progress has been made in the design and execution of ground-water investigations for site assessments, the selection of remediation options and long-term monitoring of corrective action operations. Improved reliability and cost-effectiveness in ground-water studies can result from the recognition of both the complexity of the subsurface environment and the need for simple, integrated monitoring approaches. Continued progress in the methodologies of investigation will be enhanced by improvements in conceptualization of subsurface conditions.

KEYWORDS: ground-water contamination, background levels, error-control, monitoring network design

INTRODUCTION

There has been phenomenal growth in the number and complexity of ground-water investigations in the last two decades. Large regulatory programs have been established to monitor, assess and prevent or remediate ground-water contamination. Advances in contaminant detection and assessment in support

Dr. Barcelona is Director of the Institute for Water Sciences and Professor of Chemistry at Western Michigan University, Kalamazoo, MI 49008. Mr. Helfrich is a Senior Environmental chemist with Waste Management of N. A., Inc., 3303 Butterfield Road, Oak Brook, IL 60521.

of regulatory programs have been remarkable. Soil, soil gas, and ground-water sampling and analysis techniques are now routinely applied to contamination investigations at the part-per-billion level. These concentration levels were rarely of concern 20 years ago. It is now common for legal issues surrounding zoning decisions, property-transfers and evaluations to hinge on the results of subsurface investigations.

Regulatory and legal interests have been emboldened by the advances in investigational techniques to expect high levels of reliability in all ground-water investigations. Rigid guidelines for study designs, quality assurance/quality control measures and contract compliance procedures have often taken precedence over technical experience, professional judgement and flexibly implemented ground-water investigations [1]. Expectations for future progress in ground-water investigations may be tempered by the current regulatory and legal atmosphere. However, there have been exciting improvements in scientific understanding which provide a basis for cautious optimism.

The design, execution and interpretation of results from ground-water investigations can be made more cost-effective and realistic by thorough consideration of two broad topics. These considerations include: determinations of background levels of chemical constituents (e.g. for identifying contamination and clean-up levels), and the recognition of sources of error and variability in chemical distributions.

BACKGROUND CHEMICAL CONSTITUENT LEVELS

Background levels of chemical constituents in ground water, soil, and aquifer materials are important in nearly all ground-water investigations. Naive expectations for non-detectable contaminant levels and zero-risk clean-up goals are simply not supported by the scientific literature. In addition, it is clear that left-censoring of data near the limits of detection or quantitation reduces the size of datasets available to judge background or upgradient environmental quality conditions [2,3].

In the truest sense, naturally occurring species can or may be in excess of average crustal abundance, drinking water or other environmental quality standards frequently used to represent background subsurface contamination. The most commonly measured parameters in this category are total dissolved solids (TDS), sodium, potassium, hardness, chloride, fluoride, sulfate, nitrate, manganese, chromium, iron, gross alpha, gross beta, and total radium. Synthetic organic compounds (e.g. volatile solvents, fuels, pesticides, etc.) on the other hand, are chemicals which may have been introduced into the ground water environment as a result of human activities.

The distinction between naturally-occurring and anthropogenic contaminants is somewhat blurred in the case of some chemical parameters. For example, nitrate and chloride can be introduced from fertilizers, sewer lines, septic systems, animal feedlots and road-salt, just to name a few sources. A variety of natural and man-induced sources of organic compounds may also influence "background" conditions. Surveys of background water quality may be biased significantly if obvious sources of these contaminants are not recognized.

Soils and Aquifer Materials

The geologic and hydrologic framework of the site of an investigation together provide the basis for the interpretation of sampling and analytical results. The crustal abundances of inorganic chemical constituents in soils and aquifer materials influence water quality to the extent that the solid mineral phases equilibrate with ground-water. The approach to equilibrium can vary from site to site within aquifers depending on hydrodynamics and geochemical conditions [4,5].

There have been few comprehensive reviews of background levels of inorganic constituents in soils and aquifer materials [6]. Of particular note to ground-water studies is the recent report by Kesler-Arnold and O'Hearn [7] on metallic elements and cyanide in glacial sediments in Michigan. These workers presented data for 24 metals and cyanide in 548 background soil and sub-soil (<100 ft;33m) samples from 44 sites. Focussing on constituents which are cited in the Natural Interim Primary Drinking Water Regulations, it is clear from the values in Table 1 that the sand, silt and clay sample types all had relatively high observed frequencies of detection. For example, arsenic (As) was detected at low average concentrations in 76, 100 and 92% of these sample types, respectively. Cyanide (CN^-) was detected at 8 and 19% frequencies, respectively in sand and clay fractions, albeit at ppb levels. Virtually all of the average concentrations of inorganic chemical constituents were statistically different in clay versus sand samples. It is useful to note that these authors observed statistically significant differences (i.e. at the 90% confidence level, student-t test) in results from similar sample types from separate lobes of glacial materials reflecting differences in source materials and post-depositional environments. A hasty comparison of the average values for glacial materials with the earth crust averages [8] in Table 1 might suggest that Michigan has surficial geological materials with elevated arsenic (As), cadmium (Cd), and silver (Ag). However, one must recognize that the mean values for the glacial materials span a considerable concentration range and that subsurface materials in ground-water systems were not even considered in the crustal values for these elements. It is, therefore, important to recognize that actual crustal abundances of metals and other species may be expected to range significantly between distinct geographic areas, geologic frameworks and hydrologic settings. Estimates for soil background clean-up levels [9], contributions of natural suspended or colloidal materials in ground-water [10] or interrelations between sediment composition and leachable/mobile metals, etc. (e.g. Toxicity Characteristic Leaching Procedure, [ref. 11,12]) should consider background levels very carefully. These

Table 1 -- Mean Background Concentrations of Metals and Cyanide in Glacial Drift Materials (mg/kg) (from Reference 7).

Chemical Constituent	SAND	SILT	CLAY	Average Earth Crust**
As	**4.3**	**5.9**	**8.4**	**1.5**
Ba	36	39.8	71.1	----
Cd	**0.80**	**0.81**	**1.0**	**0.11**
Cr	6.3	12	17	100
Pb	**9.8**	**7.1**	**14**	**14**
Hg	0.04*	0.05*	0.06	0.05
Ag	**0.74***	**1.3***	**1.3**	**0.07**
Fe	8,640	10,920	15,840	41,000
Mn	290	241	425	950
CN^-	0.33*	0.16*	0.13*	----

*Frequency of detection less than 25%.
**From Reference 8, Bowen (1979).

arguments apply equally to organic chemical constituents though we do not have the benefit of a large dataset of background concentrations [13,14,15]. The **expectation** here is that ranges of site-specific background levels for solid-associated chemical constituents will be developed. These ranges will be used more frequently for the identification of contaminated conditions, determination of clean-up levels, and estimation of the potential significance of leachable species. It would be particularly useful if laboratory-based leaching procedures were refined to be more representative of actual ground-water conditions. The current usage of acetic acid or acetate buffers certainly mobilizes both carbonate bound and surface-sorbed metal species [12].

Ground Water

The strong regulatory focus of ground-water investigations has often resulted in neglecting background water chemistry and major ionic-solute composition in favor of the concentrations of organic microcontaminants. Consideration of the actual condition of the resources we are trying to protect should be kept in mind, particularly in our interpretations of trends in water quality, estimates of background conditions and potential influences of remedial actions. Recent reports which highlight important aspects of ambient ground-water quality include reviews of federal and state agency data [16, 17, 18, 19].

Inorganic and radiological chemical constituents: One of the most extensive surveys of background water quality in the U.S. is the report commissioned by the Edison Electric Institute (EEI), titled "Natural Quality of Groundwater in the United States Compared to Drinking Water Standards" [16]. The authors of this report examined data from the published literature as well as from the U.S. Geological Survey's WATSTORE database to picture water quality data on national, regional (Northern Midwest, North Atlantic Coastal Plain) and state (WI, KY, NE) scales. The survey included data from wells less than 2500 feet (~763m) deep and excluded data from wells that were impacted by known point sources of contamination. Water quality data were compared to Maximum Contaminant Levels (MCLs) and Secondary Maximum Contaminant Levels (SMCLs) cited in the Safe Drinking Water Act (SDWA) [20]. Due to the extensive nature of the report only a brief summary of their findings can be presented here.

Based on the analysis of more than 122,000 wells in the WATSTORE data base, the EEI report concluded that 57% of the nation's ground water resources in the upper 2500 feet exceed at least one SDWA MCL or SMCL. Water from drinking water wells was generally of higher quality than water from wells used for other purposes. The authors concluded that this was not surprising since drinking water wells are typically developed in higher quality aquifers than non-drinking water wells. A conservative estimate of at least one SDWA exceedance was in the range of 29 to 35 percent. An estimate more representative of the entire ground water supply (i.e. including non-drinking water wells) put the range at 50 to 65 percent.

Nationwide, TDS and total iron were the most common parameters to exceed a drinking water standard. Based solely on data from drinking water wells the EEI report estimated that about 33% of drinking water supplies exceeded the SMCL for TDS of 500 mg/L, whereas that number increased to 50-65% based on all ground water data. Data for other general water quality parameters and observed exceedances are summarized in Table 2. Clearly, a substantial fraction of drinking water wells may be expected to violate one or more general water quality standard.

In the studies reviewed [16], nitrate was the parameter that exceeded an MCL most frequently. Sixty-eight percent of shallow wells less than 100 feet deep exceeded the MCL of 10 mg/L. Excluding wells <100 feet deep, on the basis that they are most easily impacted by human activities, the frequency with which nitrate exceeded the MCL was reduced to 7-10 percent of the nationwide wells.

The EEI report [16] also summarized data from a nationwide survey of public water supplies [17] to assess the distribution of radionuclides (gross-alpha, gross-beta, total radium, radium-226 and radium-228) in ground water supplies. Based on their analysis of the data the authors reported that total radium (radium-226 plus radium-228) exceeded the MCL of 5 pCi/L in 2.6% of ground water supplies, gross-alpha exceeded the MCL of 15 pCi/L in 2.2%, Ra-226 and Ra-228 exceeded the MCL of 4 mrem/year in 1.4% and 0.8%, of the samples, respectively. The gross-beta levels exceeded the MCL of 4 mrem/year in only a small number of the nation's ground water supplies. These data are considered conservative underestimates because among other reasons, private wells and small public systems were not included in the survey. The radiological data reflect geologic and hydrologic factors more so than land use or sources of anthropogenic contamination.

Organic compounds: The Illinois Environmental Protection Agency (IEPA) has been conducting an ongoing study of public water supply (PWS) wells in the state since 1985 [18]. More than 75% of Illinois' PWS wells had been tested for inorganic parameters and volatile organic compounds (VOCs) by 1989. Statistics on inorganic parameters (summarized above) were compiled for more than 2400 wells and for VOC's on all 2600 wells.

The IEPA survey included VOC determinations in samples from all PWS wells and results for pesticides in 400 high-risk wells. The data are indicative of the potential industrial, commercial or agricultural impacts that may be found in many ground-water systems. One hundred and fifteen PWS wells had detectable levels of VOCs, which represents 4.6% of the total; 41 of those wells, or 1.6% had VOC concentrations above an SDWA MCL or proposed MCL level for one or more VOC. In almost all cases of confirmed VOC detections, one or more potential sources of contamination could be identified within 1000 feet of the well head. Six of the 400 high-risk wells sampled for pesticides had confirmed detections. One of the six has one or more pesticide that could exceed a USEPA proposed MCL. Again, a site survey of the confirmed

Table 2 -- SUMMARY OF BACKGROUND GROUND-WATER QUALITY AND EXCEEDANCES OF DRINKING WATER STANDARDS

PARAMETER	STANDARD	PERCENT EXCEEDANCES		REFERENCE
		Drinking Water Well	All Wells	
Total Dissolved Solids	500 mg/L	33%	50-65%	16 (National)
			10-25%	16 (Wisconsin)
		45%		18 (Illinois)
Fe	0.3 mg/L		48%	16 (National)
			25-75%	16 (Wisconsin)
		62%		18 (Illinois)
$SO_4^{=}$	250 mg/L	9%	21%	16 (National)
		12%		18 (Illinois)
Cl^-	250 mg/L	2%	14%	16 (National)
Mn	0.05 mg/L		25%	16 (Wisconsin)
			30%	18 (Illinois)
NO_3^-	10 mg/L		68%*	16 (National)*
			<10%	16 (Wisconsin)
		<1%		18 (Illinois)
		18-22%		21 (Iowa)

*Includes wells less than 100 ft. (~31m) deep.

detections revealed potential point sources of contamination within 1000 feet of the well head.

The PWS ground-water quality situation is similar in California where detection frequencies of VOC's (i.e., perchloroethylene, trichloroethylene, dibromochloropropane, chloroform, etc.) ranged from ~ 2 to 3.5% of all wells sampled [19]. The state had sampled 52% of the large (i.e., >200 service connectors) PWS wells. A recent discussion of the relative frequencies of detection of selected VOC's in PWS wells broaches some interesting insights for the reader [22]. In this instance, the authors point out the low detection frequency of benzene in California public water supply wells, neglecting the fact that few production wells are screened at the water table where fuel-related contaminants are more likely to be detected.

These data illustrate that the ambient quality of ground water is highly variable and frequently fails to meet current drinking water quality criteria. Some of the apparent contamination may be attributed to human activities, as in the case of nitrate, VOC's and pesticides. However, many of the reported SDWA standard exceedances are due to natural conditions. As the authors of the EEI report (1987) conclude: "... the natural quality of almost 60% of the nation's ground water resources above depths of 2500 feet would not meet all federal drinking water standards". This statement contrasts with the common perception that the nation's ground water resources are naturally pristine and of high quality.

The **expectation** with regard to background ground-water quality is that more realistic and site-specific determinations will be made in future investigations. It is hoped that ground-water cleanup goals set at or below existing drinking water standards for state or national programs, will be used flexibly.

ERRORS, VARIABILITY AND NETWORK DESIGN

There are numerous elements of ground-water monitoring network design which would benefit from a closer examination of the conceptual basis for a particular investigation rather than simply meeting regulatory compliance guidelines. One might expect that with the growing number of experienced ground-water scientists and engineers that more monitoring network designs will be based on thorough preliminary geologic, hydrologic and geophysical measurements together with initial ground-water flow modeling results. There are several recent publications which should be consulted for details [23,24,25]. In this paper, we will briefly note items of concern with respect to minimizing errors and improvements in both network design and sampling protocols.

Sample Location and Frequency: Ideally, one would like to provide discrete sampling locations in specific geologic formations where ground-water flow patterns had been characterized. Then the number of samples collected per year would be planned so as to minimize redundant sampling and allow a high probability of contaminant detection. In order to achieve this goal, it will be necessary to develop a better understanding of contaminant fate and transport in ground-water environments. With the aid of integrated hydrogeologic data collection (i.e. geophysical surveys, aquifer solid characterizations, application of geostatistical techniques to mapping, etc.) and flow modeling it should be possible to target potential sampling locations (i.e. wells, multi-level sampling devices, etc.) to intercept likely paths of contaminant transport. Some very useful geostatistical tools [26] and mathematical developments are being made available to improve decision-making once the hydrologic ground-work has been established [27].

Of particular note here to considerations of sampling location are the increasing reports of short-range spatial variability in both vadose and saturated zone geochemistry and resultant vertical gradients in chemical constituent concentrations [28-31]. Chemical constituent concentrations changes in soil gas or ground-water have been measured to be greater than an order of magnitude over a few meters change in depth. The **expectation** here is that more discrete, short (i.e. <3m) screened vertical well completions or other sampling points will be utilized to improve the three dimensional coverage of hydrogeochemical data, particularly where non-aqueous fluids are concerned. The importance of reducing "well-averaging" due to the use of excessively purged or long-screened interval monitoring wells has been recognized as an order of magnitude level error by Robbins [32]. His treatment of this problem is highly recommended to the reader.

Chemical concentration variability along a ground-water flow path is also of concern and may be related to travel times and the probability of detecting contaminant concentrations within a range of the annual means. While it turns out that quarterly sampling frequency is a good starting point for monitoring designs [33] for temporal trend analyses, the estimation of a cost-effective, reliable frequency for trend detection or estimation of mean values remains largely an empirical task. Working with an "over-determined" (i.e. biweekly frequency) dataset for ground-water quality at three sites in a shallow sand and gravel aquifer, it was possible to define an "optimal" sampling frequency which resulted in maintaining a specific information content with minimal sampling effort. The information content was defined as the ratio of the effective sample (n_{eff}) size to the total sample size (n) for each of 24 water quality parameters. In most cases, the frequency maintaining an information content of 0.9 (i.e. less than 10% loss of information due to redundant data collection) ranged from one to five months at the three sites. Figure 1 contains plots of the values of n_{eff}/n versus ground-water path length corresponding to the "optimal" sampling frequency for selected water quality parameters at an uncontaminated location.

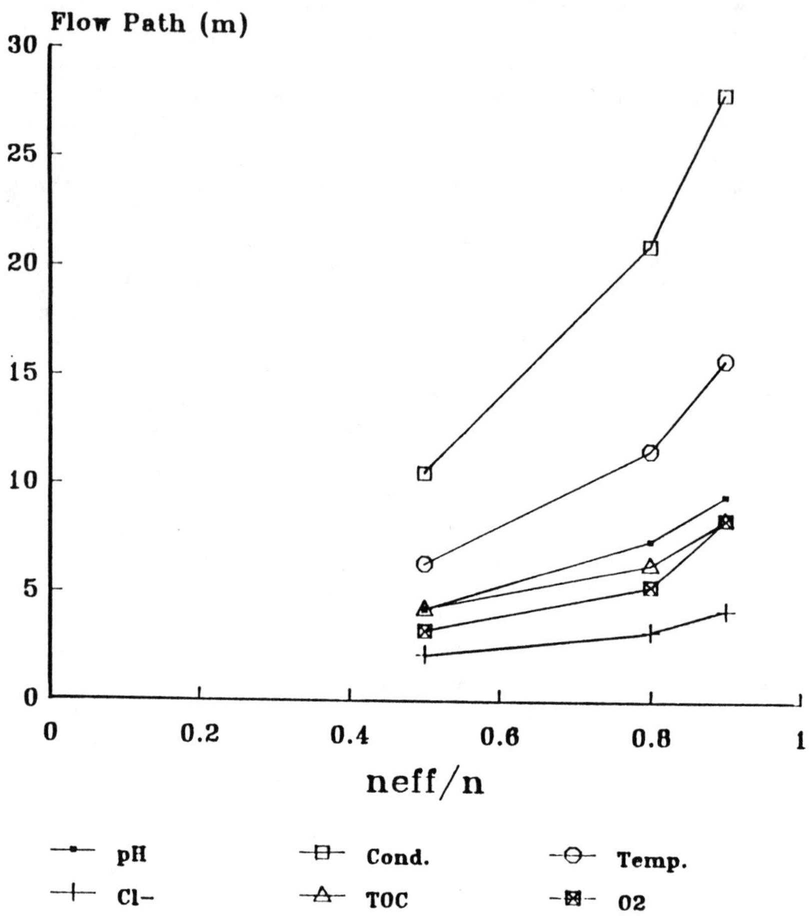

FIGURE 1

Plots of ground-water flow path (in meters) versus the ratio of effective independent sample size to total sample size.

With the exception of specific conductance, chloride and temperature at the upgradient contaminated site (i.e. variables with very strong components of temporal variability), samples taken between two and thirty meters in a ground-water flow path would correspond to a reasonably cost effective sampling frequency. In this instance, chloride variability was probably related to seasonal irrigation practices in the region.

The methodology to estimate optimal sampling frequencies and relate them to hydrogeologic processes should improve in the future. The **expectation** here is that future investigations of ground-water quality will evaluate the approximate sampling frequency and the duration of the program to better define temporal trends in chemical constituent concentrations.

Analyte selection: Most often the suite of chemical constituents of interest in an investigation are either mandated by the regulatory program or can be selected from the characteristics of the materials or wastes handled at a particular facility. In a number of studies, the volatile organic compounds have been found to be good indicators of organic contamination of ground-water [34,35]. It is clear that exhaustive analytical determinations of specific, non-volatile organic compounds (e.g. Appendix IX, etc.) often yield very little useful chemical information for the expense [36]. It is recommended that one include some target contaminant constituents (e.g. VOC's, total organic carbon), major ionic constituents (e.g. Cl^-, $SO_4^=$, Na, K, etc.) and geochemical indicators (e.g. CO_2, CH_4, O_2, FeII) in many initial ground-water investigations. The **expectation** here is that, based on a sound hydrogeologic framework, and initial flow or transport modeling results (i.e.calibrated for Cl^-, Ω^{-1}, etc.), preferred locations for chemical sampling could be identified. Then a reasoned, limited suite of likely contaminants and breakdown products could be selected for long-term monitoring if this purpose is relevant to the investigation. Redundant sampling of specific organic compounds (at or near detection levels) for long periods of time is simply too wasteful as compared to the improved information return from a simpler suite of target analytes determined at more sampling points.

Well development: With the increasing emphasis on the collection of improved geologic and hydrologic information in ground-water investigations has come the realization that the hydraulic connection between the sampling point and the geologic formation is very important. Pumping monitoring wells for development of the gravel pack, removal of fines or removal of stagnant water prior to sampling have become as important as routine slug or pump-testing to determine drawdown, yield and aquifer properties.

The removal of fines created during drilling has taken on particular importance given the potential for colloidal metal (e.g. Al or Fe) or organic transport in gravels or fractured materials [10,37] and the confusion about the need for sample filtration [38]. What is obvious is that extra care is needed in the construction, design and pumping of wells to minimize the creation and entrainment of fine particulates which are unlikely to be mobile under natural gradient conditions. Just as vigorous high-rate pumping and surging of wells during development mobilizes particulates, one should expect further development, turbidity and potentially biased water samples to result from high velocity pumping during purging or sampling (39). The **expectation** here will be that future investigations will more carefully design, construct and develop wells to minimize the collection of otherwise immobile fine materials. Also, considerable care will be taken to prevent well/gravel pack damage by high-rate pumping after development.

Well purging: The pumping strategy which is appropriate for individual monitoring wells should include the development operations described above as well as pumping during well purging and sampling. The primary consideration in purging is the isolation (via packers) or the removal of stagnant water in the casing, screen and borehole adjacent to the screen (40,41,42) which has been defined as the purge volume (43). The stored water is considered unrepresentative of ground-water in the formation of interest. Criteria for sample representativeness vary substantially. They may be characterized as: 1) pumping a specific number of purge volumes, 2) pumping the well until water quality indicators (i.e. pH, conductance, temperature, dissolved oxygen, etc.) stabilize satisfactorily, and, 3) pumping the well at a uniform rate for a time period sufficient to permit a high percentage of formation water to enter the well's screened interval (42,44). A number of workers have documented water quality changes during purging in an attempt to arrive at a generalized, consistent sampling recommendation [43,44,45,46,47,48].

The results of these investigations have disclosed three important points. The hydraulics of monitoring wells are incompletely understood (49). Order of magnitude concentration changes in general water quality parameters, metals and specific organic compounds occur in pumping idled monitoring wells between 1 and 10 purge volumes. Well design (i.e. screen placement length, slot size) construction, development, performance, rate of purge pumping and placement of the pumping device can introduce order of magnitude level changes in concentration during purging which may negate the impact of sample location, frequency and analyte selection as major network design variables. The latter theoretical and experimental gains have been made by Robbins et al. in a series of papers [50,51] which began with considerations of well-screen averaging and mass-continuity balances during purging [32].

The **expectation** here is that carefully integrated hydrogeology-based sampling protocols will be developed in the future which incorporate more vertically discrete well completions and the use of models to improve our concept of sample representativeness. In the meantime, empirical evidence suggests that low-rate pumping (so as not to cause substantial draw down) from the mid-screen portion of a well and removing a consistent number of purge volumes from each well prior to sampling is a safe approach. Clearly, the necessity of purging has been demonstrated and purging procedures should be documented (i.e. hydraulic basis, estimated purge volume, purging indicator parameter measurements) for each water sample.

Sampling device selection: There have been a number of studies which have either discussed or experimentally evaluated the attributes of ground-water sampling mechanisms [52-62]. Many of these evaluations have been based on the precision and recovery of volatile organic compound determinations on replicate samples collected in the field or laboratory. The volatile compounds are particularly sensitive to losses by degassing at reduced pressures from suction devices (e.g. peristaltic or centrifugal pumps, etc.) or turbulence in some mechanical sampling devices (e.g. gas lift, bailers, etc.). In general, positive displacement pumps (e.g. bladder pumps, etc.) which operate reproducibly over a range of lift, hydraulic head and depth conditions have been shown to provide accurate reproducible sampling performance under testing conditions.

In practical routine field use, there are additional considerations which bear on the selection of a sampling device once sufficient accuracy and precision have been demonstrated. Mainly these considerations have to do with the reduction of operator dependent error (e.g. random error in bailing, excessive sample handling/transfers, etc.), cost, ease of repair and cleaning. Dedicated sampling systems suitable for both purging and sampling are clearly preferred from the points of view of convenience, cost and consistency of results.

The **expectation** here is that along with the use of more discrete well completions with short screen lengths across a site that the use of dedicated, reproducible, sampling devices for both purging and sampling will improve the comparability and consistency of results. Since well design, pumping, and purging effects can all lead to order of magnitude errors in the determination of chemical constituents, there is much to be said for the selection of dedicated, simple sampling devices which provide long-term control over human errors.

Field determinations: Field operations at ground-water investigation sites may include sample handling, measurement of well purging parameters (i.e. pH, conductance, temperature, dissolved oxygen) or alkalinity, sample filtration, spiking samples with standards and decontamination procedures [62-66]. Most of these operations must indeed be accomplished at the well head for the samples and subsequent measurements to be valid. They have been greatly facilitated by in-line well-purging parameter measurement systems [31,63,67], in-line filtration devices and increasing attention to the documentation of sampling details. The **expectation** here is that sampling and analysis steps which include sample pumping, collection, handling and field determinations involve errors that can and will be controlled through attention to error and simple sampling protocols [33]. In this way better estimates of spatial and temporal variability can be obtained at known levels of confidence. This will greatly improve the approximation of actual contaminant distributions, transport and fate in ground water.

Data analysis and interpretation: Improved data quality and confidence come from both the application of informed technical judgement and normal attention to quality assurance and quality control. The serious levels of error, or apparent variability, which have been identified with well location, design and pumping clearly call for quality assured decision-making. As the cadre of experienced ground-water investigators has grown, the interpretation of water quality data has improved proportionally. Improved software and statistical tools for handling ground-water quality datasets will undoubtedly be developed to aid in the integration of chemical and hydrogeologic information. The **expectation** here is that investigators will be given increased latitude in network design and operation to maximize the information return from ground-water investigations. Since the bulk of the scientific and engineering data that will be collected on ground-water systems will be done in regulatory programs, it is hoped that flexibility will be granted in these efforts.

CONCLUSIONS

Our understanding of subsurface geochemistry and hydrogeologic controls on contaminant behavior has improved substantially in recent years. It is anticipated that, in the next decade, the purposes and designs for ground-water investigations will incorporate more in-depth considerations of natural variability in chemical constituent concentrations and of the need for careful error identification and control. The increased use of integrated hydrogeologic and geochemical data collection strategies should lead to more meaningful performance criteria for valid monitoring efforts and results. It may be expected that ground-water investigations will expand in the future to adopt more long-term monitoring approaches to both contamination identification and remediation.

ACKNOWLEDGEMENTS

The authors appreciate the help of Ms. Bonnie Dube, Ms. Janene Hoover, Ms. Linda Jones, and Ms. Kim Finkbeiner in the preparation of the manuscript.

REFERENCES

[1] Freeze, R.A. and Cherry, J.A. (1989). What has gone wrong? Ground Water, 27, 4, 458-464.

[2] Gilliom, R.J., Hirsch, R.M., and Gilroy, E.J. (1984). Effect of Censoring Trace-Level Water Quality Data on Trend Detection Capability. Environmental Science and Technology, 18, 530-536.

[3] Gilbert, R.O. (1987). Statistical Methods for Environmental Pollution Monitoring. Von Nostrand Reinhold, N.Y.

[4] Stumm, W. and Morgan, J.J. (1981). Aquatic Chemistry, 2nd edition, John Wiley and Sons, N.Y. 780pp.

[5] Katz, B. and Choquette, A.F. (1991) Aqueous Geochemistry of the Sand and Gravel Aquifer, Northwest Florida, Ground Water, 29, 1, 47-55.

[6] Forstner, U. and Wittmann, G.T. Metal Pollution in the Aquatic Environment. Springer-Verlag, N.Y., N.Y 1983.

[7] Kesler-Arnold, K.A., and O'Hearn, M. (1990). Background Concentrations of Metals and Cyanide in Lower Michigan Soils, p. 33-47, Chapter 4 in 44th Purdue Industrial Waste Conference Proceedings, J.M. Bell (ed.), 1990. Lewis Publishers, Inc., Chelsea, MI 48118.

[8] Bowen, H.J.M. (1979) Environmental Chemistry of the Elements. Academic Press, New York, 333pp.

[9] Moore, J.N. and Luoma, S.N. (1990) Hazardous wastes from large-scale metal extraction. Environmental Science and Technology, 24, 9, 1278-1285.

[10] McCarthy, J.F. (1989) Mobility of Colloidal Particles in the Subsurface: Chemistry and Hydrology of Colloid Aquifer Interactions. DOE/ER-0425, U.S. Department of Energy, Washington, D.C.

[11] Chemical and Engineering News (1990) "Hazardous Wastes: EPA Adds 25 Organics to RCRA List." March 12, 1990 American Chemical Society, p.4.

[12] Forstner, U. (1990). Inorganic Sediment Chemistry and Elemental Speciation. Chapter 3, p. 61-105 in Sediments: Chemistry and Toxicity of In-Place Pollutants. R. Bauds, J.P. Giesy, H. Muntau (eds.), Lewis Publishers, Inc., Ann Arbor, MI.

[13] Cheng, H.H. (1990) Organic Residues in Soils: Mechanisms of Retention and Extractibility. International Journal Environmental Analytical Chemistry, 39, 165-171.

[14] Dragun, J. (1988). The Soil Chemistry of Hazardous Materials. Hazardous Materials Control Research Institute, Silver Spring, MD.

[15] Slaine, D.D. and Barker, J.F. (1990). The Detection of Naturally Occurring BTX During a Hydrogeologic Investigation. Ground Water Monitoring Review, 10, 2, 89-94.

[16] Edison Electric Institute (1987). Natural Quality of Groundwater in the United States Compared to Drinking Water Standards. Prepared by Roy F. Weston, Inc., September 1987. R. F. Weston, Inc. Weston Way, West Chester, PA 19380.

[17] Horton, T.R. (1985). Nationwide Occurrence of Radon and other Natural Radioactivity in Public Water Supplies. USEPA 520/5-85-008, 208p.

[18] Illinois Environmental Protection Agency (1990). Illinois Groundwater Protection Program: A Biennial Report Prepared by the Interagency Coordinating Committee on Groundwater, February, 1990.

[19] California Dept. of Health Services (1986). Organic Chemical Contamination of Large Public Water Systems in California, Calif. DHS, Sacramento, CA, April 1986.

[20] U.S. Environmental Protection Agency (1986). Safe Drinking Water Act - 1986 Amendments. USEPA 570/9-86-002. Office of Drinking Water, Washington, D.C.

[21] Morris, R.L. and Johnson, L.G. (1969). Pollution Problems in Iowa, in P.J. Horick, (ed.), Water Resources of Iowa, Iowa Acad. Sci. University Printing, Iowa City, IA, pp89-110.

[22] Hadley, P.W. and Armstrong, R. (1991). "Where's the Benzene?" Examining California Ground-Water Quality Surveys. Ground Water, 29, 1, 35-40.

[23] Nielsen, D. (1991). (ed.), Ground-Water Monitoring. Lewis Publishers, Chelsea, MI.

[24] American Petroleum Institute (1989). "A Guide to the Assessment and Remediation of Underground Petroleum Releases." 2nd ed., API Publication #1628, American Petroleum Institute, Washington, D.C., 1989.

[25] Keith, L.H. (1988) editor, "Principles of Environmental Sampling", American Chemical Society Professional Reference Book, ACS Washington, D.C., 458 pp.

[26] USEPA (1989). Geostatistical Environmental Assessment Software (Geo EAS), EPA-Environmental Monitoring Systems Laboratory. Office of Research and Development, Las Vegas, NV.

[27] Freeze, R.A.,Massman, J., Smith, L., Sperling, T., and James, B. (1990). Hydrogeological Decision Analysis, 1.A. Framework. Ground Water, 28, 5, 738-766.

[28] Suchomel, K.H., Kreamer, D.K. and Long, A. (1990) Production and Transport of Carbon Dioxide in a Contaminated Vadose Zone: A Stable and Radioactive Carbon Isotope Study. Environmental Science and Technology, 24, 12, 1824-1831.

[29] Kaback, D.S., Bergren, C.L., Carlson, C.A., Carlson, C.L. (1990) Testing A Ground Water Sampling Tool: Are the Samples Representative? Presented at National Water Well Assoc.-Assoc. of Ground-Water Scientists and Engineers. Outdoor Action Conferences May 14-17, 1990, Las Vegas, NV.

[30] Barcelona, M.J., Holm, T.R., Schock, M.R., George, G.K. (1989). Spatial and Temporal Gradients in Aquifer Oxidation-Reduction Conditions. Water Resources Research, 25, 5, 991-1003.

[31] Walton-Day, K., Macalady, D.L., Brooks, M.H., Tate, V.T. (1990). Field Methods for Measurement of Ground Water Redox Chemical Parameters. Ground Water Monitoring Review, 10, 4, 81-89.

[32] Robbins, G.A. (1989) Influence of Using Purged and Partially Penetrating Monitoring Wells on Contaminant Detection, Mapping and Modeling. Ground Water, 27, 2, 155-162.

[33] Barcelona, M.J., Lettenmaier, D.P. and Schock, M.R. (1989) Network Design Factors for Assessing Temporal Variability in Ground-Water Quality. Environmental Monitoring & Assessment, 12, 149-179.

[34] Plumb, R.H. (1987). A comparison of Ground-Water Monitoring Data from CERCLA and RCRA Sites. Ground Water Monitoring Review, 7, 4, 94-100.

[35] Feld, J., Connelly, J.P. and Lindorff, D.E. (1987). Ground-Water Sampling: Addressing the Turbulent Inconsistencies p. 237-255 in Proceedings of the NWWA Outdoor Action Conference, May 1987, Las Vegas, NV. National Water Well Association, Dublin, Ohio.

[36] Plumb, R.H. (1991). The Occurrence of Appendix IX Organic Constituents in Disposal Site Ground Water, Ground Water Monitoring Review, 11 (in press).

[37] Gschwend, P.M. and Reynolds, M.D. (1987). Monodisperse Ferrous Phosphate Colloids in an Anoxic Ground-Water Plume. Journal of Contaminant Hydrology, 1, 309-315.

[38] Puls, R.W., and Barcelona, M.J. (1989). Filtration of Ground-Water Samples for Metals Analysis. Hazardous Waste and Hazardous Materials, 6, 4, 385-393.

[39] Puls, R.W., Eychaner, J.H., and Powell, R.M. (1991) Colloidal-Facilitated Transport of Inorganic Contaminants in Ground Water: Part 1. Sampling Considerations (in press).

[40] Barcelona, M.J., Gibb, J.P., Helfrich, J.A., and Garske, E.E. (1985). Practical Guide for Ground-Water Sampling. Illinois State Water Survey Contract Report 374, Champaign, IL, 94pp.

[41] Robin, M.J.L. and Gillham, R.W. (1987). Field Evaluation of Well-Purging Procedures, Ground Water Monitoring Review, 7, 3, 85-93.

[42] Keely, J.F. and Boateng, K. (1987). Monitoring Well Installation Purging and Sampling Techniques - Part 1: Conceptualizations: Ground Water, 25, 3, 300-313.

[43] Robbins, G.A., Bristol, R.D., Hayden, J.M. and Stuart, J.D. (1989). Mass Continuity and Distributions: Implications for Collection of Representative Ground Water Samples from Monitoring Wells, p. 125-139. In Proc. of Conf. on Petroleum Hydrocarbons and Organic Chemicals in Ground Water: Prevention, Detection and Restoration. Nov. 15-17, 1989, Houston, TX. National Water Well Assoc., Dublin, Ohio.

[44] Gibs, J. and Imbrigiotta, T.E. (1990). Well-Purging Criteria for Sampling Purgeable Organic Compounds. Ground Water, 28, 1, 68-78.

[45] Barcelona, M.J. and Helfrich, J.A. (1986). Well Construction and Purging Effects on Ground-Water Samples. Environmental Science and Technology, 20, 11, 1179-1184.

[46] Pennino, J.D. (1988). There's No Such Thing as a Representative Ground Water Sample. Ground Water Monitoring Review, 8, 3, 4-9.

[47] Smith, J.S., Steele, D.P. Malley, M. J. and Bryant, M.A. (1988). Chapter 17, p255-260 in Principles of Environmental Sampling, L.H. Keith (ed.), American Chemical Society Professional Reference Book ACS, Washington, D.C.

[48] Schuller, R.M., Gibb, J.P., Griffin, R.A. (1981). Recommended Sampling Procedures for Monitoring Wells. Ground Water Monitoring Review, 1, 1, 42-46.

[49] Unwin, J.P. and Maltby, V. (1988). Investigations of Techniques for Purging Ground-Water Monitoring Wells and Sampling Ground Water for Volatile Organic Compounds in "Ground Water Contamination, Field Methods", A.G. Collins and A.I. Johnson (eds.), American Society for Testing and Materials, STP #963, pp. 240-252.

[50] Robbins, G.A., Martin-Hayden, J.M. (1991). Mass Continuity Modeling of Monitoring Well Purging. Journal of Contaminant Hydrology (in press).

[51] Robbins, G.A., Martin-Hayden, J. M., Bristol, R.D., and Stuart, J.D. (1991). A Field Study of Mass Continuity Influences on the Characterization of Subsurface Gasoline Contamination. Journal of Contaminant Hydrology (in press).

[52] Barcelona, M.J., Helfrich, J.A., Garske, E.E., and J.P. Gibb (1984). A Laboratory Evaluation of Ground Water Sampling Mechanisms. Ground Water Monitoring Review, 4, 2, pp. 32-41.

[53] Nielsen, D.M. and Yeates, G.L. (1985). A Comparison of Sampling Mechanisms for Small-Diameter Ground Water Monitoring Wells. Ground Water Monitoring Review, 5, 2, pp. 83-99.

[54] Sonntag, W.H. (1987). Comparative Test of Two Sampling Devices for Obtaining Purgeable Organic Compounds from Ground Water Wells. U.S. Geological Survey Open-File Report, 87-109, p. E-7.

[55] Pearsall, K.A. and Eckhardt, D.A. (1987). Effects of Selected Sampling Equipment and Procedures on the Concentration of Trichloroethylene and Related Compounds in Ground Water Samples. Ground Water Monitoring Review, 7, 2, pp. 64-73.

[56] Pohlmann, K.F. and Hess, J.W. (1988). Generalized Ground Water Sampling Device Matrix. Ground Water Monitoring Review, 8, 4, pp. 82-84.

[57] Imbrigiotta, T.E., Gibs, J., Fusillo, T.V., Kish, G.R. and Hochreiter, J.J. (1988). Field Evaluation of Sampling Devices for Purgeable Organic compound in "Ground Water Ground-Water Contamination, Field Methods" ASTM STP 963, A.G. Collins and A.J. Johnson, (eds.). American Society for Testing and Materials, Philadelphia, PA, p. 258-273.

[58] Barker, J.F., and Dickhout, R. (1988). An Evaluation of Some Systems for Sampling Gas-Charged Ground Water for Volatile Organic Analysis. Ground Water Monitoring Review, 8, 4, pp. 112-120.

[59] Rehm, B.W., Stoltzenburg, T.R., and Nichols, D.G. (1985). Field Measurement Methods for Hydrogeologic Investigations. A Critical Review of the Literature. EPRI Proj. Report EA-4301, Project #2485-7, October 1985 by Residuals Management Technology, Inc., Electric Power Research Institute, Palo Alto, CA.

[60] Unwin, J.P. (1984). A Laboratory Study of Four Methods of Sampling Ground Water for Volatile Organic Compounds. National Center for Air and Stream Improvement, Technical Bulletin. No. 441, August 1984, 18pp.

[61] Blegen, R.P., Pohlmann, K.F. and Hess, J.W. (1987). Bibliography of Ground-Water Sampling Methods. EPA Report EPA/600/X-87/327, September 1987, USEPA-EMSL, Las Vegas, NV.

[62] Pankow, J.F. (1990). Minimization of Volatilization Losses During Sampling and Analysis of Volatile Organic Compounds in Water. Chapter 5, p. 73-86. In Significance and Treatment of Volatile Organic Compounds in Water Supplies. N.M. Ram, R.F. Christman, and K.P. Cantor (eds.), Lewis Publishers, Chelsea, MI, 558pp.

[63] Garner, S., (1988). Making the Most of Field-Measurable Ground Water Quality Parameters. Ground-Water Monitoring Review, 8, 3, 60-66.

[64] Dodge, C.J. and Francis A.J., (1988). Anoxic Collection and Analysis of Subsurface Water Samples. DOE/ER-0262, DE 86008700 U.S. Department of Energy report by Brookhaven National Laboratories, February, 1986, 18pp.

[65] Clark, L. and Baxter, K.M., (1989). Ground Water Sampling for Organic Micropollutants: UK experience. Quarterly Journal of Engineering Geology, 22, 159-168.

[66] Mickam, J.T., Bellandi, R., and Tifft, E.L. (1989). Equipment Decontamination Procedures for Ground Water and Vadose Zone Monitoring Programs: Status adn Prospects. Ground Water Monitoring Review, 9, 2, 100-121.

[67] Garske, E.E. and Schock, M.R. (1986). An inexpensive flow-through cell and measurement system for monitoring selected chemical parameters, Ground Water Monitoring Review, 6, 3, 79-84.

Peter T. Regan[1], Jonathan A. Conte, and Michael M. Mitchell

THE NEED FOR IMPROVED USE OF GEOLOGIC DATA IN GROUND WATER INVESTIGATIONS

REFERENCE: Regan, P. T., Conte, J. A., and Mitchell, M. M., "The Need for Improved Use of Geologic Data in Ground Water Investigations," Current Practices in Ground Water and Vadose Zone Investigations, ASTM STP 1118, David M. Nielsen and Martin N. Sara, Eds., American Society for Testing and Materials, Philadelphia, 1992.

ABSTRACT: In our view, the predominant cause of many poor feasibility studies, irrational risk assessments, and failed remediations has been the lack of a sound understanding of the physical characteristics of the aquifer system. Specifically, we refer to the lack of attention to detailed, geologic descriptions of aquifer materials during drilling and the subsequent interpretation of the geologic factors that control the three-dimensional shapes of permeability fields. The environmental industry has generally underutilized the potential benefits of better prediction of aquifer geometries and, therefore, more cost-effective exploitation of geologic conditions. We urge a return to the roots of geologic practice that historically formed the basis for ground water investigations.

In this paper, we propose a retrograde and not entirely innovative method that is based on classical geologic principles. This method depends on the process of developing a detailed evaluation of the physical features of the aquifer skeleton. Hydrogeologic investigations must be based on detailed geologic descriptions and interpretations of complete sections of all boreholes to define stratigraphy and interpret depositional environments. In cases where

[1]Mr. Regan is a Senior Hydrogeologist for ERM-Southwest, Inc. in New Orleans, LA. Mr. Conte and Mr. Mitchell are Senior Hydrogeologists for NUS Corporation in Oak Ridge, TN.

structural features control ground water flow, detailed measurements of orientations and densities of fabric elements are essential. These interpretive tools provide the physical foundation that will support the remarkable equation-solvers (computer models) which have become so popular. We suggest that cost-effective solutions to problems of contaminant migration in flow systems that may be dominated by advective transport require better prediction of the spatial distribution of permeability fields between widely spaced data points. Better prediction can only be accomplished through refined geologic interpretations based on detailed and reliable field data.

KEY WORDS: hydrogeology, aquifer skeleton, sedimentary facies.

A recent report by the Environmental Protection Agency [1] concluded that aquifer restoration, particularly active remediation by pump-and-treat methods, is more difficult to accomplish than originally supposed by the regulatory and engineering communities. The EPA study evaluated factors that might affect the success or failure of ground water extraction remedies. The two most commonly cited factors in the report are contaminant characteristics and site hydrogeology.

The physical and chemical behaviors of contaminants in the subsurface are poorly understood. Recent field-scale studies at intensively instrumented sites [2] have substantially improved our appreciation of the complexity of subsurface contaminant migration. Continuing efforts at these sites will result in more detailed understanding of physicochemical and biochemical reactions and processes that control the movement of anthropogenic constituents in the subsurface. Detailed investigations and remedial efforts underway at the large DOE and DOD facilities in the United States will also add to this database.

The EPA study concluded that another primary contributor to the success of ground water extraction systems is detailed definition of site hydrogeology. Although comprehension of the site hydrogeology would seem to be a prerequisite for aquifer restoration, the EPA study pointed out that data collection during site investigations was frequently insufficient to properly assess and exploit hydrogeologic conditions for later remedial measures. This is disturbing because so much effort is spent on this very subject during site investigations. Furthermore, hydrogeologic conditions beneath many, if not most, sites are readily comprehensible and can be successfully exploited if properly defined.

We agree with the EPA conclusion that inadequate characterization of hydrogeologic conditions often dooms effective aquifer remediation. A number of factors explain this trend. Perhaps the most important is the fact that aquifer restoration is still a young discipline that lacks a pool of experienced workers. Another is the tendency of managers and clients to attempt to save money by avoiding costly "science projects". Furthermore, the engineering community may underestimate or overlook the importance of a detailed characterization of the geology of an affected aquifer.

As the ground water industry matures, the great cost to define and remediate ground water contamination will force industry performance to improve rapidly. We believe that one way to improve the quality of remedial investigations is to better define the hydrogeologic conditions at sites with ground water contamination. This improvement can be accomplished without substantial increase in cost. Improved understanding of hydrogeologic conditions is also essential for application of promising remedial techniques such as enhanced biodegradation and surfactant flushing.

THE NATURE OF GROUND WATER INVESTIGATIONS

Investigations of ground water contamination are normally conducted with one of three goals in mind. The first is to determine whether contamination is present in the subsurface. This is the first step in all investigations and typically consists of a limited scope of work including collection of a small number of soil and ground water samples for laboratory analyses. Semi-quantitative data, such as regional aquifer studies from the USGS or potentiometric data from three or four wells on-site, are used to develop a preliminary conceptual model of the aquifer beneath the site.

The second goal, once contamination is confirmed, is to provide enough information to determine whether the contamination threatens human health or the environment. This requires both a greater quantity and quality of data so that the conclusions of a risk assessment have a sound technical basis. The databases for risk assessment are normally very large and include highly accurate chemical data. Detailed information regarding the hydraulic properties of the aquifer is developed and reduced for modeling the threat of contaminants to potential receptors. Less commonly, important geochemical data such as redox conditions or organic carbon fraction are collected to better describe the processes that affect the migration of the unsavory constituents in the aquifer. Unfortunately, the techniques and assumptions of risk assessment are far more subjective than indicated by the precision of the data used in the determination of risk.

The third goal of a ground water investigation, if the risk assessment indicates its necessity, is to gather sufficient data to effectively remediate the aquifer. Whether the ultimate goal is containment of the contaminant plume or removal of the constituents, the solution normally involves pumping contaminated water from the aquifer and treating the water prior to discharge. In most cases, the ultimate goal of ground water remediation is some degree of restoration.

To adequately restore the aquifer, the designer must understand a number of important chemical and physical attributes. The types and concentrations of the chemicals of concern must be known with some certainty. The geochemistry of the aquifer and the confining units must be described in sufficient detail to understand the chemical reactions that control the rate of contaminant migration. The hydraulic properties of the aquifer, including the confining units, if any, must be estimated with some precision. The ground water flow paths, not just the general flow direction based on hydraulic head, must be carefully defined.

CONTAMINANT MIGRATION IN AQUIFERS

Much of the research in contaminant hydrogeology in the past several years has focused on contaminant transport. Although discussion regarding certain details continues, most workers agree that advective transport of constituents is the dominant mechanism of migration. Advective transport is the process by which constituents are carried along with ground water as it flows through an aquifer. This transport is largely a physical process whereby the energy of the flowing water causes molecules of dissolved constituents, whether naturally occurring or man-made, to migrate predominantly in the direction of ground water flow.

If advective transport is the dominant mode of contaminant migration, then this precept should be integrated with a thorough understanding of site hydrogeology. In particular, because the greatest mass of contaminant in an aquifer will generally be located in areas of highest hydraulic conductivity, a detailed understanding of this distribution will result in more accurate assessments of ground water contamination. In turn, this will allow more successful exploitation of the natural features of the aquifer during remediation.

A number of other important processes help to control contaminant transport in the aquifer. Most of these are geochemical in nature and tend to retard the migration of constituents relative to the rate of ground water flow. These processes include the adsorption of contaminants to solid particles in the aquifer, the desorption of the contaminants from the solid particles, the dissolution of contaminants from a free-phase liquid, and the diffusion of contaminants from areas of higher concentration to areas of lower concentration. All of these physiochemical processes, as well as others that are not as well understood, are important factors that influence contaminant migration and continue to be the focus of both field and laboratory research.

THE CONCEPTUAL HYDROGEOLOGIC MODEL

A conceptual hydrogeologic model consists of four basic components: the aquifer skeleton, the aquifer boundaries (geologic boundaries), the aquifer geochemistry (both solids and fluid), and aquifer-surface water interaction (hydraulic boundaries). Discussions of the geochemistry of the aquifer and the aquifer-surface water interaction, which represent very significant components of the conceptual model, are beyond the scope of this paper. Instead, we focus on the physical features of the aquifer that define the distribution of hydraulic conductivity and, ultimately, advective transport of contaminants.

The Aquifer Skeleton

The aquifer skeleton refers to the earth materials through which ground water flows. The skeleton includes the soil column, near-surface sediments, and the underlying bedrock including weathered and fractured rock. The heterogeneity of earth materials causes large variations in hydrologic properties in aquifers, particularly hydraulic conductivity. The variety in spatial trends of the earth materials due to changes in thickness or areal extent of individual geologic units further complicates an aquifer. Spatial trends of hydraulic properties of an aquifer, therefore, can be directly related to the geology of the aquifer. Geologic features such as porosity, grain-size distribution, degree of lithification, and fracture interconnection control hydraulic properties such as hydraulic conductivity and storativity. The resulting aquifer skeleton is a complex package of flow fields characterized by unique geometries and hydraulic properties.

Aquifer Boundaries

Ground water/surface-water interactions represent important hydrologic boundaries in an aquifer system. Here, we distinguish hydrologic boundaries such as streams and geologic boundaries such as the presence of confining units. These can be particularly important in contaminant hydrogeology especially in shallow aquifers. The shapes and hydrologic properties of geologic boundaries, both lateral and vertical, exert considerable influence over, and in some cases, completely control, the ground water flow fields. The physical characteristics of fine-grained materials, which represent barriers to flow, are particularly important.

Many investigations forego evaluation of these basic characteristics and extrapolate geologic data between widely spaced boreholes by 'connecting the dots'. The availability of computer programs to draw cross-sections has exacerbated this tendency among the geologists and engineers charged with the interpretation of borehole data. Three-dimensional aspects of geologic features often are virtually ignored. Because the geologic heterogeneities that control hydrologic properties are often oversimplified, the complexity of the hydrogeologic systems can be severely underestimated. This has led to unrealistic expectations among the engineering and regulatory communities, as well as the general public, regarding the feasibility of easily restoring contaminated aquifers.

CONSTRUCTION OF THE CONCEPTUAL HYDROGEOLOGIC MODEL

The initial step in the construction of a conceptual model is a recognition of an area's broad geologic features. This is easily accomplished by reviewing relevant geologic maps and reports. For problems involving contaminant hydrogeology, the task is simplified because many contaminated aquifers are close to the ground surface. An understanding of gross features such as the presence of thick glacial sequences in the northern part of the country or thick fluvial packages in the Mississippi Valley simplifies this task further.

Once the general geology of an area is known, the investigator should begin to develop a list of assumptions regarding the geology of an aquifer. This thought process is part of a method known as the method of multiple working hypotheses and is essential to the progress of any scientific investigation [3]. All too often, however, this critical step is eliminated in favor of reducing budgets or shortening schedules. However, a few hours of reading or map study during the planning stages of an investigation will result in a better planned, more successfully implemented field program, and will eventually shorten aquifer remediation.

With several hypotheses about the possible geologic features that might be found during the drilling program, the investigator begins to develop the conceptual hydrogeologic model. For example, if the site is located in a modern fluvial environment, the geologist should anticipate that the boreholes will contain a wide range of sedimentary units of varying shapes and grain sizes. Gravel units will generally be overlain by sandy units which, in turn, may be overlain by silts and clays. The sandy units might contain beds that are cross-stratified, indicating current direction. The clayey units might be finely laminated with thin interbeds of silty material.

Based on world-wide studies of modern fluvial environments, sedimentologists have constructed and refined a large number of models to explain the common occurrences of such features. These models are known as facies models.

Using these models, sedimentologists have found that the sizes and shapes of sand deposits in fluvial environments, for instance, are limited by the physical processes that are active in the depositional environment. By understanding how the sand bodies are deposited by rivers, the facies models can be used to predict the extent, thicknesses and orientations of sand units from a limited amount of data. Furthermore, the physical relationships between sand and clay bodies, which has been referred to as facies architecture, can be used to refine the initial interpretations. The concept of sedimentary facies only recently has been applied to contaminant hydrogeology [4].

The geologists and geological engineers who study rock mechanics and structural geology have discovered that structural fabrics also can be classified by genetic relationships. The concept of structural facies, which implies that fabric elements of particular type and orientation are formed under similar conditions, has been used extensively to map complex structural features in great detail. For example, the orientation of cleavage in a highly deformed rock mass can be used to infer the orientation of the stress field that caused the rock strain. Genetically related structural fabrics have been mapped over great distances. Structural geologists in the Rocky Mountains have interpreted the presence of secondary deformation of structural fabrics to infer and locate folds or faults at depth. In most cases, the structural facies contain a limited number of fabric elements and a limited number of orientations. This concept can be exploited by hydrogeologists in the same fashion as sedimentary facies.

Development of the Facies Model Through Geologic Logging

The conceptual hydrogeologic model should be developed during the drilling phase of the investigation. In our view, all boreholes must be continuously sampled to allow detailed geologic logging of the entire section. In our experience, continuous sampling of boreholes does not add significant cost to the project budget. For example, most drillers who use the hollow-stem auger technique now have access to the five-foot split-barrel samplers. If not, Shelby tubes or split-spoon samplers are very useful for collecting continuous samples. Under certain circumstances, such as sampling in fractured rock aquifers, continuous sampling will add a significant cost to a project. However, these situations involve far more complicated hydrogeologic settings, and for that reason, the extra expense is not only warranted, but is essential for successfully interpreting the hydrogeologic system.

The logging of geologic samples from drilling operations has been notoriously poor in the environmental industry. Paltry descriptions of the dominant grain size, along with notations of color, are the industry standard. This style of logging is completely inadequate to develop an appropriate understanding of the geology of an aquifer. Other methods, such as the Unified Soil Classification System (USCS) are also inadequate for hydrogeologic work. The engineering community has inappropriately applied the USCS, which is used to describe various geotechnical properties of soils such as bearing capacity and plastic limits, to geologic materials for the purposes of describing ground water flow. We suggest that borehole logging should always include the classical geologic descriptions that form the basis for interpreting depositional environments.

Descriptions of geologic samples should be rigorous, complete and useful. The drilling of boreholes is a very expensive proposition and the data generated during drilling must be developed to its fullest capacity. Samples are rarely stored and, once discarded as is the norm, cannot be retrieved without incurring additional expense. As Compton so eloquently pointed out in 1962, the sampling location cannot be revisited to test every new idea [5].

We are presenting the argument that sedimentary and structural facies models can be used to extrapolate between data points in a logical and consistent fashion. To accomplish this, the geologic logs generated during drilling and sampling of boreholes must be as complete and descriptive as possible. Because the drilling is used to sample geologic media, the classical methods of geologic description and logging are not only appropriate, they are essential to the correct interpretation of the geology of an aquifer.

A number of books and papers have been published on the subject of geologic descriptions and logging. All are very useful and can be applied to a number of geologic situations. However, we ascribe to the philosophy of Compton as opposed to a strict set of procedures or rules [5].

Geologic descriptions of cores should be made in the field as the drilling progresses. The descriptions form the basis for interpretation of geologic facies. Compton [5] points out that sound descriptions cause the geologist to think critically and to refine his or her hypotheses as the work proceeds in the field. In general, most of the descriptions will be factual data or observations of specific items noted in samples. Of particular importance are those features that might indicate the origin of the geologic unit. No matter how small the investigation, the descriptions must be thorough and complete for the entire borehole. If certain features are noted in a series of boreholes, this should be noted in the field. Consistency in logging procedures and descriptions among different workers at the same site is critical.

The specific items that should be included in the geologic descriptions of boreholes will vary depending on the hydrogeology of the area. Table 1 is a general list of observations modified after Compton [5]. The list is not intended to be comprehensive nor will all items apply in all situations. However, it does provide an indication of the geologic features that will aid in the interpretation of sedimentary facies and, subsequently, in the development of a conceptual hydrogeologic model. A similar list of items can be prepared prior to mobilization to the field to ensure consistent logging.

Depending on the nature and extent of the hydrogeological investigation, field descriptions should be augmented with laboratory analyses of selected samples. In particular, standard geotechnical analyses such as grain-size analyses and Atterberg limits can be used to characterize geologic units in a semi-quantitative fashion. These data are often useful for geological correlation across large sites and to characterize stratigraphic variations between boreholes. Simple geochemical parameters such as pH, Eh, cation exchange capacity, and total organic carbon can also be used to describe the geochemistry of geologic units for the purposes of correlation. These data are also important for analysis of contaminant retardation in an aquifer.

Interpretation of Geologic Data

The collection of geologic data in the field consists of careful observations and detailed recording of those observations. This is a rather straight-forward process that can be accomplished successfully by all scientists and engineers. The interpretation of that data, however, is much more difficult and requires special training and experience.

The interpretation of geologic data requires the conscious integration of technical skill and creativity. The data must be reduced to classify them into functional groupings. Geologists are often referred to as "lumpers" or "splitters" depending on their tendencies to group data. Accordingly, geologic interpretations are non-unique and imprecise. This explains one of the primary difficulties in the

Table 1 - Outline of Geologic Observations

1. Name and location of borehole.

2. Specific depths (typically marked on the logging form).

3. Description of sediments/rocks, with most abundant variety described first:

    a. color, fresh and weathered.
    b. degree of induration.
    c. grain sizes; principal size, ranges of sizes.
    d. degree of sorting.
    e. shapes of grains.
    f. orientations or fabric of shaped grains.
    g. nature and amount of matrix or cement.
    h. nature and amount of pores.
    i. mineralogy of grains.

4. Characteristic structures of each unit:

    a. range of thicknesses of different beds.
    b. shapes of beds.
    c. primary sedimentary structures including grading, laminations, cross-bedding, erosional features, etc.
    d. secondary structures such as weathering.
    e. frequency and thicknesses of fractures or joints.

5. Biological features:

    a. distribution of fossils, including bioturbation.
    b. position and condition of fossils (i.e., fragmented, rounded).

6. Nature of contacts:

    a. sharp or gradational, with descriptions and dimensions of gradations.
    b. descriptions of possible geologic unconformities.

application of extremely precise, yet rigid computer models to inherently equivocal interpretations.

In our view, the interpretation of geologic data must be done by those knowledgeable in field geology, stratigraphy, and structural geology. The appropriate use of geologic descriptions from boreholes is central to our theme that the geology of an aquifer exerts considerable control on ground water flow. Often, correlation between boreholes seems impossible because the sections are so different. Drawing widely spaced lines on a stratigraphic cross-section to indicate lithologic correlation or facies boundaries through areas where no data exist requires unusual skill or great arrogance (or perhaps both). However, the hydrogeologist has assistance in this kind of situation. The geologic literature contains hundreds of papers and books devoted to the subjects of sedimentary and structural facies.

The hydrogeologist often has one additional tool for use. In many parts of the country, outcrops of the aquifer materials are available for detailed study. If not, then test pits can be dug by a backhoe to provide exposures for examination. Study of the exposures can reduce the guesswork associated with the extrapolation of geologic data between the widely-spaced boreholes. Instead of relying solely on the geologic literature, the hydrogeologist can reduce the number of potential interpretations by carefully measuring, describing, and photographing the actual geology of the aquifer. This is an inexpensive and easy way to significantly improve the quality of the conceptual hydrogeologic model.

Refinement of the Conceptual Hydrogeologic Model

Many ground water investigations are designed to be completed in phases, with each subsequent phase building upon the previous work. In such cases, the conceptual model will be refined as the investigation progresses. The model can be used to predict specifically the geologic and hydrogeologic features anticipated in subsequent boreholes. In this fashion, the model can be improved based on the new findings or removed from further consideration if it proves inadequate. In either case, the hydrogeologist will understand the ground water flow system in far more detail than previously.

The conceptual model can also be improved using standard techniques of aquifer testing. Multiple-well pumping tests can be used to measure or evaluate the distribution of hydraulic conductivity as indicated by the geologic data. Piezometers can be located appropriately based on the facies models as well as on hydrologic parameters such as saturated thickness and expected hydraulic conductivity. Single-well tests can be used to characterize individual facies in a similar fashion to the geotechnical analysis of samples.

How Much Data is Enough?

No two people will provide the same answer to this question. However, it is clear that the quality of the conceptual model of an aquifer system depends on the geologic and hydrogeologic data collected during field investigations. More data, of the appropriate kind, will result in a better understanding of the aquifer. However, data alone cannot solve problems. The informed interpretation of data, whether enough or not, will result in better solutions to problems of ground water contamination.

The improved use of geologic data that we have presented here does not require additional boreholes, exotic tools, or computers with extended memories. The methods of describing geologic features encountered in boreholes are not difficult and, in fact, are taught to nearly all undergraduate geology students. Furthermore, more detailed descriptions will not add any cost to a job. Data interpretation will require more time than has traditionally been spent on such matters. However, this investment is inconsequential when compared with the cost of purchasing and installing a ground water extraction and treatment system.

USE OF THE REFINED HYDROGEOLOGIC MODEL

The refined conceptual model has two important uses. The first is related to the use of the model in risk assessment. EPA guidance stipulates that a conservative approach will be used in risk assessment to ensure that errors, if they are made, favor a higher degree of protection to human health and the environment. While we agree with this approach, we also believe that risk assessment should be based on the best science that is possible. For instance, many geochemical reactions involving man-made chemicals in the subsurface are extremely complicated and remain poorly understood. An improved understanding of the hydrogeology of a site will yield a more informed risk assessment. Improving our definition of the hydrogeology of a system will decrease the uncertainty inherent in risk assessments.

The second use of an improved conceptual hydrogeologic model is in ground water remediation. Many early attempts at aquifer rehabilitation have been poorly designed, overly expensive, and inadequate. Some systems have been grossly over-engineered by including far too many extraction wells. Many in situ bioremediation projects have failed because the hydraulics of the ground water system were not clearly defined. Promising new technologies, such as soil washing and surfactant flushing, may fail if they are applied in geologic settings that are not completely defined. An improved understanding of ground water flow paths, which we believe are intimately connected to the geology of the aquifer, will improve our ability to remediate contaminated aquifers.

CONCLUSIONS

Ground water assessment and remediation have improved over the past several years. Important new tools have been developed and applicable old tools have been borrowed from other industries to enhance proven methods of remedial investigations. However, as the remediation industry matures and these and other new techniques are developed, we urge a return to the roots of geologic practice that historically formed the basis for ground water investigations. In porous media, hydrogeologic investigations should be based on detailed geologic descriptions and interpretations of depositional environments. Cost-effective solutions to problems of contaminant migration in flow systems dominated by advective transport can only be accomplished through refined geologic interpretations based on detailed and reliable field data.

ACKNOWLEDGEMENTS

The authors acknowledge with appreciation the help of Jeff Melby, Mary Louise Conte, and four anonymous reviewers in the preparation on this paper.

REFERENCES

[1] U.S. Environmental Protection Agency, Evaluation of Ground Water Extraction Remedies, EPA/540/2-89/054, September 1989.

[2] Cherry, John A., Editor, 1983, Migration of Contaminants in Ground Water at a Landfill: A Case Study, Journal of Hydrology, Volume 63, pp. 1-198.

[3] Chamberlin, T. C., "The Method of Multiple Working Hypotheses", Journal of Geology, Vol. 5, 1897.

[4] Anderson, Mary P., "Hydrogeologic Facies Models to Delineate Large-Scale Spatial Trends in Glacial and Glaciofluvial Sediments", Geological Society of America Bulletin, Vol. 101, p. 501-511, April 1989.

[5] Compton, R. R., Manual of Field Geology, John Wiley and Sons, New York NY, 1962.

Alva M. Parsons and Paul A. Davis

A PROPOSED STRATEGY FOR ASSESSING COMPLIANCE WITH THE RCRA GROUND-WATER MONITORING REGULATIONS

REFERENCE: Parsons, A. M. and Davis, P. A., "A Proposed Strategy for Assessing Compliance with the RCRA Ground Water Monitoring Regulations," Current Practices in Ground Water and Vadose Zone Investigations, ASTM STP 1118, David M. Nielsen and Martin N. Sara, Eds., American Society for Testing and Materials, Philadelphia, 1992.

ABSTRACT: To satisfy the Resource Conservation and Recovery Act (RCRA) ground-water monitoring regulations, a hazardous waste management facility must have a ground-water monitoring system consisting of at least one upgradient and three downgradient wells and show that the downgradient wells are capable of immediately detecting a statistically significant amount of contamination at the water table. Because the regulations are subjective, it is often difficult for the owner/operator and the regulator to assess whether a monitor-well network satisfies the regulations. A probabilistic strategy is presented which satisfies the regulations and attempts to minimize subjectivity in evaluating the performance of a monitor-well network.

The proposed strategy is based on a determination of the likely ground-water flow paths through both the unsaturated and saturated zones and ground-water travel times. The approach involves three stages of analysis: 1) optimization of monitor well location, 2) evaluation of the sampling interval, and 3) assessing the monitor-well network performance through time.

KEY WORDS: ground water, ground-water monitoring, monitor-well network, RCRA regulations, quantification, strategy

A. M. Parsons is a hydrologist and P. A. Davis is a division supervisor at Sandia National Laboratories, Environmental Impact and Restoration Division 7723, Albuquerque, NM 87185-5800.

1. INTRODUCTION

Current RCRA ground-water monitoring regulations are subjective, making it difficult for the owner/operator and the regulators to resolve differences as to whether a monitor-well network satisfies the regulatory requirements. Sandia National Laboratories (SNL) has developed a probabilistic approach to meeting the RCRA regulations which attempts to minimize the subjectivity in evaluating the performance of a monitor-well network by prior agreement to a set approach, and by defining and quantifying the performance measures to evaluate the monitor-well network.

This strategy was developed to determine the adequacy of an existing monitor-well network at a specific SNL waste site located in Albuquerque, New Mexico. The intent was to provide a mechanism which allowed the regulators and SNL to assess whether the number and locations of existing wells met the regulations. Quantification of the well-network problem also required clarification of regulatory requirements and definitions. A discussion of these requirements and definitions along with the proposed strategy for assessing compliance with the RCRA ground-water monitoring regulations is presented below. We are currently in the development stages of implementing this strategy at a site, however the strategy is not ASTM or U.S. EPA approved at this time.

2. REGULATORY REQUIREMENTS AND DEFINITIONS

To satisfy the RCRA ground-water monitoring criteria (40 CFR 265.91), "a ground-water monitoring system must be capable of yielding ground-water samples for analysis and must consist of: 1) Monitor wells (at least one) installed hydraulically upgradient (i.e., in the direction of increasing static head) from the limit of the waste management area. Their number, locations, and depths must be sufficient to yield ground-water samples that are: i) representative of background water quality in the uppermost aquifer near the facility; and ii) not affected by the facility; and 2) Monitoring wells (at least three) installed hydraulically downgradient (i.e., in the direction of decreasing static head) at the limit of the waste management area. Their number, locations, and depths must ensure that they immediately detect any statistically significant amounts of hazardous waste constituents that migrate from the waste management area to the uppermost aquifer." Additionally, the RCRA closure permit regulations (40 CFR 270.14c) require the "identification of the uppermost aquifer and aquifers hydraulically interconnected beneath the facility property, including ground-water flow direction and rate, and the basis for such identification (i.e., the information obtained from hydrogeologic investigations of the facility area)."

These regulations contain several terms which need to be defined prior to developing a strategy for assessing compliance. Also, the U.S. EPA regulations do not account for many of the hydrologic complexities associated with the arid

Southwest. In the Southwest, it is not uncommon to find waste sites located hundreds of meters above unconfined alluvial aquifers (> 1000 meters thick) as is the case at the SNL site [1]. Thus, additional clarification of U.S. EPA definitions was required to develop our proposed strategy for application to the SNL site. The definitions employed in developing this monitor-well network optimization strategy are as follows:

Ground Water. Although 40 CFR 260.10 and ASTM define ground water as subsurface water in a zone of saturation, flow paths of water below the land surface in both the unsaturated and saturated zones are considered herein. Ground-water flow through the unsaturated zone must be considered to determine the location of a potential plume arriving at the water table. Therefore, ground water is defined as all water below the land surface, with the saturated zone water being the only currently regulated portion of the ground water.

Uppermost Aquifer. In an aquifer system of substantial thickness it may be neither feasible or necessary to consider the entire aquifer thickness when determining monitor well locations. Therefore, the uppermost aquifer is defined within by the flow paths between the waste site and the monitor wells. That is, the uppermost aquifer is that portion of the aquifer which could plausibly be contaminated by a release from the waste site.

Immediate Detection. "Detection" in RCRA terms involves the determination of statistical differences between contaminant concentrations in monitor wells compared to background wells when sampled at an appropriate frequency to detect a contaminant plume. A strict interpretation of immediate detection would require constant monitoring at the exact point where the contaminant reaches the water table. Instead, immediate detection is defined within as the detection of a contaminant release prior to significant deterioration of the aquifer water quality. The term "significant deterioration" is defined below.

Hydrogeologic Characterization. This term is not defined by the U.S. EPA, and is defined here as sufficient data collection and analysis to allow for the determination of ground-water flow direction and rate. Specifically, the distribution of hydraulic conductivity and hydraulic gradient are the data required for the purpose of well location. For the determination of the sampling interval, the distribution of hydraulic conductivity, the hydraulic gradient, and the effective porosity are required.

Ground-Water Flow Direction. The U.S. EPA requirements imply that the direction of ground-water flow and the direction of the hydraulic gradient are coincident. However, the direction ground water travels (the path any dissolved contaminant would follow) is controlled by the distribution of hydraulic conductivity and the hydraulic gradient. The latter definition is used here.

Ground-Water Flow Rate. Although not defined by the U.S. EPA, we propose to use the Darcy flow velocity (a volume flux defined as the volume of discharge per unit bulk area per unit time) for the purpose of well location and the seepage velocity (discharge rate per unit of pore space in a unit bulk area) for the purpose of determining the sampling interval.

The following terms are not specifically found in the regulations, but are used in developing this strategy (Figure 1):

Detection Boundary. The boundary surrounding the waste management area defined by the monitor-well network.

Significant Deterioration. A contaminant release which affects a significant portion of the aquifer.

Significant Portion of the Aquifer. This term is defined as that portion of the aquifer beyond the recovery boundary of the monitor wells. That is, any aquifer location outside the capture zones of the monitor wells. The capture zone is that area surrounding a well within which contaminants could be removed by pumping the existing monitor wells.

Recovery Boundary. The boundary defined by the capture zones of the monitor wells is herein referred to as the recovery boundary.

Finally, the assumption is made that the U.S. EPA required ground-water monitoring network is to be designed to detect a continuous type of release from the waste site. Otherwise, assuming a pulse release would require continuous sampling.

3. APPROACH TO ASSESSING COMPLIANCE

The proposed strategy for assessing compliance with the RCRA regulatory requirements is to: 1) determine if the monitor-well network is optimally located such that the wells have a high probability of detecting the likely flow paths through the saturated zone, 2) evaluate the sampling interval, and 3) assess the performance of the monitor-well network through time. Following is a description of each of these steps.

3.1. Determine Optimal Well Location

Determination of the optimum well location is based on an assessment of likely flow paths through both the unsaturated and saturated zones (Figure 2). The approach used to simulate the likely paths and evaluate the well locations is described in the following sections. Similar approaches have been used elsewhere to optimally locate a monitor-well network [2,3].

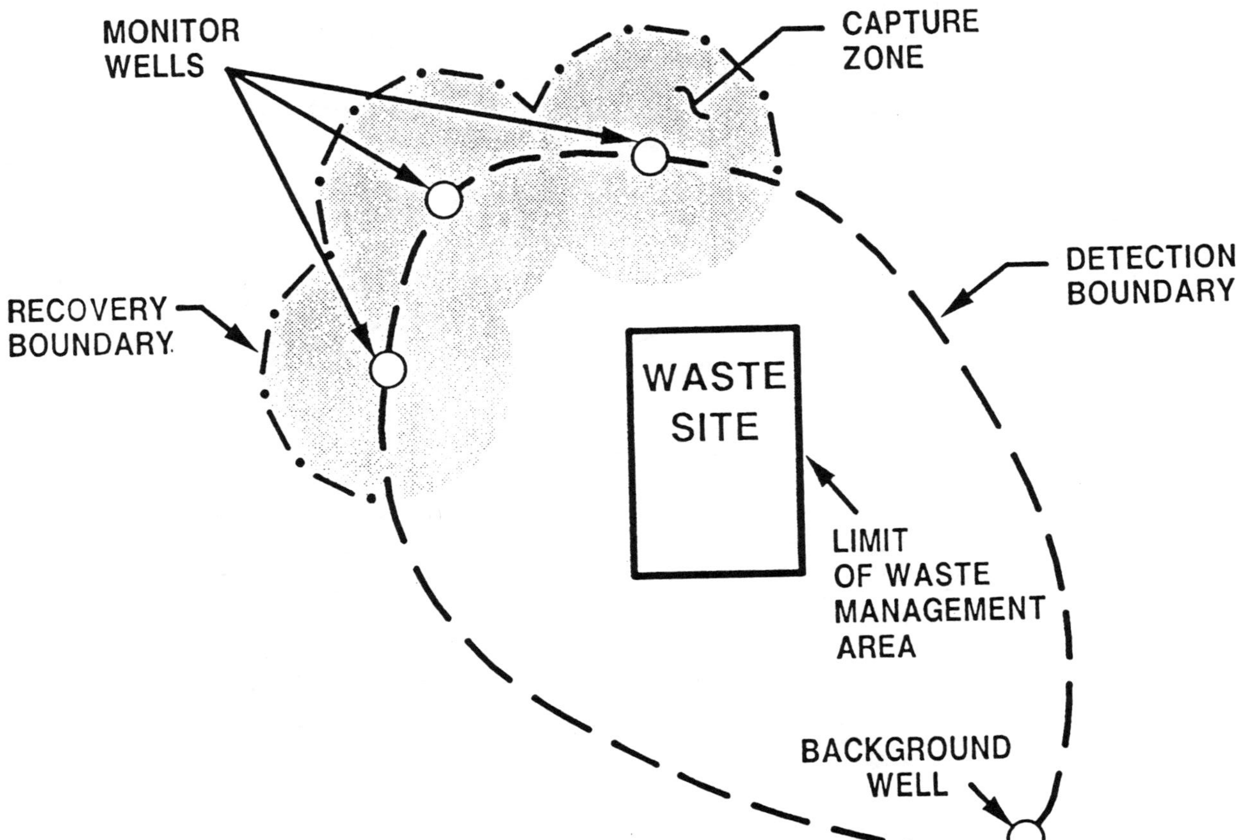

Figure 1. Illustration of some terms used to develop this strategy.

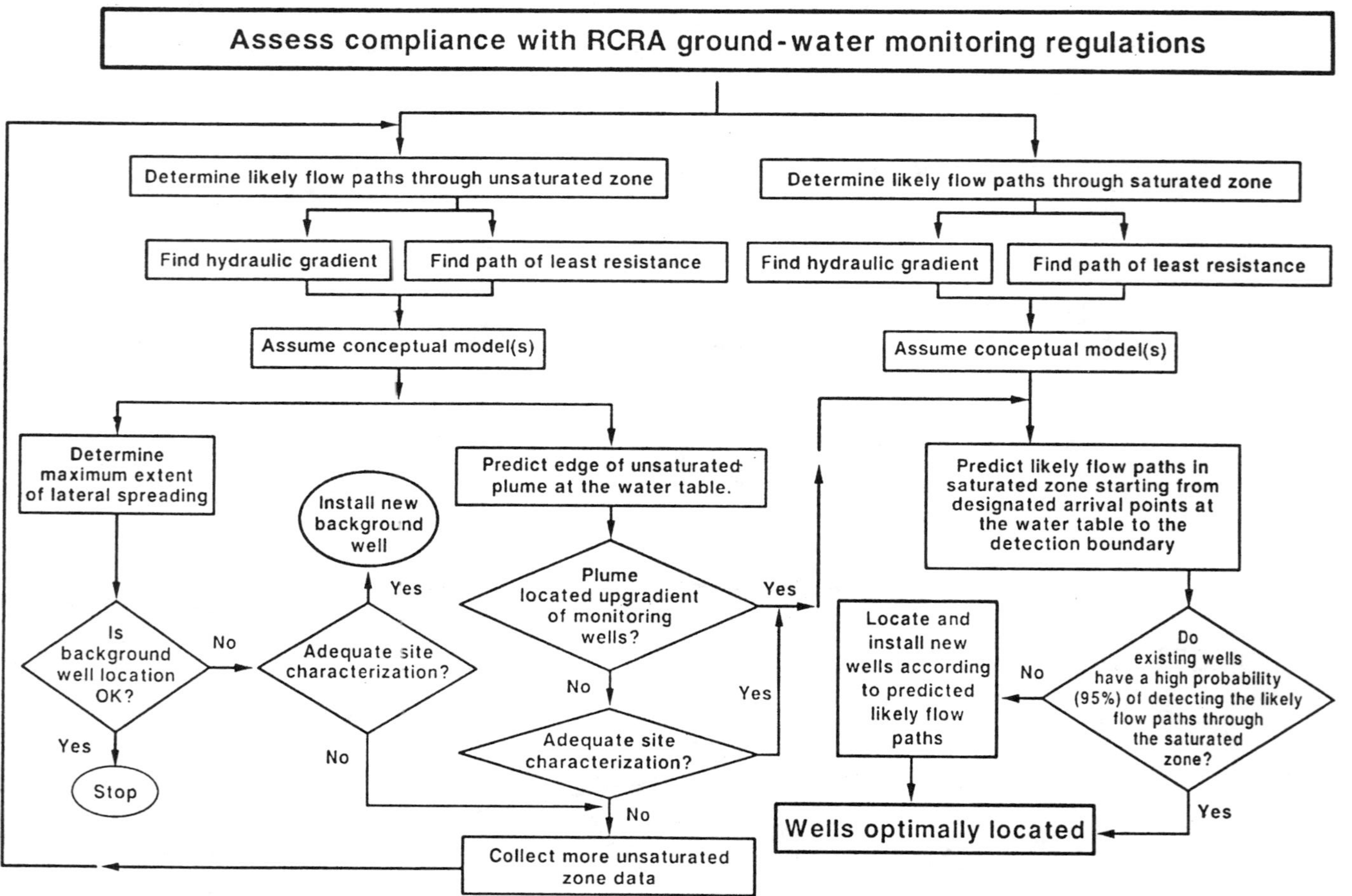

Figure 2. Approach to optimizing the monitor-well locations.

3.1.1. Determine Likely Flow Paths Through the Unsaturated Zone: The likely flow paths in the unsaturated zone, just as in the saturated zone, are controlled by the hydraulic gradient and the distribution of hydraulic conductivity (Figure 2). In the unsaturated zone, the hydraulic gradient is a function not only of elevation, but also of the distribution of the negative pressure head or suction. The distribution of the negative pressure head may be directly measured or inferred from a measured moisture content and the associated moisture retention curve (moisture content versus negative pressure head relationship). Alternatively, for the case of steady-state vertical flow, the hydraulic gradient can be assumed to be equal to unity.

The distribution of hydraulic conductivity in the unsaturated zone is governed by spatial variability of the geologic media and varies according to the negative pressure head. Hydraulic conductivity as a function of the negative pressure head can be determined from the saturated hydraulic conductivity, moisture retention curve, and porosity by assuming that Mualem's pore structure model applies [4].

Hydraulic gradient and hydraulic conductivity data are never complete or unambiguous. Therefore, the next step in the strategy involves making additional assumptions that describe the thickness and extent of the geologic layers, spatial variation of hydraulic properties, as well as the initial conditions and boundary conditions at the site. The sum of all of the assumptions defines the conceptual model, both here and for the saturated zone. Acceptance of the conceptual model is the key step in this strategy and should involve collaboration with the regulators.

Finally, numerical simulations which are consistent with the chosen conceptual model will be used to simulate the water movement from the waste site through the unsaturated zone to determine possible plume locations at the water table. Currently, MEAN2D [5], VAM2D [6], and DCM3D [7] are codes which could be used for the unsaturated zone numerical simulations. Uncertainty in the input parameters (hydraulic conductivity, hydraulic gradient, and porosity) will be propagated through to the model output by performing multiple conditional simulations using Latin Hypercube sampling techniques [8] to reduce the number of required simulations.

The simulated locations of the edge of the plume at the water table will be used for two purposes: 1) to determine if the location of the background well is outside the range of potential contamination, and 2) to define the starting points at the water table that will be used in simulations of the likely flow paths in the saturated zone (see Section 3.1.2). The first point requires further explanation.

To evaluate the potential for background well contamination, the numerical simulation which produces the maximum amount of lateral spreading in the unsaturated zone will be used (Figure 2). If this simulation indicates that the edge of the plume could be located between the edge of the waste site and the background well, then the background well is appropriately located to obtain samples of the ambient aquifer water quality and the background well evaluation is completed. On the other hand, the edge of the simulated plume at the water table could be located at or beyond the background well location (Figure 3). In this case, the monitor-well network may not be capable of determining a statistically significant amount of contamination (i.e., there is no statistical difference between the concentration of contaminant in the upgradient and downgradient monitor wells).

However, realizing that considerable uncertainty could exist in the unsaturated zone data leads to additional considerations prior to concluding that the background well is incorrectly located. Evaluation of simulation results will indicate whether or not additional unsaturated zone data could reduce the uncertainty in the simulated plume locations. If this is true, these data will be collected and the unsaturated zone analysis will be repeated. However, if the level of uncertainty in the unsaturated zone analysis can not be reduced, a new background well will be installed, the location of which is based on the simulated maximum extent of lateral spreading (Figure 2).

3.1.2. Determine Likely Flow Paths Through the Saturated Zone: Just as for the unsaturated zone, the likely flow paths in the saturated zone are controlled by the hydraulic gradient and the path of least resistance (distribution of hydraulic conductivity). In the saturated zone, the hydraulic gradient is determined from the distribution of hydraulic head. This gradient may have both vertical and horizontal components. The horizontal gradient is based on differences in water levels from wells spacially separated, screened within the same vertical horizon of the aquifer while the vertical gradient is determined by the difference in head between wells at the same location but screened at different depths within the aquifer. We assume that information about the distribution of hydraulic conductivity can be determined from aquifer test data. For this problem, assessing the adequacy of an existing monitor-well network, we propose that these data should be collected by conducting several multi-well aquifer tests.

Following aquifer testing, assumptions will be made to define a conceptual model of the ground-water flow system. This conceptual model will provide the basis for the simulation of the likely flow paths in the saturated zone. Again, this is a key step in the strategy, and it should involve collaboration with the regulators.

Simulations of likely flow paths in the saturated zone will be based on numerical modeling of the ground-water flow field using the pre-defined conceptual model and will be designed to produce results that are consistent with the measured parameters (i.e., hydraulic head and hydraulic conductivity). Possible computer codes to use for the saturated zone analyses include GEOINVS [9] and USGS3D [10]. To account for uncertainty in hydrologic parameters, multiple conditional simulations will be performed, resulting in many likely flow paths from the simulated plume arrival points at the water table (see section 3.1.1) to the detection boundary.

Assessment of the existing monitor-well network is then based on the simulated plume locations relative to the well locations. If a monitor well is not

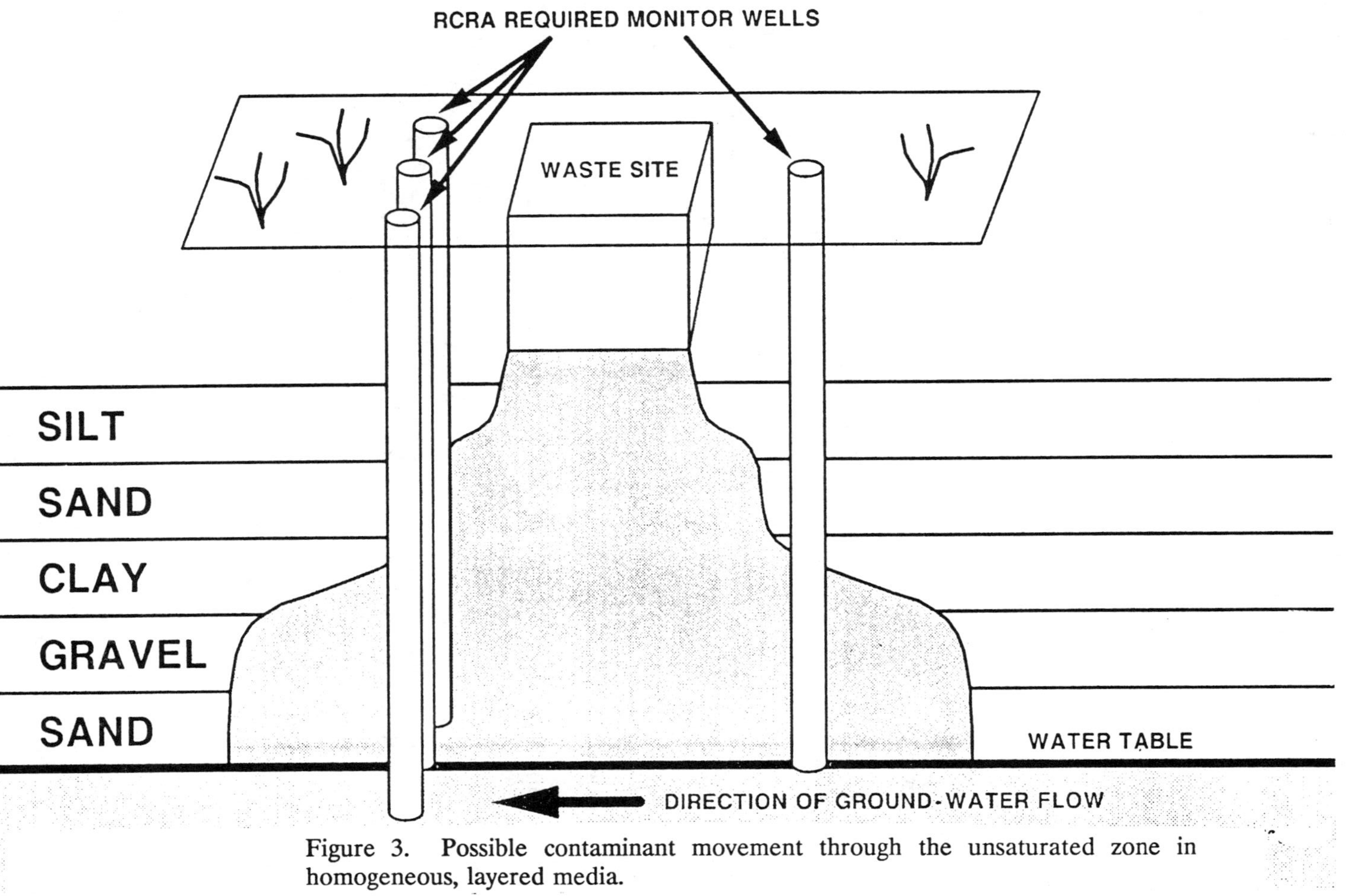

Figure 3. Possible contaminant movement through the unsaturated zone in homogeneous, layered media.

located within the simulated plume, it is assumed to fail in detecting the plume. The total number of successes and failures based on all of the simulations is then calculated. In this manner, the probability of detecting the likely flow paths in the saturated zone can be determined with the existing monitor wells (Figure 4). We propose that the minimum acceptance criteria for the monitor-well network performance be set at 95% detection, however the exact value could be based on other criteria including regulatory input and cost analysis.

The results may indicate that new monitor wells are required. In this case, the simulated likely flow paths provide the information necessary to determine the number and location of additional wells (Figure 2). That is, the results show which location(s) have the highest probability of detecting the likely flow paths which were undetected by the existing network.

3.2. Sampling Interval Evaluation

The first part of this strategy results in an optimal monitor-well network with respect to location. Once this is accomplished, the next step is to determine the optimum sampling interval or frequency. Here the purpose is to detect the movement of contaminant prior to significant deterioration of the aquifer water quality. The RCRA regulations state that the wells must be sampled, at a minimum, quarterly for the first year and semi-annually thereafter. Whether or not the minimum required sampling times are adequate can only be determined by an analysis of likely contaminant travel times. Therefore, the sampling interval is based on the regulatory requirements as a minimum frequency and on expected contaminant travel times for the maximum frequency (Figure 5).

A conservative approach was taken in order to determine an adequate sampling frequency. The contaminants were assumed to be moving at the average ground-water velocity (i.e., no retardation or degradation occurs) in the direction of ground-water flow. It was also assumed that there is no dispersion other than that caused by the defined spatial variability of hydraulic conductivity. Both of these assumptions lead to conservative results; shorter ground-water travel times and more frequent sampling.

The appropriate sampling interval was assumed to be less than or equal to the shortest travel times across any of the capture zones of the monitor wells. That is, we assumed that the contaminant to be detected lies at the edge of a capture zone at the time of the previous sampling so that the contaminant was not detected during the last sampling event. The contaminant should be detected before it travels past the outer edge of the capture zone of the monitor wells. By not allowing "significant deterioration of the aquifer" to occur, the existing monitor-well network could aid in containing the plume.

This minimum travel time can be calculated from combining particle tracking with the previous saturated zone likely flow path simulation results (Section 3.1.2). The only additional data required for this step are the effective porosities which

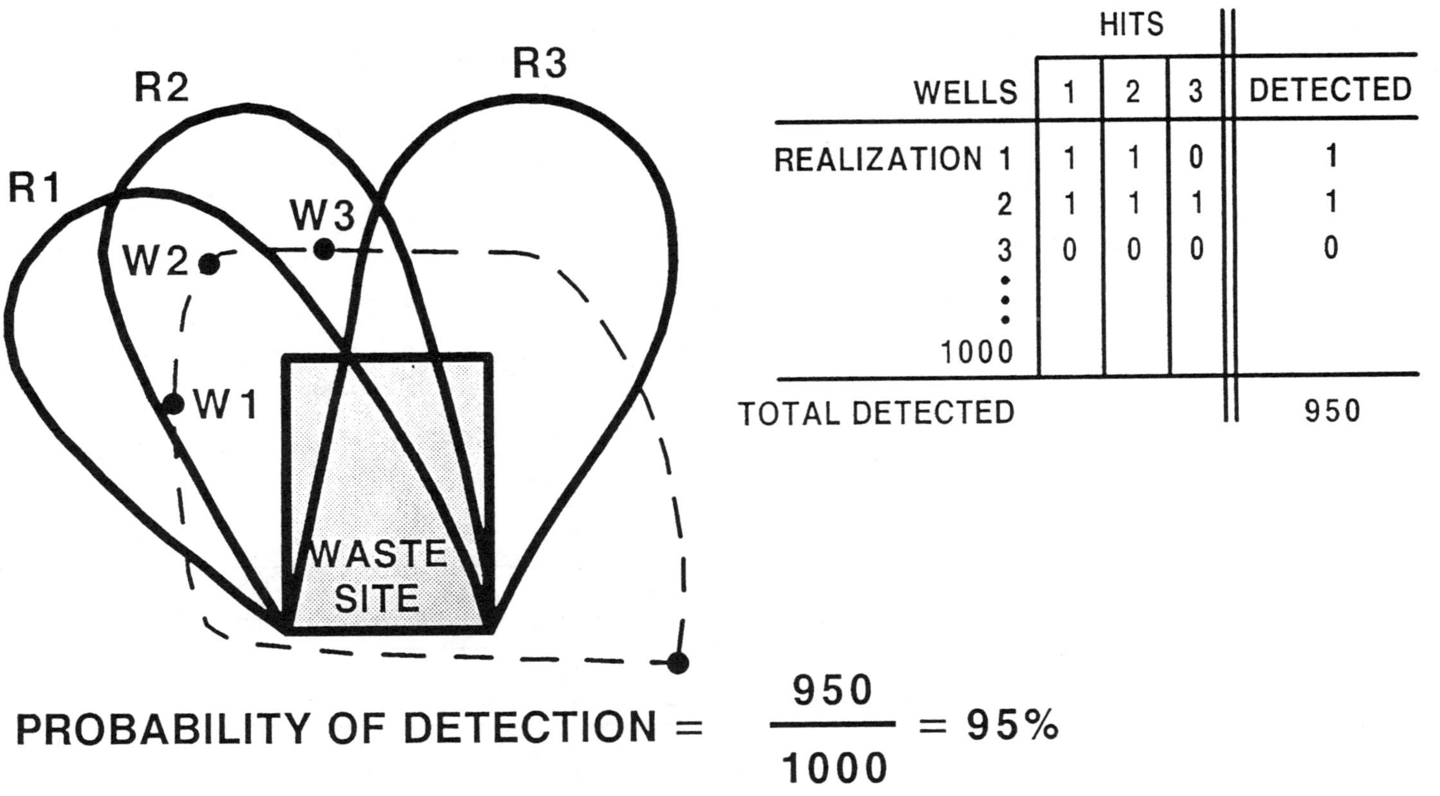

Figure 4. Approach to calculating the probability of detecting likely flow paths in the saturated zone. Realizations depict possible plume locations.

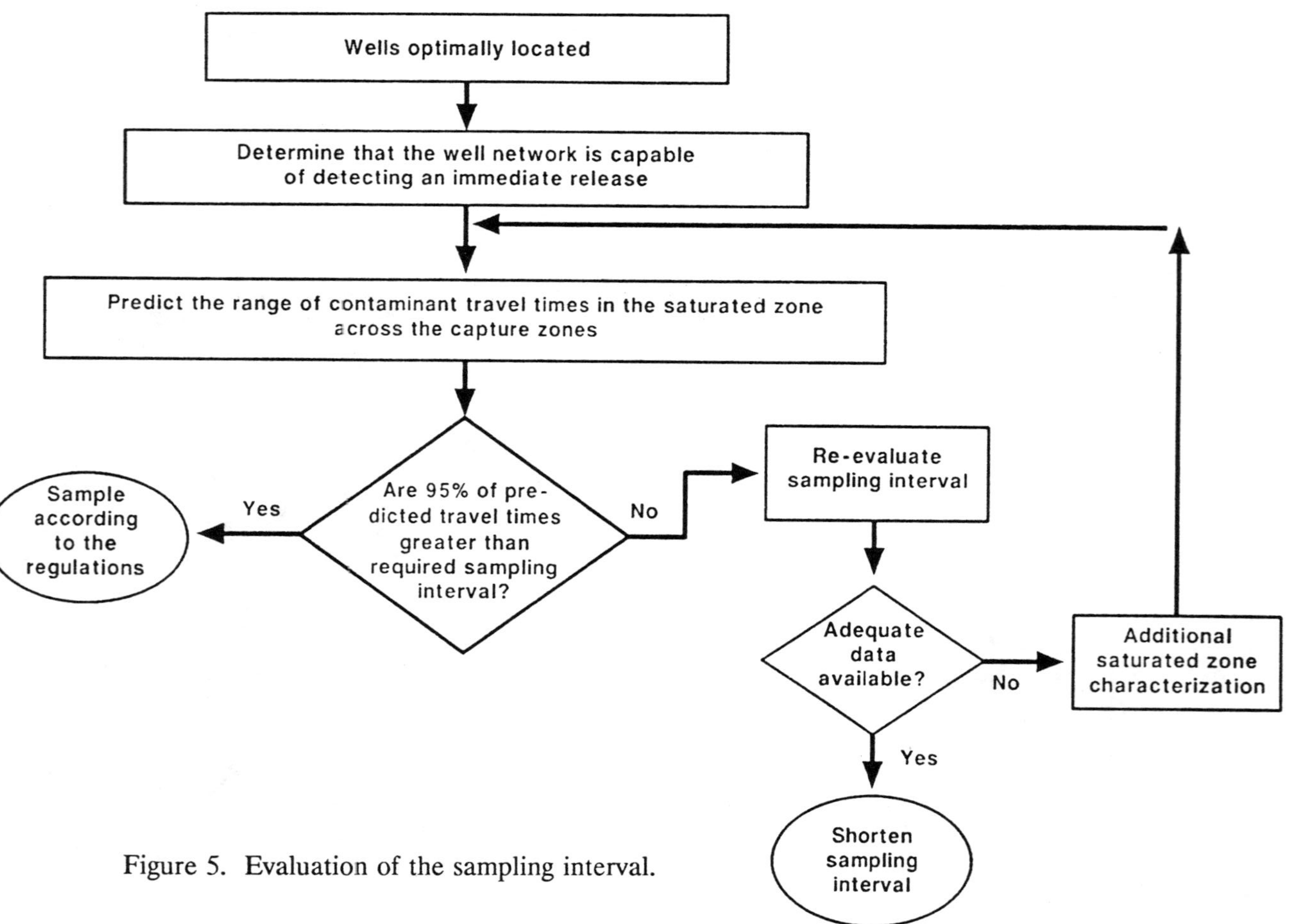

Figure 5. Evaluation of the sampling interval.

can be estimated from the literature according to the known lithology of the aquifer, geophysical logging results, or core laboratory testing. Uncertainty in the input parameters (hydraulic gradient, hydraulic conductivity, and effective porosity) will be accounted for by conducting multiple simulations of the travel time across the capture zones of each well, resulting in a range of possible contaminant travel times (sampling intervals).

If at least 95% of the estimated contaminant travel times across the capture zones are greater than the required regulatory sampling interval, the minimum regulatory ground-water sampling schedule will be followed. If not, then it is necessary to re-evaluate the sampling interval. The regulatory ground-water sampling schedule will continue to be followed during the period of re-evaluation. Prior to shortening the sampling interval to less than the U.S. EPA requirements (sampling more frequently), evaluation of the travel time results will indicate whether or not the collection of additional and/or more site specific data would reduce the uncertainty in the simulated travel times. If this is true, additional data will be collected and the travel time analysis will be repeated. However, if uncertainty can not be further reduced, then the sampling interval should be shortened (Figure 5).

3.3. Assessing Changes in the Monitor-Well Network Through Time

The monitor-well network that results from the previous steps will have been based on present and past hydrologic conditions, including consideration of seasonal fluctuations that are documented in the existing records. However, hydrologic conditions may change in the future due to such factors as long-term environmental changes, or long- and short-term human-induced changes. U.S. EPA requires that the monitor-well network will continue to be able to "immediately" detect a release from the waste site even if such changes exist. Our approach to insuring the adequacy of the monitoring network with time is based on understanding the cause of the hydrologic change and assessing potential changes in the flow-path directions and ground-water travel times (i.e., the possible contaminant travel paths) resulting from the change.

This approach is not solely based on changes in the hydraulic gradient. Instead, determination of the cause of change is necessary for defining a course of action. Specifically, we want to be able to distinguish between one time perturbations to the system versus permanent or long-term changes to the system. For example, performing an aquifer test at the waste site could change the direction of the hydraulic gradient by pumping a well or wells for a short time. After pumping was discontinued, the hydrologic system would return to its original gradient. In this case, a contaminant would only travel a very short distance under the altered gradient while over a larger time (i.e., the sampling interval), the contaminant would travel to the monitor wells. Thus, it is unnecessary to drill additional wells. On the other hand, long term or permanent changes to the hydrologic system which adversely affect the performance of the monitor-well network may have to be accounted for by drilling one or more additional monitor wells. If we are unable to determine the cause and the nature of the change then additional wells will be installed. The following procedure will be followed in assessing the adequacy of the monitor-well network with time (Figure 6).

The performance of the monitor-well network will be evaluated after each sampling event as described in Section 3.1.2. First, if the probability of detecting a plume at the existing wells drops below 95% (see Section 3.1.2), the frequency of water-level monitoring will be increased while possible causes for the change in the monitor-well network performance are investigated. Then, if the existing wells still do not have a 95% probability of detecting the likely flow paths through the saturated zone after two consecutive sampling intervals, a decision will be made as to whether a new monitor well is required. A new monitor well or wells will be installed if the cause of the change is unknown or if the change is known to be permanent.

If the cause of the change is known to be temporary relative to the sampling interval, then the only action taken would be to continue water-level monitoring to assure that the change was temporary. The remaining case is that the cause of the change is not permanent but extends past two sampling intervals. Then the decision on whether or not to drill a new well should be based on the expected plume location after two sampling intervals. New wells would be added only if the likelihood of detecting this plume with existing wells would be less than 95% (Figure 6).

4. DISCUSSION

Using current U.S. EPA guidance [11], neither the owner/operator nor the regulator have a clear method of determining the adequacy of a monitor-well network with respect to the number of wells, well location, and sampling frequency. The owner/operator often feels that the regulator has a very limited basis for requesting additional wells while the regulator feels that U.S. EPA has not provided sufficient guidance for evaluating how many wells are necessary. Both parties may acknowledge that the process is very subjective from start to finish. This proposed strategy provides a framework for the owner/operator and the regulator to agree upon the required number and location of monitor wells, the sampling frequency, and the monitor-well network performance through time.

This strategy attempts to minimize the current subjectivity associated with monitor-well network evaluation, but subjectivity in the evaluation is in no way eliminated. Instead, the subjective aspects of the evaluation process are moved up front where they can be jointly addressed by the owner/operator and the regulator. Specifically, the subjective components arise in the conceptual model development stage and in determining the adequacy of the data. These two components are currently part of every step in the monitor-well network evaluation problem, beginning with the first well drilled and continuing each time new data (i.e., quarterly or semi-annual water-level measurements) are collected.

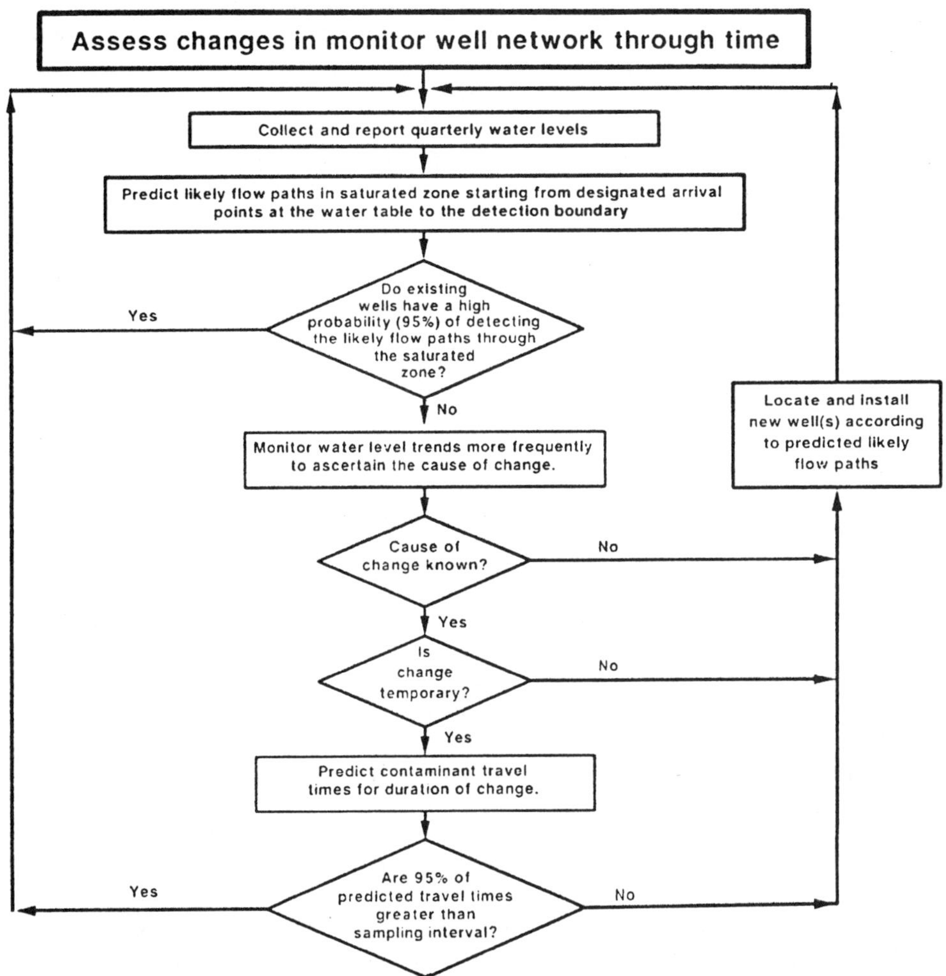

Figure 6. Approach to assessing changes in the monitor-well network performance through time.

In any case, the owner/operator and the regulator must resolve differences of opinion over these issues. Using this proposed strategy, the subjective decisions are made only at the onset of the problem. From the point of agreement on the conceptual model, data adequacy, and performance measures, the numerical model results will indicate whether or not the monitor-well network is adequate.

This proposed strategy is based on the acknowledgement of uncertainty both in our understanding of the ground-water flow system and the parameters that describe the system. In addition, this approach provides a framework for combining the U.S. EPA ground-water monitoring requirements with the requirements for aquifer characterization.

5. SUMMARY

To satisfy the RCRA ground-water monitoring regulations, a waste site must have a ground-water monitoring system consisting of a minimum of one upgradient and three downgradient wells and demonstrate that the downgradient wells have an acceptable probability of detecting a statistically significant amount of contamination at the water table until at least 30 years after site closure.

A strategy has been presented here for assessing compliance with these RCRA ground-water monitoring regulatory requirements at a specific SNL waste site. The proposed strategy provides a mechanism which allows the owner/operator and the regulators to determine the adequacy of an existing monitor-well network. Additional clarification of some U.S. EPA regulatory definitions was required to develop this strategy for application to a waste site at SNL because the U.S. EPA regulations do not account for many of the hydrogeologic complexities associated with the arid Southwest. We are currently in the development stages of implementing this strategy at a site, however the strategy is not ASTM or U.S. EPA approved at this time.

Multiple interactions between the owner/operator and the regulator are required to implement this proposed approach. The strategy attempts to minimize the current subjectivity of the monitor-well network evaluation process by prior agreement to a set approach and by defining the performance measures to evaluate the monitor-well network. In addition, conceptual model development is a key step which should involve collaboration with the regulators.

This strategy is based on the determination of likely ground-water flow paths through both the unsaturated and saturated zones and ground-water travel times. The general approach involves: 1) determining that the monitor-well network is capable of detecting 95% of the likely flow paths in the saturated zone and that the background well is located such that it monitors ambient conditions, 2) assuring that no contaminant travels beyond the capture zones of the monitor wells in between sampling events, and 3) assessing whether the monitor-well network performs adequately through time.

ACKNOWLEDGEMENTS

The authors would like to thank Allan Gutjahr, Tito Bonano, and Roger Cox for their review comments used in developing this strategy. Thanks goes to Steve Conrad and Robert Knowlton for reviewing this paper.

REFERENCES

[1] Bjorklund, L.J. and B.W. Maxwell, "Availability of Ground Water in the Albuquerque Area, Bernalillo and Sandoval Counties, New Mexico," New Mexico State Engineer Report 21, 1961.

[2] Meyer, P.D. and Brill, Jr., E.D., "A Method for Locating Wells in a Groundwater Monitoring Network Under Conditions of Uncertainty," Water Resources Research, Vol. 24, No. 8, 1988, pp. 1277-1282.

[3] Massman, J. and Freeze, R.A., "Groundwater Contamination from Waste Management Sites: The Interaction Between Risk-Based Engineering Design and Regulatory Policy, 1. Methodology," Water Resources Research, Vol. 23, No. 2, 1987, pp. 351-367.

[4] Mualem, Y., "A New Model for Predicting the Hydraulic Conductivity of Unsaturated Porous Media," Water Resources Research, Vol. 12, No. 3, 1976, pp. 513-522.

[5] Polmann, D.J., Vomvoris, E.G., McLaughlin, D., Hammick, E.M., and Gelhar, L.W., "Application of Stochastic Methods to the Simulation of Large-Scale Unsaturated Flow and Transport," NUREG/CR-5094, U.S. Nuclear Regulatory Commission, Washington, DC, 1988.

[6] Huyakorn, P.S., Kool, J.B., and Robertson, J.B., "Documentation and User's Guide: VAM2D - Variably Saturated Analysis Model in Two-Dimensions (Version 5.0 with Hysteresis and Chained Decay Transport)," NUREG/CR-5352, U.S. Nuclear Regulatory Commission, Washington, DC, 1989.

[7] Updegraff, C.D. and Lee, C.E., "DCM3D: A Dual-Continuum, Three-Dimensional, Ground-Water Flow Code for Unsaturated, Fractured, Porous Media," NUREG/CR-5536, U.S. Nuclear Regulatory Commission, Washington, DC, 1990 (in press).

[8] Iman, R.L. and Shortencarier, M.J., "A Fortran 77 Program and User's Guide for the Generation of Latin Hypercube and Random Samples for Use with Computer Models," NUREG/CR-3624, SAND-2365, Sandia National Laboratories, Albuquerque, NM, 1984.

[9] Bras, R., "A Description of the Computer Code GEOINVS," Informal Report, 1986.

[10] McDonald, M.G. and Harbough, A.W., "A Modular Three-Dimensional Finite-Difference Ground-Water Flow Model," Open File Report 83-875, U.S. Geological Survey, 1983.

[11] U.S. Environmental Protection Agency, "RCRA Ground-Water Monitoring Technical Enforcement Guidance Document," Report No. OSWER-9950.1, 1986.

Surface and Borehole Geophysics in Ground-Water Investigations

Richard P. Allen and Mark A. Seelen[1]

THE USE OF GEOPHYSICS IN THE DETECTION OF BURIED TOXIC AGENTS AT A U.S. MILITARY INSTALLATION

REFERENCE: Allen, R. P. and Seelen, M. A., "The Use of Geophysics in the Detection of Buried Toxic Agents at a U. S. Military Installation," Current Practices in Ground Water and Vadose Zone Investigations, ASTM STP 1118, David M. Nielsen and Martin N. Sara, Eds., American Society for Testing and Materials, Philadelphia, 1992.

ABSTRACT: In 1989 personnel at a U.S. Army base in the southeastern U.S. discovered a 1945 U.S. Army Corps of Engineers Water Distribution System Map identifying an area as "Toxic Agents Buried". The date of the handwritten inscription is unknown; however, a study of several aerial photos of this area revealed a disturbed area in approximately the same location dating back to the early 1940's. A geophysical study was conducted in this general area in an attempt to pinpoint this location so that the materials could be exhumed and properly disposed of, and to provide a preliminary indication of the distribution of any potential contaminants in shallow soils and shallow ground water.

KEY WORDS: geophysics, magnetometer, gradiometer, terrain conductivity, ground penetrating radar, chemical agents, landfill

INTRODUCTION

While base personnel were reviewing documents at a U.S. Army base in the southeastern U.S., a U.S. Army Corps of Engineers water distribution map was discovered in the base archives with the cryptic handwritten caption: "Toxic Agents Buried". The map was of a portion of the base which included several of the many test firing ranges for small arms as well as mortars and tanks. The notation was located in one of the

[1]Mr. Allen and Mr. Seelen are employed by ABB Environmental Services, Inc. (formerly the E.C. Jordan Company), P.O. Box 7050, Portland, ME 04112. Mr. Allen is a Senior Geophysicist and Mr. Seelen is a Project Manager.

buffer zones between ranges along with a small outline representing its apparent approximate location.

Experts at the base hypothesized that the entry dated to the late 1940's coinciding with demilitarization following World War II. Although the validity of the handwritten entry was unknown, a geophysical survey was planned to determine the exact burial location of these materials so that if the map entry was valid, and if the materials were present after these many years, they could be exhumed and properly disposed of. Aerial photographs from the 1940's and 1950's were examined and areas where clearings were identified on the aerial photographs were superimposed over current topographic maps of the site and the 1945 water distribution map to facilitate locating the geophysical survey grid to encompass all suspected fill areas in the immediate area (Figure 1).

SURVEY PROCEDURES

The survey grid was laid out with a compass and cloth tape prior to any geophysical measurements. The grid was used for lateral control during the geophysical measurements and would allow future investigators to reoccupy locations of interest identified from the geophysical measurements. The geophysical survey consisted of magnetometer (with vertical gradiometer), terrain conductivity, and ground penetrating radar measurements.

Magnetometer survey

The magnetic method is a versatile, relatively inexpensive, geophysical exploration technique traditionally used for evaluating both shallow and deeper regional geologic structures. Magnetometers are used in environmental studies for locating repositories of buried wastes in steel drums. Locating and quantifying these materials is essential to any remediation effort. Magnetometer surveys can provide an extra measure of safety to those personnel involved in the clean-up activities because the technique can provide indications of the waste location prior to any removal activities.

The magnetometer used for this work is of the "proton precession" type. This device utilizes the precession of spinning protons in a sample of hydrogen-rich fluid (e.g., kerosene, alcohol, or water) to measure the total magnetic field intensity.

Protons spinning in an atomic nucleus behave like tiny magnetic dipoles which can be aligned (polarized) by an external magnetic field. The protons are initially aligned parallel to the earth's field. A second, much stronger magnetic field is produced approximately perpendicular to the earth's field by introducing electric current through a coil of wire. The protons become temporarily aligned with this stronger field. When this stronger field is removed, the protons tend to realign themselves with the earth's field, causing them to

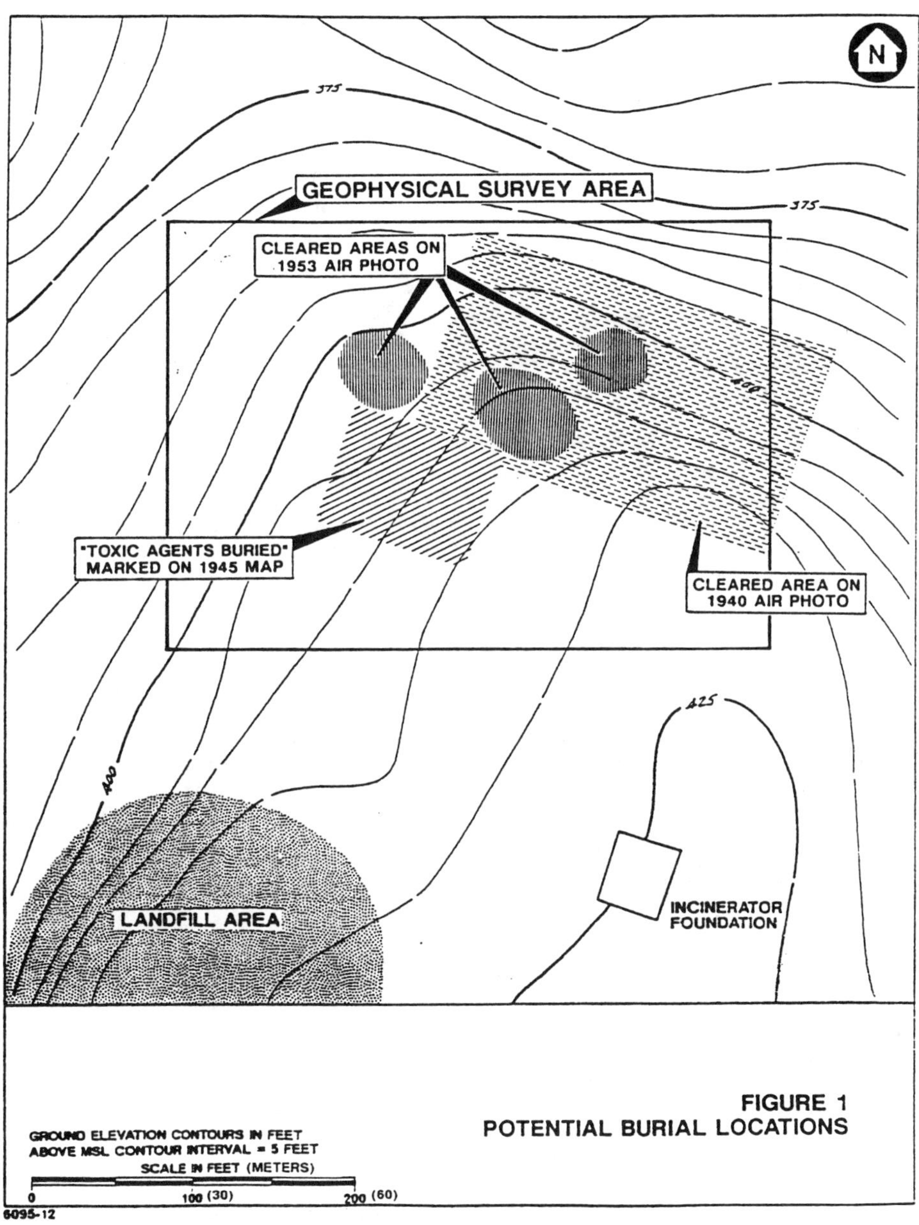

FIGURE 1
POTENTIAL BURIAL LOCATIONS

precess about this direction at a frequency of approximately 2,000 Hz. The precessing protons will generate a small electric signal in the same coil used to polarize them with a frequency proportional to the total magnetic field intensity and independent of the coil orientation. By measuring the signal frequency, one can obtain the absolute value of the total earth's field intensity with a high degree of accuracy (better than 1 part in 50,000).

At hazardous waste sites, a typical reconnaissance grid might start out at perhaps a 6-meter interval, and would be closed down to 1.5 to 3 meters in areas where more detail is desired. Vertical gradient measurements involve the simultaneous acquisition by two sensors of two values of the total field. The sensors are mounted on a staff that is held vertically during a measurement. A known distance (commonly 0.5 or 1 meter) separates the sensors on the staff. Vertical gradient measurements tend to be more sensitive to the presence of near-surface ferrous objects than total field values alone. There are commercially available magnetometers that record field data in memory, and the data can be transferred each day at the completion of field activities onto a personal computer for processing and evaluation.

A proton precession vertical gradiometer was used for this magnetometer survey. The gradiometer had two sensors separated by one-half meter mounted on a 2.4 meter staff which is held vertically during a measurement. Magnetic measurements were made every 3 meters (north-south) along traverses separated by 6 meters (east-west). The gradiometer has a memory which records the field data as a series of data blocks, each of which contains the date and time of each measurement, the "x" and "y" coordinates, the total field value for each of two sensors mounted on the vertically-oriented sensor staff, a comments column to briefly note landmarks and relevant objects observed in the field for each measurement location, and several parameters which provide the user with an indication of data quality. The vertical gradient is simply the difference between the total field value recorded by the upper sensor and that recorded by the lower sensor divided by the distance between them, the result expressed in gammas per meter. A total of 381 magnetic stations were established during this portion of the investigation.

Terrain Conductivity Survey

Terrain conductivity surveys have traditionally been used in mineral exploration for tracing conductive ore bodies (i.e., massive sulfides). They are also used for tracing conductive contaminant plumes in ground water. For example, leachate from municipal landfills tends to be much more conductive than naturally occurring ground water. Accordingly, the shape and relative extent of a plume can be studied with terrain conductivity surveys.

Since the instrument does not require electrical contact with the ground, data acquisition is more rapid than conventional, galvanic, earth resistivity surveys. However, quantification of conductivity data to yield a layered-earth solution is less reliable than with conventional earth resistivity.

The instrumentation consists of a transmitter and receiver. When the transmitter is energized by an alternating current, a primary magnetic field is produced which induces small electric currents to flow in the earth which, in turn, produce a secondary magnetic field. This secondary magnetic field is related to both the transmitter/receiver separation and the operating frequency of the transmitter. The ratio of the secondary field to the primary field is proportional to the terrain conductivity. This ratio is converted into conductivity values in units of millimhos per meter.

The volume (and therefore the depth) of earth materials sampled increases with increasing separation between the transmitter and receiver. The separation can be either fixed or operator-selectable, depending upon the configuration of the specific instrument in use.

Terrain conductivity measurements were made along the same measurement grid used for the magnetometer survey, above. The particular conductivity meter utilized here is sensitive to depths of up to approximately 4.5 meters. A data logger facilitated data transfer and processing and enabled both the quadrature phase and inphase components of the instrument to be recorded simultaneously. The quadrature phase is a measure of the conductivity of the soil and rock materials beneath the transmitter-receiver array of the instrument. The inphase component is more sensitive to metallic objects around the instrument and yields results similar to those from a metal detector.

Ground Penetrating Radar

The GPR technique uses high frequency electromagnetic waves to determine the presence of subsurface objects and structures. Energy is radiated downward into the subsurface from an antenna that is pulled slowly across the ground at speeds varying from about 0.25 to 5 mph, depending on the amount of detail desired and the nature of the target. The electromagnetic wave energy is reflected from surfaces where there is a contrast in the electrical properties of subsurface materials. These surfaces may be naturally occurring geologic horizons (e.g., soil layers, changes in moisture content, voids and fractures in bedrock) or manmade (e.g., buried utilities, tanks, drums). The reflected energy is processed and displayed as a continuous strip chart recording of distance versus time (where time can be thought of as proportional to depth).

The depth of penetration of any GPR system is highly site-specific, and depends (among other factors) on: (1) the soil types at the site (well-sorted granular materials are best), (2) moisture conditions (dry is best), and (3) the frequency of the antenna (the lower the frequency, the deeper the penetration, and the less the resolution capability).

Typical applications for GPR include delineating the boundaries of buried hazardous waste materials and the perimeters of abandoned landfills; finding steel reinforcement bars and voids in concrete structures; recording the depth of geological interfaces and bedrock; and locating and mapping buried utilities.

The GPR system used for this work was adjusted for a penetration depth of approximately 3 meters (70 nanoseconds of two way travel time). This is based on the assumption that the soil materials at the site would have a velocity of approximately 7 nanoseconds per 0.3 meters. Traverses were made on a 1.5-meter grid within the two portions of the study area selected for this technique. The antenna must be in reasonably good contact with the ground surface to obtain good quality data, requiring that the areas to be surveyed be cleared of brambles, bushes, and small trees prior to any survey work. A total of 285 meters of GPR profiling (22 separate traverses) was completed during the investigation.

RESULTS

The geophysical survey was completed over a three-acre area in three field days and resulted in the selection of a primary target area for further investigation. Selection was based on magnetic and conductivity anomalies, and the presence of a GPR anomaly which appeared to be a filled excavation with an unknown reflective target at its center at a depth of approximately one meter.

The results of the magnetometer survey are presented as Total Magnetic Field contours, Figure 2. The contour interval is 10 gammas. The high density of contours in the southern portion of the study area is due to the remaining foundation of an outbuilding and the presence of surface metallic debris including metal cans, miscellaneous trash, and concrete rubble with reinforcing rod. A magnetic anomaly ("A") in the southwestern portion of the study area is believed to be due to the presence of numerous discarded dry cell batteries observed to be partially buried at the surface. A second magnetic anomaly ("B") in the west central portion of the study area occurs coincident with a nearly round surface depression 12 to 15 meters across.

The results of the terrain conductivity survey are presented as conductivity contours in millimhos per meter (Figure 3). Note the high density of contours in the southern portion of the study area as with the magnetic data. An anomaly ("A") in the southwestern portion of the study area is coincident with the magnetic anomaly and is believed to be due to the dry cell batteries. A second anomaly ("B") is coincident with the magnetic anomaly in the west-central portion of the study area. A third anomaly ("C") is located in the east-central portion of the study area. The source of this anomaly is not known.

Anomalies "B" and "C" were surveyed with GPR. The GPR profile data did not reveal any distinct features of interest in the vicinity of anomaly "C". However, several profiles in the vicinity of anomaly "B"

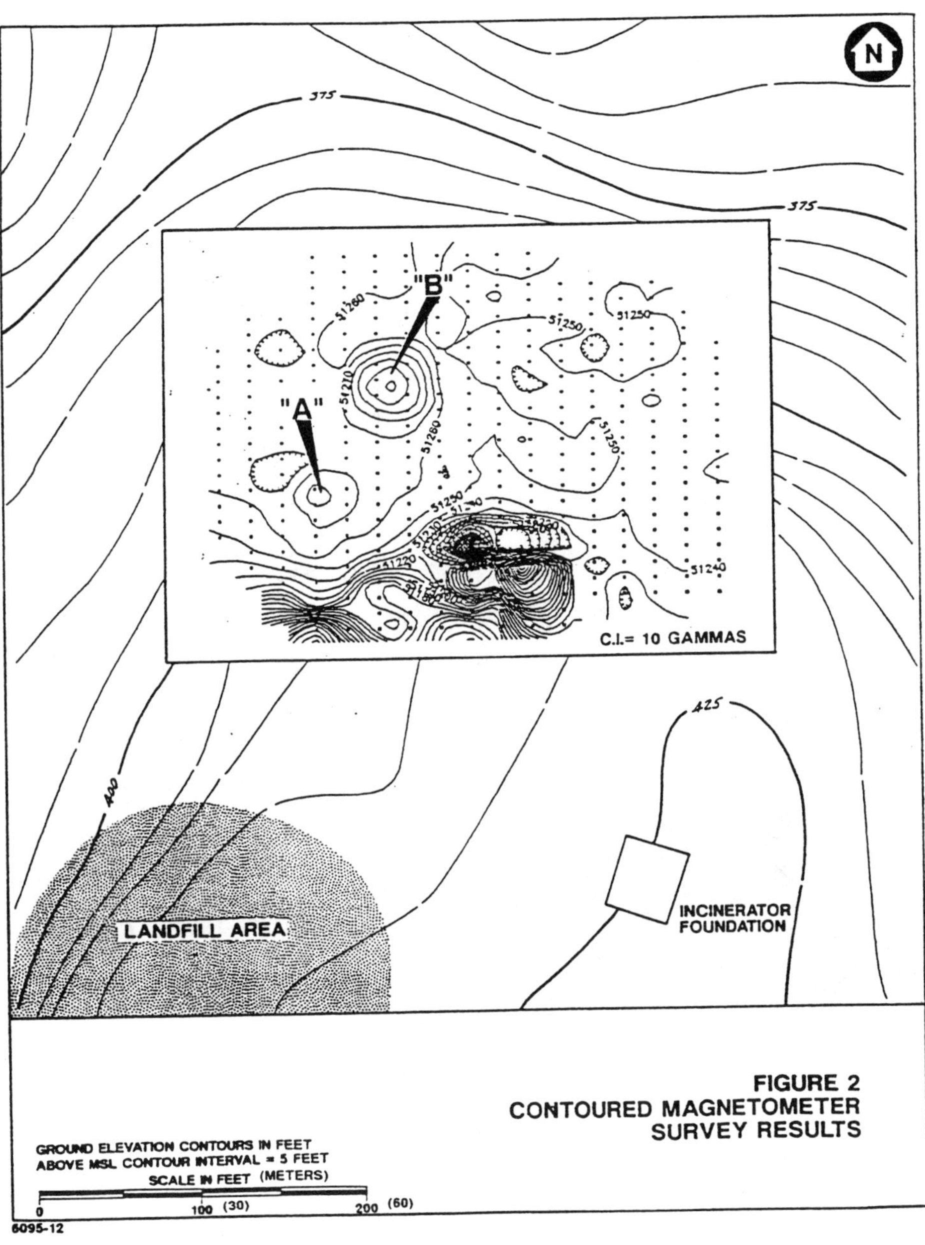

FIGURE 2
CONTOURED MAGNETOMETER
SURVEY RESULTS

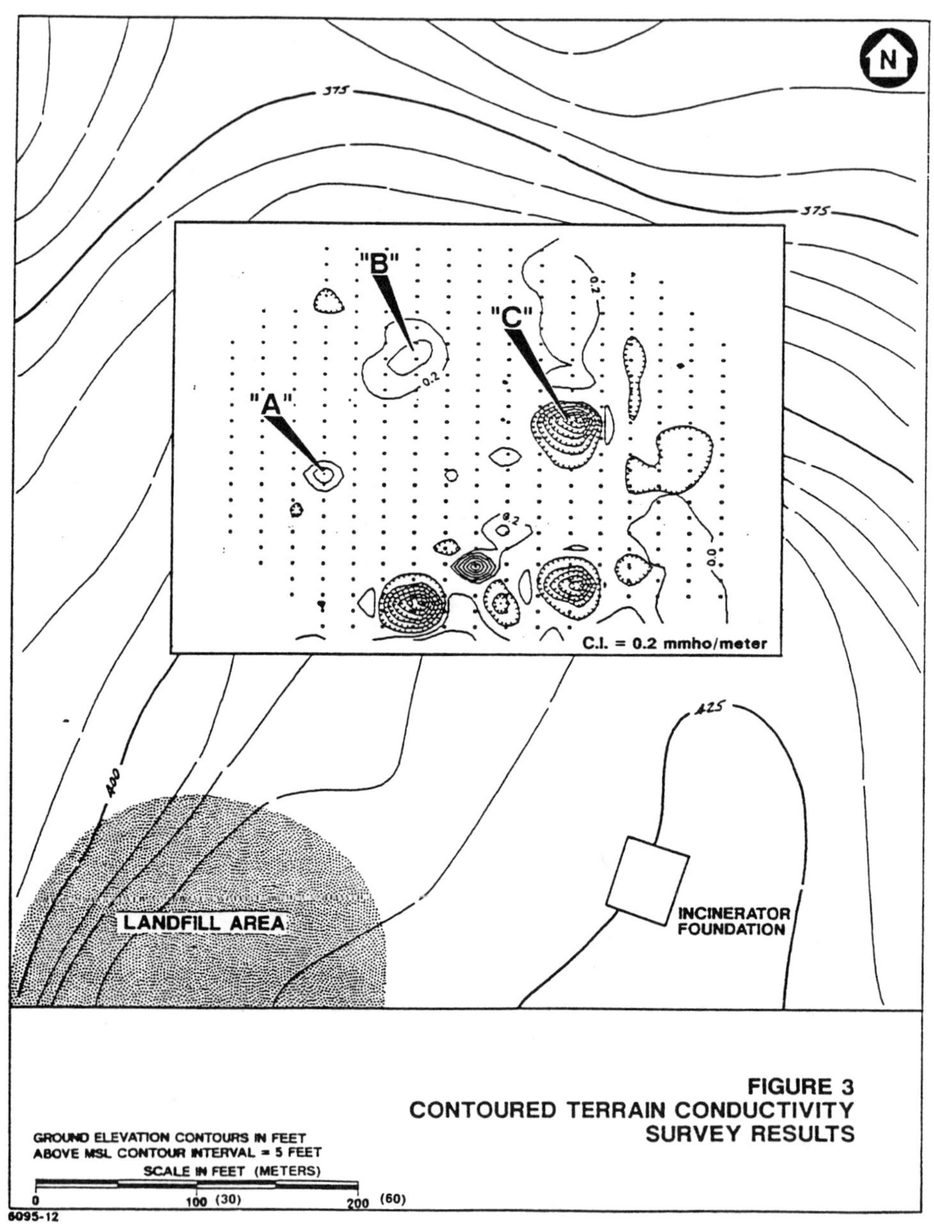

FIGURE 3
CONTOURED TERRAIN CONDUCTIVITY
SURVEY RESULTS

revealed a feature which appeared to be a filled excavation, and at least one profile revealed the presence of a possible metallic object or objects at the base of the excavation (estimated at a depth of 1.2 to 1.5 meters - Figure 4). Because of the coincidence of geophysical anomalies from all three geophysical techniques in Location "B", it was targeted for further evaluation and subsequent excavation.

WASTE REMOVAL

A backhoe-dug test pit at geophysical anomaly "B" resulted in the excavation of an old road sign at a depth of 0.3 to 0.6 meters, a 1.2 meter long iron pipe at 1 meter, and eight badly rusted steel drums at 3.7 meters with incinerated contents believed by U.S. Army personnel to have originally been mustard and decontaminating agents. In addition, approximately 100 sealed glass vials were exhumed along the north wall of the excavation at a depth of 1 meter. The vials turned out to contain dilute solutions of the chemical agents phosgene, lewisite, chloropicrin, adamsite, and mustard, and are thought to have been components of World War II era chemical agent identification kits.

U.S. Army EOD (Explosives and Ordnance disposal) and chemical disposal personnel removed and disposed of the toxic materials. Subsequently, five monitoring wells were installed (four downgradient and one upgradient) to determine the impact, if any, of the previously disposed materials on ground-water quality.

The application of geophysical techniques greatly facilitated locating the "Buried Toxic Agents" area, and was instrumental to the success of this investigation.

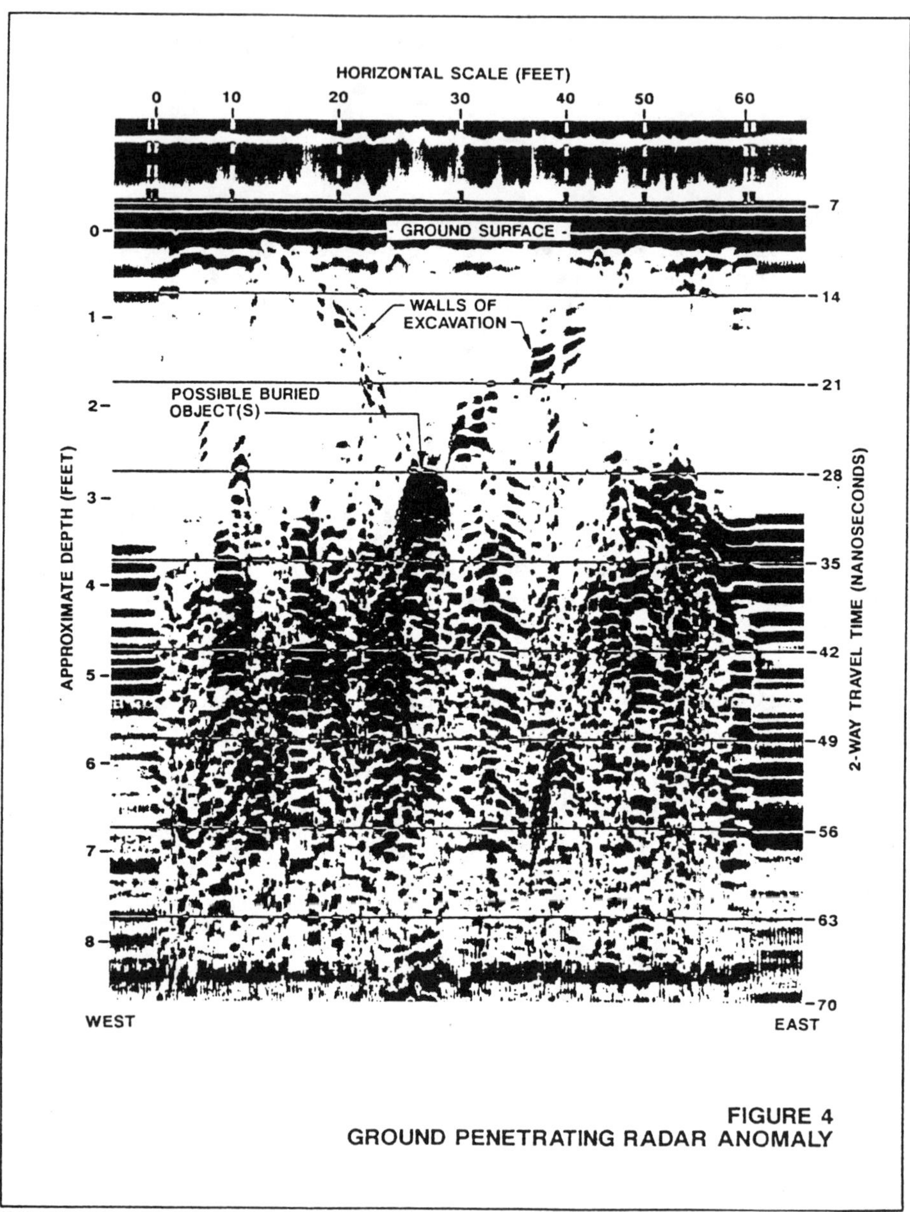

FIGURE 4
GROUND PENETRATING RADAR ANOMALY

6095-12

W. Edward Wightman, Brian C. Martinek, and Dale Hammermeister

GEOPHYSICAL METHODS USED TO GUIDE HYDROGEOLOGICAL INVESTIGATIONS AT AN UMTRA SITE NEAR GRAND JUNCTION, COLORADO

REFERENCE: Wightman, W. E., Martinek, B. C., and Hammermeister, D., "Geophysical Methods Used to Guide Hydrogeological Investigations at an Umtra Site Near Grand Junction, Colorado," Current Practices in Ground Water and Vadose Zone Investigations, ASTM STP 1118, David M. Nielsen and Martin N. Sara, Eds., American Society for Testing and Materials, Philadelphia, 1992.

ABSTRACT: Surface geophysical surveys played a key role in identifying a suitable hydrogeologic location for a Department of Energy Uranium Mill Tailings Remedial Action (UMTRA) disposal cell near Grand Junction, Colorado. Drilling and excavation work conducted as part of a first phase of site characterization activities had determined that a previously identified 35 hectare (90 acre) disposal cell location was underlain by shallow ground water (6 to 12 meters deep) suitable for domestic use. This shallow ground water was found to be confined to a complex system of alluvial-colluvial filled paleochannels eroded into the upper surface of the underlying Mancos shale bedrock. Geophysical methods were then used to cost-effectively direct a second phase of drilling and excavation activities which delineated an adjacent disposal cell location that was free of shallow ground water. Since this new disposal cell location was within the original designated site area of approximately 240 hectare (600 acres), considerable cost and time savings were realized by this surface geophysics directed field characterization program.

KEYWORDS: surface geophysics, electromagnetic conductivity, uranium mill tailings

INTRODUCTION

The Uranium Mill Tailings Remedial Action (UMTRA) program is a federal program administered by the Department of Energy that is designed to relocate uranium mill tailings in several western states to engineered, environmentally stable containment cells. These cells are to be constructed at a number of carefully selected sites in these states. The selected sites are required to meet specific geologic and hydrogeologic criteria that, together with a properly designed containment cell, insure that the environmental impact of the tailings will be minimized. One of these criteria is that hydrogeologic conditions at the site are such that the containment cell can be constructed, and the tailings placed, above the water table.

The subject of this study is the 240 hectare (600 acre) Cheney Reservoir (Cheney) UMTRA site at which a containment cell is to be constructed for uranium mill tailings from the Grand Junction,

Dr. Wightman and Mr. Martinek are Principal Scientists at Geraghty & Miller, Inc., 1099 18th Street, Suite 2100, Denver, CO 80202; Dr. Hammermeister is a Senior Hydrologist at Daniel B. Stephens & Associates, 4415 Hawkins, N.E., Albuquerque, NM 87109.

Colorado area. The Cheney site is located approximately 24 kilometers (km) south of Grand Junction (Figure 1). Design plans for this site called for the construction of a 35 hectare (90 acre) containment cell at the location shown on Figure 1 (original cell location) which would extend to a depth of 5 to 10 meters. Once constructed, the cell would be filled with uranium tailings and subsequently capped. Initial site characterization, which consisted of drilling and trenching at this original cell location, indicated that ground water was not present to depths of 15 meters and hence engineering designs for the cell were prepared. However, during the drilling conducted as part of engineering design, shallow ground water was unexpectedly encountered in some parts of this original cell location at depths of 5 to 10 meters.

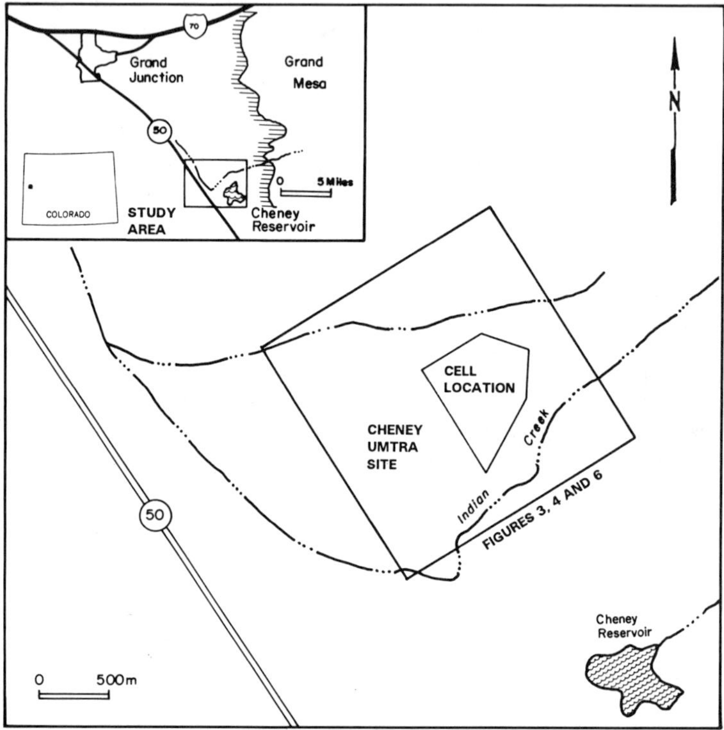

FIGURE 1. SITE LOCATION MAP

The discovery of shallow ground water within the cell area could, depending on its extent, preclude the cell, or the entire Cheney site, from meeting Department of Energy siting or design criteria. The cost-related implications of this discovery were significant and ranged from the requirement for cell design modifications to abandoning the Cheney site altogether. The costs of these options ran from several hundred thousand to several million dollars. In addition, the tight schedule for building the cell, and existing commitments for contractors and equipment, meant that delays would also be costly. It was therefore imperative to quickly, yet confidently, identify the extent of the occurrence of the shallow ground water both in the original cell location, and over the remainder of the Cheney site if the original cell location proved unsuitable. It was also apparent that this could not be done over such a large area quickly enough and in sufficient detail solely by drilling.

Based on these considerations and favorable site conditions, geophysical methods were selected to evaluate the extent of the shallow ground water in the original cell location. Later, the survey was expanded over the remainder of the 240 hectare site in an attempt to identify a more suitable location for the construction of the cell. The geology of the Cheney site, the geophysical method selection, and the results and interpretation of these two surveys are discussed below.

PHYSICAL SETTING AND GEOLOGY

The Cheney site is located at an elevation of about 1,550 meters and is about 10 km west of the Grand Mesa, which is at an elevation of about 3,400 meters. The site generally slopes gently to the west with a uniform gradient. Its northern boundary is marked by a gentle valley and the southern boundary by a sharp valley with 5 to 10 meters of relief. Both of these valleys contain stream beds that are dry for much of the year. Rainfall in the area is less than 25 centimeters (cm) per year and vegetation comprises desert grasses and shrubs.

The geology of the site consists of up to about 15 meters of alluvium and colluvium resting on the Mancos shale. The alluvium and colluvium consists mostly of mixtures of silty gravel with cobbles and boulders derived from the basalt rock that caps Grand Mesa several kilometers east of the site [1]. Silty surface soils, derived from the erosion of the Mancos and Mesa Verde formations, have developed on the gently sloping surface pediment. Underlying the surface soils are a chaotic mixture of particle sizes ranging from graded gravel and sand to clay. Basalt boulders as large as 1 meter in diameter were encountered in test pits and trenches. Locally, there are cemented zones of caliche directly overlying perched, or formerly perched, water-bearing zones. Thick gypsiferous deposits also overlie some of these same shallow perched or formerly perched zones in the middle and upper portions of the overburden.

The Mancos shale bedrock consists of finely bedded shale to depths of approximately 15 to 25 meters [1]. Below these depths, the shaley bedding appears only intermittently, with most of the formation appearing as massive claystone. The degree of weathering varies from severely to slightly weathered in the upper 15 meters of bedrock. Stress relief fracturing and surfical erosion are determining factors in the depth of the weathered zone. In the weathered rock, paleochannels have been incised from 1.5 to 7.5 meters below adjacent bedrock surfaces. Debris flows have filled most of these paleochannels. Fractures in the weathered zone are commonly filled with gypsum, and iron and manganese stains are prominent on partings. In the underlying unweathered bedrock, most fractures are unfilled and unstained, although some joints may have clay or calcite filling.

Ground water is typically encountered at a depth of about 215 meters in the area, in the Dakota Sandstone Formation underlying the Mancos shale. The shallow perched ground water encountered during the engineering design phase drilling was observed at depths of as little as 5 meters. Information on the occurrence of this ground water was, prior to the commencement of geophysical surveys at the site, limited to that obtained from about 35 drillholes and three trenches that had been excavated to the top of the shale. The trenches were about 150 meters long and 10 meters wide. The perched water was observed directly above the Mancos shale in alluvial and colluvial overburden that filled topographic lows on the surface of the shale. Thus, the occurrence of the water appeared to be controlled by the surface configuration of the Mancos shale, probably occurring in channels on top of the shale. Apart from these localized occurrences of water, the overburden and shale were visually dry.

GEOPHYSICAL METHOD SELECTION

Preliminary tests at the site were conducted using seismic techniques along with surface conductivity measurements and in situ measurements of the resistivity of the alluvium and the unsaturated and saturated shale. Based on these results, along with site conditions, electromagnetic conductivity methods using a Geonics EM34 instrument (EM34) were tentatively selected to delineate areas of shallow ground water at the site. Since the electrical conductivity of a material is strongly influenced by the presence of water, the selection of this method was based on the hypothesis that the conductivity of wet shale/alluvium would be higher than dry shale/alluvium and, moreover, that there would be sufficient conductivity contrast between the wet and dry areas to discriminate between them in the conductivity data.

Electromagnetic conductivity instruments, such as the EM34, measure the conductivity of a volume of the earth using electromagnetic waves [3],[4]. The conductivity measured is influenced by the composition and porosity of the soil and the conductivity of the fluids within the soil pores [2] as well as the degree of saturation. The EM34 consists of separate hand-held transmitter and receiver coils and power source. During the operation of the EM34, the transmitter coil is energized by a low frequency alternating current that radiates an electromagnetic field into the earth. This primary field induces eddy currents in the ground below the instrument. The receiver coil on the EM34 detects both the primary field and the secondary magnetic field resulting from the eddy currents. The ratio between the primary field and the quadrature (out-of-phase with the primary field) component of the secondary field is converted to conductivity which is displayed by the EM34. This reading is a bulk measurement of the conductivity beneath the instrument and is made up of the cumulative response to subsurface conditions extending from the surface to the effective depth of penetration of the instrument. More specific details of the instrument response can be found in Technical Note TN-6 [3].

To test the hypothesis that shallow ground water is a cause of an increase in conductivity that can be detected from surface measurements, conductivity readings were taken along a traverse beside one of the trenches excavated during the engineering design phase work. Tests were conducted using both the horizontal and vertical dipole modes and with spacings between the transmitter and receiver coils of 10, 20, and 40 meters (see "Data Collection"). Ground water occurred in this trench within a topographic low on the surface of the shale. The depth to the wet shale, which occurred near the center of the trench, was approximately 12 meters. To either side of this, the shale became shallower and dryer. The water in the trench was about 1 meter deep.

Contrary to what was expected, there was not an increase in conductivity over the wet shale. Instead, there were low conductivity values where the shale was relatively deep regardless of whether it was wet or dry, and higher conductivity values where the shale was shallow. These results indicated that the shale, even where visually dry, was significantly more conductive than the overburden and that the occurrence of the shallow ground water had an insignificant effect on the bulk conductivity. This relationship is shown schematically on Figure 2. The conductivity values shown in Figure 2 are representative of those found at the site. The reason for the unexpected results is due to the relatively high conductivity of the shale, presumed to result from the significant amounts of gypsum which could be observed. Under these conditions, only very small amounts of moisture in the rock pores are needed to produce high conductivity. Very little change in conductivity occurs as the rocks become completely saturated. With additional surface geophysical measurements, it was determined that this

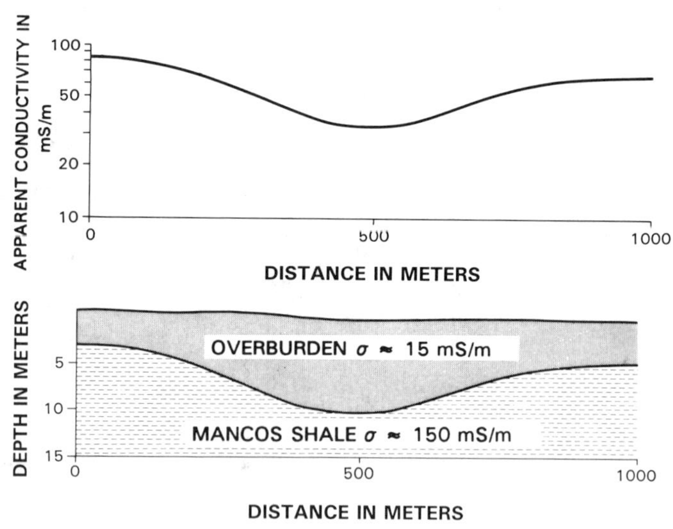

FIGURE 2. HYPOTHETICAL RELATIONSHIP OF CONDUCTIVITY TO DEPTH TO TOP OF SHALE

relationship was consistently observed across the site. Thus, while conductivity readings could not be used to detect the areas of shallow ground water directly, they could be used to map the topography of the shale.

DATA COLLECTION

The primary target for the geophysical survey was the location of channels on the surface of the shale in the region of the original cell location. Once these had been defined, and proved by drilling, the survey was extended over the remainder of the site to determine if a more suitable cell location could be identified.

As part of the final design of the survey, the instrument configuration and line and station spacings were selected. The instrument configuration variables consist of coil orientation and spacing. Conductivity readings are made with the EM34 coils coplanar and oriented either with their planes parallel to the ground surface (vertical dipole mode) or at right angles to it (horizontal dipode mode). In addition, readings can be taken with three different spacings between the transmitter and receiver coils. Available coil spacings are 10, 20, and 40 meters. The frequencies of the transmitted electromagnetic waves at these coil spacings, which are automatically selected by the instrument, are 6.4 kilohertz (Khz), 1.6 Khz, and 0.4 Khz, respectively. Together, the variable orientation of the coils and the transmitter-receiver spacings result in a total of six different reading configurations. These coil configurations control both the depth, and volume, of the earth investigated and the response of the EM34 to lateral variations in conductivity [3], [4].

The optimum coil separation and operating mode for the field survey of the Cheney site were selected based on the results of several test lines. These test lines were run using the EM34 at both 10 and 20 meter coil spacing using both the horizontal and vertical dipole modes in areas with drillhole or trench control. The horizontal dipole mode provided much smoother data, and gave a larger amplitude response over the shale undulations, than the vertical dipole mode. Of the two coil spacings, the 20 meter spacing was chosen because the maximum exploration depth required over the site was uncertain and the greatest depth penetration could be achieved using the 20 meter spacing.

Using the horizontal dipole mode and 20 meter coil spacing, conductivity readings were taken at nearly 5,000 stations across the site along lines spaced about 60 meters apart (Figure 3). In the eastern area of the site, in the vicinity of the original cell location and where rapid changes in the conductivity were observed, the station spacing was 6 meters, while in the western area a 12 meters station spacing was sufficient.

DATA INTERPRETATION

General

A contour map of the conductivity values obtained over the whole site together with the original cell location is shown on Figure 4. A small regional trend of increasing conductivity values to the east has been removed from the contoured data using a first order trend surface. The relationship between conductivity and the depth to the Mancos shale that was observed during the preliminary testing was the basis for interpreting the contoured conductivity data. Areas of high conductivity occur where the shale is shallow and the overburden thin; areas of low conductivity occur where the shale is deep and the overburden thick. The arrows on the map show the major paths taken by surface water across the site, which usually occurs during summer thunderstorms and during spring snow melt. The higher conductivity in these channels results from topographic effects caused by stream valleys cutting into the overburden. Since the Mancos shale is nearer to the surface in these valleys, the measured

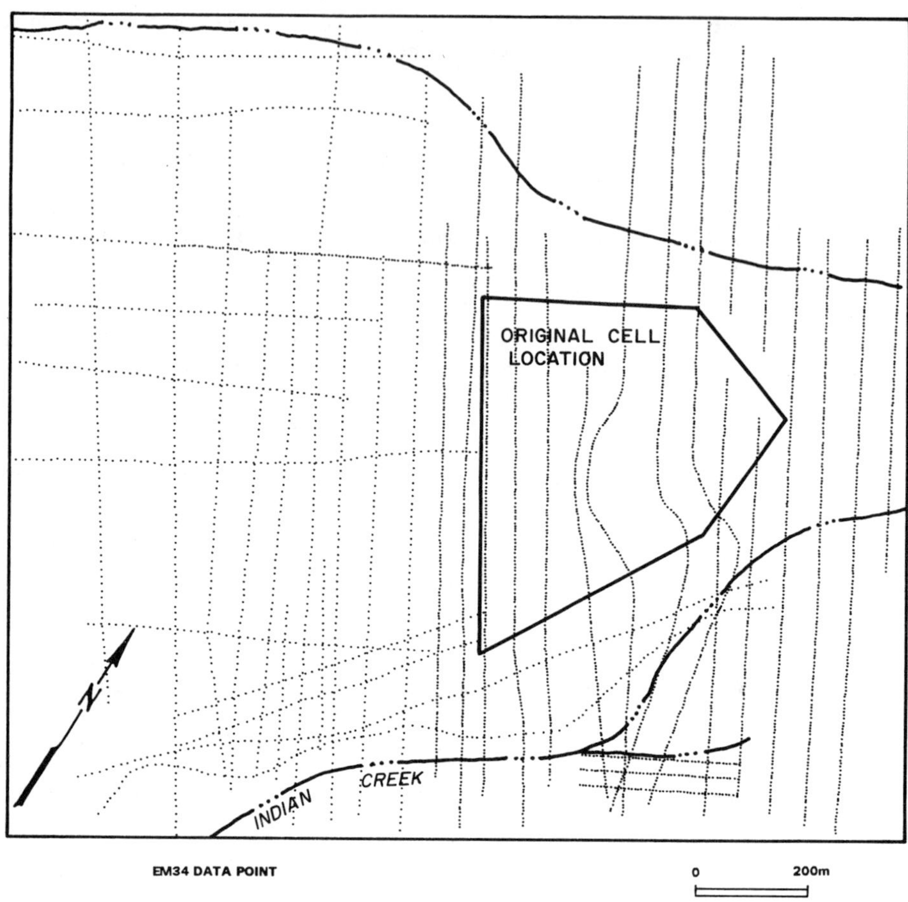

FIGURE 3. GEOPHYSICAL SURVEY LINES AND STATIONS AT CHENEY SITE

conductivity is higher. The valley in the south of the area is the most deeply incised and hence has the highest conductivity values.

Paleochannel in Cell Area

The geophysical results in the cell area (Figure 4) show a conductivity high in the center of the cell and a southwesterly trending, linear, low conductivity feature to the west of this high. This conductivity low extends southwestward, beyond the western margin of the cell, until it is eclipsed by the high conductivity values in the southern stream valley. The pattern of low conductivity values was interpreted to indicate a relatively large depth to the top of the shale reflecting a paleochannel in this area.

This interpretation is consistent with the results of the engineering design phase drilling which indicated that Mancos shale was at depths as great as 14 meters just south of the center of the western edge of the original containment cell. Additional drilling undertaken based on this geophysical interpretation confirmed that this is a topographic depression on the Mancos shale, and that this depression contains ground water to the east of the western boundary of the original cell. A section

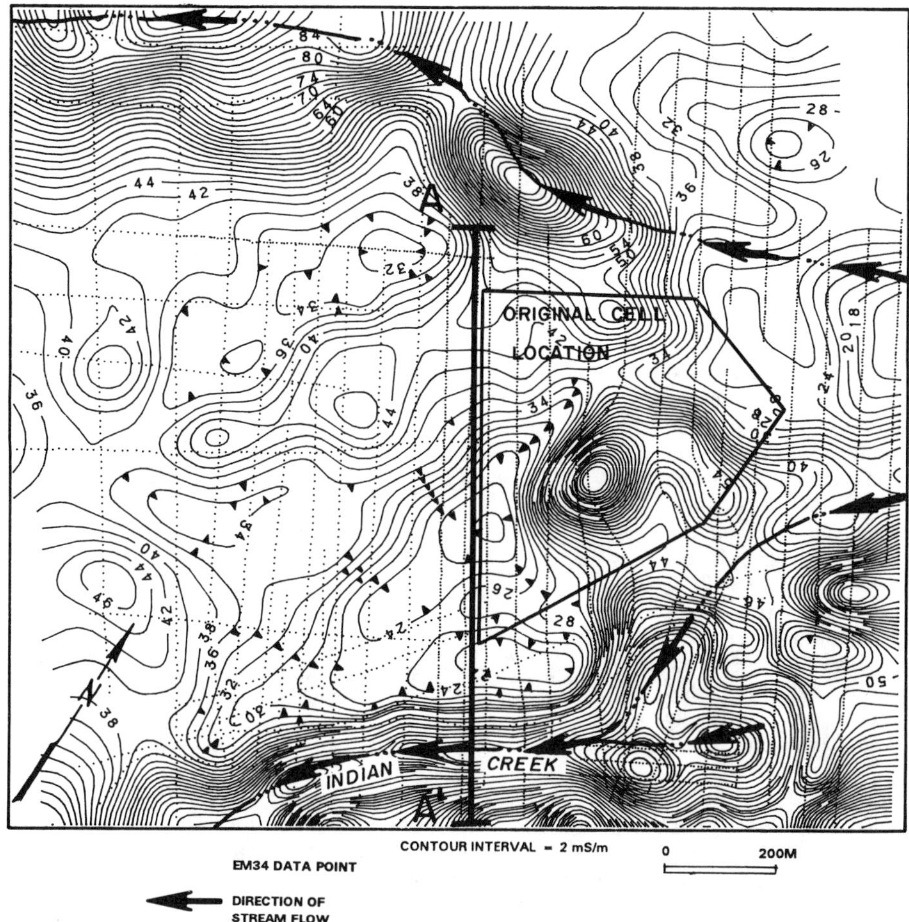

FIGURE 4. CONTOURED EM34 CONDUCTIVITY DATA

through the feature is shown on Figure 5 (the location of the section line, marked A A', is shown on Figure 4 and is approximately topographically flat). The conductivity high in the center of the cell results from the shale being at a depth of only about 3 meters, while in the depression the depth to the shale is as great as 15 meters. It was also confirmed with subsequent drilling and trenching over the entire cell area that this was the main feature controlling the occurrence and flow of ground water. This feature, along with the interpreted ground-water flow, is shown on Figure 6.

Conditions at the Remainder of the Site

As the geophysical survey was expanded to the remainder of the site following the confirmatory drilling described above, it was observed that the variability of the conductivity values to the west of the original cell location (Figure 4) were, on average, much lower than those in the cell area. This is shown on Figure 7 in which histograms of the conductivity values from the two regions called the West Area and East Area are compared. A wide range of conductivity values, with two predominant peaks at about 25 millisiemens per meter (mS/m) and 55 mS/m is shown in the histogram for the East Area, whereas, as can be seen in the plot for the West Area, the conductivity values are predominantly concentrated between 25 and 30 mS/m with a much smaller peak at between 35 and 40 mS/m.

These differences in the conductivity values between the two areas are interpreted to result from two factors. These are the depth of the shale, the variability of which controls the amplitude variations in the EM34 conductivity measurements, and the shale conductivity, which is primarily a function of moisture content. Both of these factors influence the magnitude of the measured conductivity. Overburden conductivity variations were not found to be important since its conductivity is much lower than that of the shale (15 mS/m versus 150 mS/m). The lower average conductivity values in the West Area, along with their smaller amplitude variations, suggest that either the shale is deep or, if shallow, it has a relatively low conductivity and is comparatively dry. Both of these factors are favorable for the location of the containment cell.

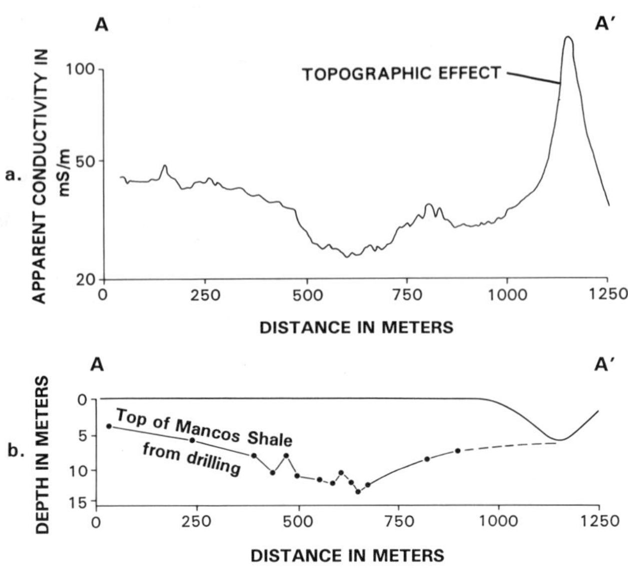

FIGURE 5. MEASURED EM34 CONDUCTIVITY DATA (a) AND DEPTH TO TOP OF MANCOS SHALE (b) ALONG PROFILE A-A' (SEE FIGURE 4 FOR LINE OF SECTION).

In order to test these interpretations, a number of preliminary test pits and boreholes were excavated in the West Area to identify depth and moisture conditions for the shale. These test pits and boreholes indicated that the depth to the shale was relatively shallow and that the shale was comparatively dry. Based on these observations and the geophysical data, it was therefore concluded that the shale was comparatively dry and shallow throughout most of the West Area. The subsequent excavation of additional confirmatory test pits found no shallow ground water or significant paleochannels in the West Area. As a result of this, the location of the containment cell was changed to a new location on the dry shale areas, shown on Figure 6.

CONCLUSIONS

With the use of the geophysical survey, in conjunction with geologic studies and confirmatory drilling, it was possible to cost effectively define the occurrence of shallow ground water at the Cheney site and identify the location of an acceptable area for a tailings containment cell on-site. Hundreds of thousands of dollars to millions of dollars were saved that might have otherwise been spent selecting an entirely new site or drastically modifying the design of the cell if constructed at the original location. These savings far outweighed the cost of the survey, which was approximately $50,000.

Key components in the success of the project are both technical and organizational. Careful pre-survey testing was necessary to develop the correct understanding of how the geophysical information could be used to investigate shallow ground water conditions at the site. Through the close interaction between hydrogeologists and geophysicists involved in the project it was possible for a geological-geophysical hypothesis to be established prior to conducting routine measurements. In addition, project managers were convinced of the importance of allocating time and money for the

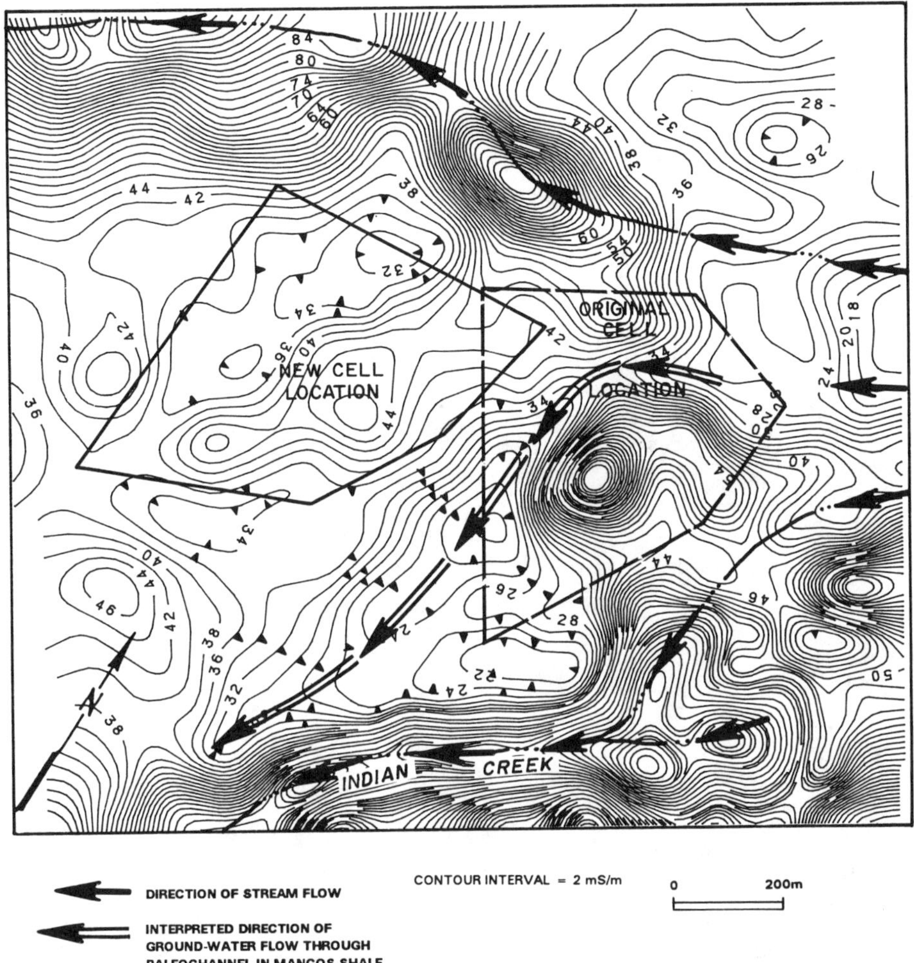

FIGURE 6. CONTOURED CONDUCTIVITY DATA SHOWING INTERPRETED GROUND-WATER FLOW PATH THROUGH ORIGINAL CELL, AND PROPOSED NEW CELL LOCATION

pre-survey testing. By integrating the geophysical survey with confirmatory drilling work, the confidence and accuracy of the geophysical interpretation was enhanced while unnecessary drilling was minimized. The resulting site characterization achieved by the conductivity survey was more comprehensive and detailed than could be achieved by drilling alone, and more accurate than could be obtained solely through the collection of geophysical data.

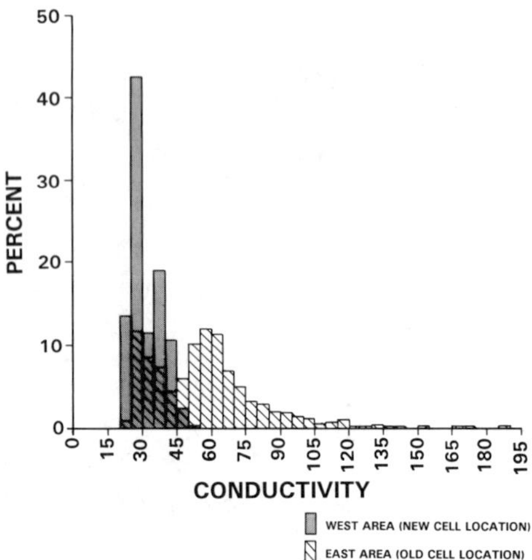

FIGURE 7. HISTOGRAMS OF MEASURED EM34 CONDUCTIVITY VALUES IN WEST AND EAST AREAS OF THE CHENEY SITE

REFERENCES

[1] United States Department of Energy, Remedial Action Plan and Site Design for Stabilization of the Inactive Uranium Mill Tailings Site at Grand Junction, Colorado, Attachment 2: Geology Report, August 1990.

[2] Keller, G. V. and Frank C. Frischknecht, Electrical Methods in Geophysical Prospecting, International Series of Monographs in Electromagnetic Waves, Volume 10, Pergamon Press.

[3] McNeill, J. D., Electromagnetic Terrain Conductivity Measurements at Low Induction Numbers, Geonics Ltd., Technical Note TN-6, 1980, 15 pp.

[4] McNeill, J. D., Electrical Conductivity of Soils and Rocks, Geonics Ltd., Technical Note TN-5, 1980, 22 pp.

Oldrich Mazac, William E. Kelly and Ivan Landa

GEOELECTRICS IN COMPREHENSIVE GROUND WATER CONTAMINATION STUDIES

REFERENCE: Mazac, O., Kelly, W. E., and Landa, I., "Geoelectrics in Comprehensive Ground Water Contamination Studies," Current Practices in Ground Water and Vadose Zone Investigations, ASTM STP 1118, David M. Nielsen and Martin N. Sara, Eds., American Society for Testing and Materials, Philadelphia, 1992.

ABSTRACT: A suite of geoelectric techniques were used to characterize an aquifer for a pollution study. Vertical electrical soundings were used to define aquifer layer resistivities and thicknesses. Porosities, hydraulic conductivities, and transmissivities were then estimated from the resistivities and thicknesses. The mise-a-la-masse method was used to define direction of flow, tracer velocity, and dispersion coefficients. Finally, borehole logging was used to obtain additional information on porosities and velocities.

KEYWORDS: ground water, geophysics, electrical resistivity, pollution

Geoelectrics can be applied to many specific hydrogeologic problems from aquifer characterization to pollution detection.
In general, a single hydrogeological measurement technique is not sufficient, and supplementary techniques are needed. Geoelectric methods may themselves be supplementary methods.

Mazac and others [1,2] have presented the principles of geoelectrics as applied in aquifer characterization studies, and protection and pollution studies respectively. In this paper a comprehensive case history is presented, in which a suite of borehole and surface geoelectric methods were employed to characterize an aquifer as part of a comprehensive pollution study.

Drs. Mazac and Landa are geophysicists at Geofyzika, N.C., Geologicka 2, 152 00 Prague 5, Barrandov, Czechoslovakia; Professor Kelly is chairman of the Civil Engineering Department, University of Nebraska, Lincoln, NE 68588-0531.

BACKGROUND

Vertical electrical soundings typically are a primary geoelectrical measurement technique. With proper appraisal, sounding results can provide a complete picture of subsurface electrical conditions subject to the interpretation limitations. Normally, soundings should be calibrated at boreholes. Maps of layer resistivities, layer thicknesses and depths, or the Dar Zarrouk parameters [3], and geoelectrical cross sections can be prepared for determining or estimating geometric and hydrogeologic characteristics of an aquifer.

If information on ground water and relevant aquifer resistivity is available, the effective porosity can be calculated [4].

Aquifer resistivity can sometimes be related to aquifer hydraulic conductivity [1], and there are several conditions in which relations may be obtained. One case is where the percent of clay varies systematically in an aquifer. In this environment, the resistivity of the matrix can be expected to decrease with clay content and a relationship between permeability and matrix resistivity can be obtained. A complicating factor is that field measurements of matrix resistivity require the measurement of layer resistivities at at least two ground water resistivities. With the aquifer resistivity known, and if aquifer thickness can be determined, transmissivity can be calculated.

Flow direction can be obtained by monitoring the movement of an injected tracer using the mise-a-la-masse method [5,6]. Tracer experiments or movement of contaminants can be interpreted to estimate dispersivities. This can be done using monitoring wells and measuring specific conductances or, under favorable conditions, surface geoelectrics [7] or the mise-a-la-masse method can be substituted for the measurement of specific conductances.

The characteristics of aquitards can also be defined by electrical measurements [3]. The vertical permeability or leakance (ratio of vertical permeability to thickness) can be determined and can be useful in both protection and pollution studies.

SETTING

A geophysical survey was conducted at a locality in southern Moravia (Czechoslovakia) to investigate the causes of pollution in two drainage areas designated areas DA1 and DA2 on Fig. 1. The aquifer is used as a potable water source and more than 40 l/s of water is pumped from the aquifer for drinking water. The aquifer consists of Quaternary sand and gravel terrace deposits along the Satava River overlying an impermeable Neocene clay.

To characterize the pollution and define the probable pollution sources, water quality measurements were made along with surveys using several geoelectrical methods. The first objective of the geolectric surveys was to define the geometry of the aquifer and its hydrogeological characteristics [8].

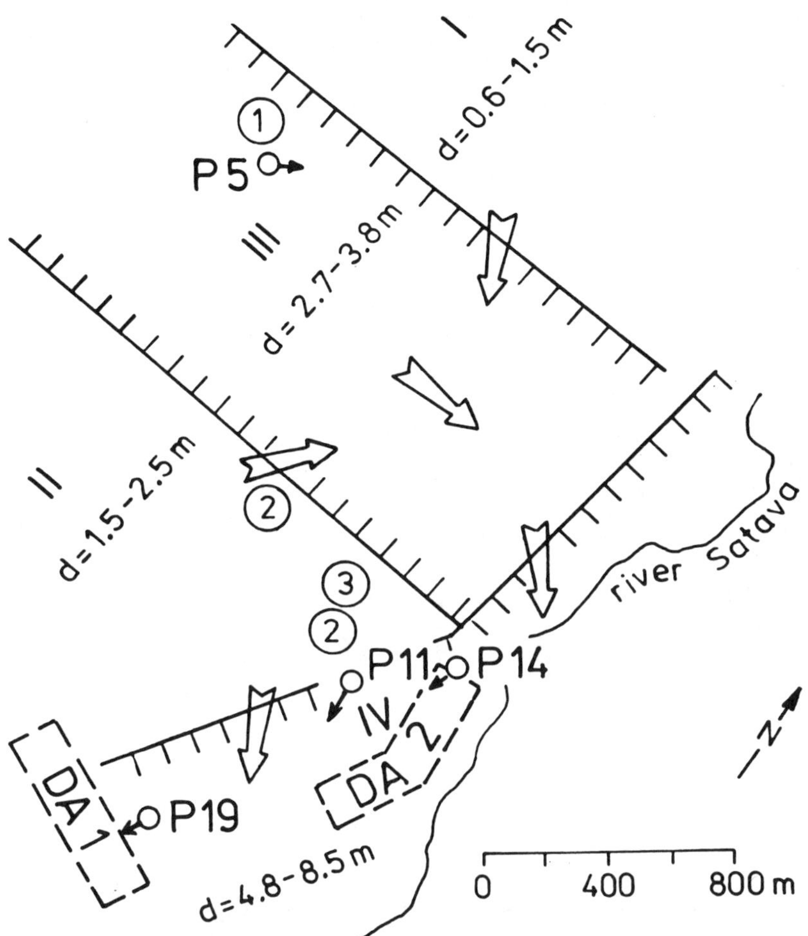

FIG. 1 -- Boundaries of hydrogeological units I, II, III, and IV based on electrically determined thicknesses.

From the vertical electrical sounding results, four areas were distinguished on the basis of the thicknesses and resistivities of the sand and gravel layers (Figs. 1,2). Thicknesses range from about 5 to more than 8 meters along the river (Area IV) to less than 1 meter in the northern part of the study area (Area I). The aquifer resistivity varies from less than 200 ohm-meters to more than 700 ohm-meters.

AQUIFER PROPERTIES

Porosity

The effective porosity P_{eff}(%) was calculated from the estimated resistivities using the relation (Worthington, 1976).

$$P_{eff} = \left[a\frac{\rho_{w,i}}{\rho_{o,i}} - a\frac{\rho_{w,i}}{\rho_s} \right]^{1/n} \qquad (1)$$

where $\rho_{w,i}$ is the groundwater resistivity (contaminated $\rho_{w,c}$ = *3–10 ohm–m*, uncontaminated $\rho_{w,u}$ *12 ohm–m*). The aquifer resistivity determined by VES is $\rho_{o,i}$ (contaminated $\rho_{o,c}$ = *170–550 ohm–m*, uncontaminated $\rho_{o,u}$ = *550 ohm–m*), and the constants *a=1* and *n=2* (Table 1) and ρ_s is the matrix resistivity. Matrix resistivity can be directly related to permeability where permeability is proportional to clay content; matrix resistivities can be calculated if measurements are made at the same point before and after contamination. The matrix resistivity is then

$$\rho_s = \frac{\rho_{o,u}\rho_{w,u}\rho_{o,c} - \rho_{o,u}\rho_{w,c}\rho_{o,c}}{\rho_{w,u}\rho_{o,c} - \rho_{o,u}\rho_{w,c}} \qquad (2)$$

The effective P_{eff} (%) can also be calculated as

$$P_{eff} = 100\,\frac{V_f}{V_r} \qquad (3)$$

where the filtration (V_f) and the actual (V_r) velocities are determined by the dilution and mise-a-la-masse geophysical methods as shown later.

TABLE 1 -- Observed data $\rho_{o,c}$, $\rho_{o,u}$, $\rho_{w,u}$, $\rho_{w,c}$ and interpreted values of ρ_s and P_{eff}.

Borehole no.	$\rho_{o,c}$ (ohm-m)	$\rho_{w,c}$ (ohm-m)	ρ_s 10^{-3} (ohm-m)	P_{eff} (%) n = 2	Remarks ρ (ohm-m)
P 1a	300	7	7.50	15.0	$\rho_{o,u}$ = 500
P 1	350	8	3.50	14.3	...
P 6	200	4	2.00	13.4	$\rho_{w,u}$ = 12
P 5	170	3	1.42	12.7	...

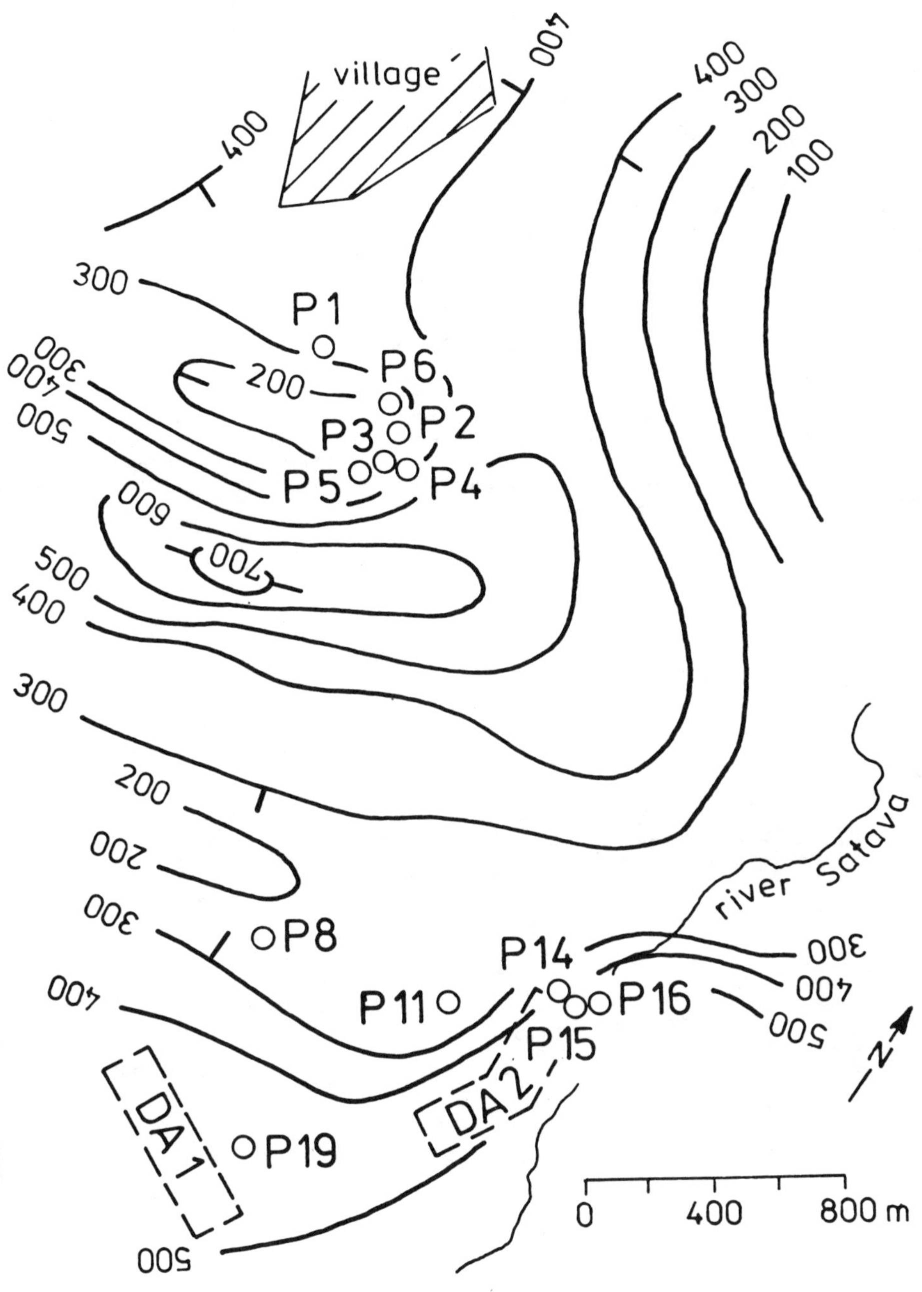

FIG. 2 -- Distribution of aquifer resistivity based on vertical electrical soundings, and location of boreholes.

Flow direction and velocity

Both the direction and the magnitude of the groundwater flow velocity can be determined by a hydrogeological variant of the mise-a-la-masse method. The mise-a-la-masse method was originally used for detecting and determining the extent of conductive ore bodies.

The hydrogeological variant of the method [9,10] involves the continuous injection of a tracer (e.g., NaCl, KCl) which forms a conductive body in an aquifer which increases in size with time in the flow direction mainly due to convective motion, partly to dispersion. The body (plume) is charged by an electrical current and the distribution in time and space of equipotential lines are observed on the earth's surface (Fig. 3).

The actual or tracer flow velocity V_r in the mise-a-la-masse method is usually determined from the velocity of the movement $V_e = \frac{\Delta a}{\Delta t}$ (Fig. 4) of the equipotential lines as defined by the maximum slope of the $a_i(t_i)$ curve where a_i is the time displacement of the equipotentials in the direction of A of the maximum change. The main advantage of the mise-a-la-masse method is that only a single borehole is needed for determining the flow velocity and direction.

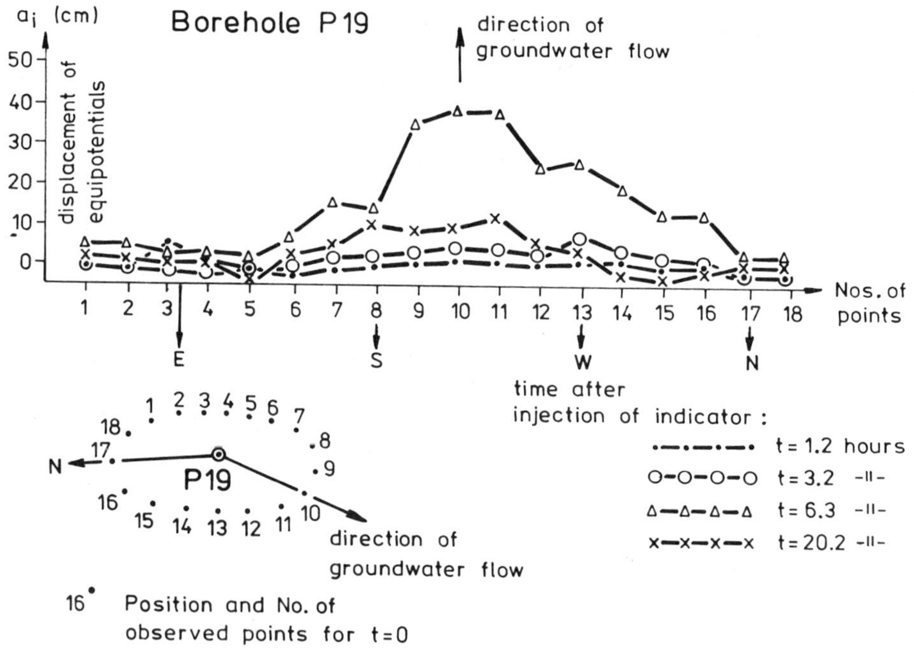

FIG. 3 -- Determination of the flow direction from the displacement of equipotential lines a_i (cm) by the mise-a-la-masse method.

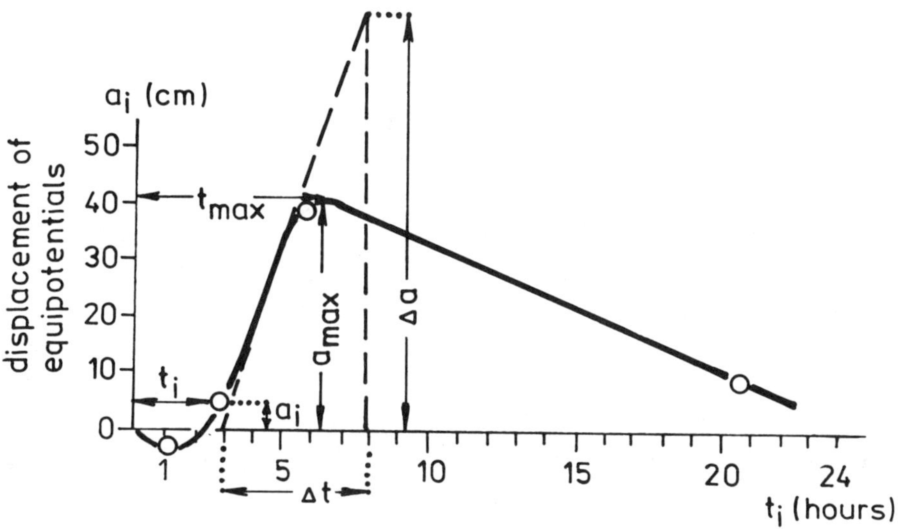

FIG. 4 -- Displacement of equipotential lines a_i (cm) in the mise-a-la-masse method (derived from Fig. 3).

Due to the finite resistivity of the charged body (plume) the velocity V_e does not correspond to the true or tracer velocity V_r. The relation between the two velocities in the direction A is determined by Eq. (4):

$$V_r = C_n V_e \tag{4}$$

where the coefficient C_n $(1 \le C_n \le \infty)$ depends on resistivity contrasts, depth to the water level and on other geometric factors [6]. If the resistivity ρ_0 of an aquifer within the plume approaches zero $(\rho_0 \rightarrow 0)$ the constant $C_n = 1$ and $V_r = V_e$ which is a very rare case. At the surveyed locality $C_n = 1.3$ *to 1.5* (Table 2).

TABLE 2 -- Observed or interpreted values of A, V_e , V_r, D_e, and D_{er}.

Borehole no.	A (degrees)	V_e 10^{-5} (m/s)	V_r 10^{-5} (m/s)	D_e (m^2/day)	D_{er} (m^2/day)
P 5	75	0.95	1.24	0.42	3.55
P 11	200	4.17	5.42	0.75	5.76
P 14	195	0.16	0.24	...	...
P 19	220	4.58	6.87	0.62	5.43

The depth penetration of the method is limited to about ten meters. The accuracy of the direction determination is higher than for the velocity determination. It is desirable to observe the move-ment of at least two different equipotentials at a distance from the borehole greater than 1.5 times the depth to the aquifer with a non-conductive casing and more than 3 times the depth in case of a conductive casing but in both cases less than seven times the depth to the aquifer. Equipotential movement should be observed for a continuous tracer input until their displacement $a_i(t_i)$ reaches a maximum (Figs. 3,4). Tracer concentrations should be in the range of 2 to 5 g/l, since higher concentrations can cause vertical segregation due to density contrast within the injected body and invalidate the geometric and resistivity assumptions. The continuous concentration is mantained constant by monitoring the groundwater resistivity

Dispersion

The electrically determined dispersion coefficient $D_e = D_{er} P_{eff} 10^{-2}$ and the actual dispersion coefficient D_{er} are, together with the parameters V_r and P_{eff}, the most important transport parameters. Traditionally, determination of dispersion coefficients is based on monitoring variations of a tracer concentration $C(x,t)$, at a fixed point x, with time t. The point (e.g, an observation well) should be situated in the direction A of the groundwater flow which can be advantageously determined by the hydrogeologic variant of the mise-a-la-masse method described above (Fig. 3). Concentration variations at the point can be determined either in an observation well by chemical, electrical, photo- and thermometric methods or geophysically using two approaches: surface vertical electrical soundings at a point situated in the direction of groundwater flow or with the mise-a-la-masse method if only one borehole is available. The former method uses an impulse method and the latter a continuous input.

Relations for determining the dispersion coefficient D_e by the mise-a-la-masse method are based on the very rough assumption that the $C(t)$ curve corresponds to the $a_i(t)$ curve (Fig. 4).

This assumption is approximate, but is analogous to the assumption that the tracer concentration C can be determined from changes in aquifer resistivity ρ_0^+ interpreted from surface geoelectrical measurements [6]. Present interpretation techniques do not enable reliable determination of the distribution curve ρ_0^+ (x,t) and consequently, the concentration curve $C(x,t)$.

The relation for calculating D_e (Table 2) is taken from classical hydrogeological methods [7, 11, 12] using $a_i(t)$ for $C(t)$, with the notation as defined in Fig. 4. Here it is assumed that the dispersion is one-dimensional.

Filtration Velocity

The apparent $V_{a,f}$ and actual V_f filtration velocities were determined by the borehole resistivity dilution method (Fig. 5) based on monitoring groundwater resistivity with the depth to the water level.

The principle of the method was developed in the late fifties [13, 14]. Typically, salts (NaCl, KCl), organic dyes or radioactive elements are used as tracers. The apparent filtration velocity $V_{a,f}$ (m/s) can be calculated using equation (5):

$$V_{a,f} = -\frac{\pi r}{2t} \ln \left[\frac{C_t - C_0}{C_1 - C_0}\right] \tag{5}$$

where r is the well diameter in meters, t is the time since injection of the tracer in seconds, C_0, C_1, and C_t are concentrations in gl^{-1} of the tracer in the groundwater, immediately after introduction of the tracer and at time t.

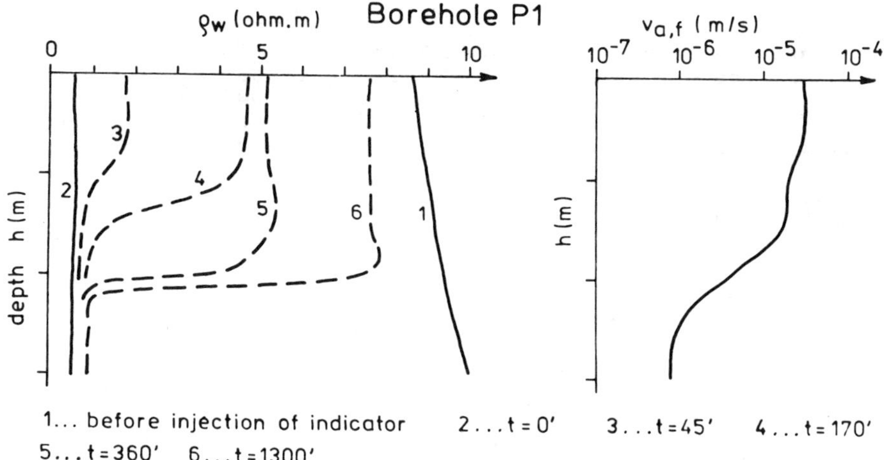

FIG. 5 -- Determination of $V_{a,f}$ by the dilution method.

When resistivity is used to measure dilution of the tracer, NaCl is commonly used as the tracer. The NaCl concentration *(C)* can then be determined from the resistivity *R (ohm-m)* values using the simple relationship

$$C_t = \frac{\alpha_t}{R_t^n}; \quad C_0 = \frac{\alpha_0}{R_0^n}; \quad C_1 = \frac{\alpha_1}{R_1^n}$$

where $\alpha_{t,0,1}$ are constants depending on groundwater temperature, $R_{t,0,1}$ are measured groundwater resistivities and $n = 1.04 \pm 0.02$ is a constant determined experimentally and valid for $0.001\ g/l \le C \le 10\ g/e$ [15].

Substituting the expressions for C_t, C_o, and C_1 and provided that $\alpha_t=\alpha_0=\alpha_1$ the expression for the apparent filtration velocity can be rewritten in the following form:

$$V_{a,f} = 3.62 \left[\frac{r}{t}\right] \log \left[\frac{\left(\frac{R_0}{R_1}\right)^{1.04} - 1}{\left(\frac{R_0}{R_t}\right)^{1.04} - 1}\right] \quad (6)$$

The true filtration velocity, V_f (Table 3) is calculated from Eq. 7

$$V_f = \frac{V_{a,f}}{\beta} \quad (7)$$

where β is the drainage coefficient [15] depending on the well configuration $(0<\beta<8)$. In the study locality $\beta=3$.

TABLE 3 -- Calculation of $V_{a,f}$ and V_f.

Borehole no.	$V_{a,f}$ 10^{-5} (m/s)	V_f 10^{-5} (m/s)
P 1	2.41	0.80
P 2	3.50	1.17
P 3	6.65	2.22
P 4	2.02	0.67

Hydraulic Conductivity

The hydraulic conductivity K_f *(m/s)* (Fig. 6) was calculated from the aquifer resistivity ρ_0 using the following relation based on 8 parametric measurements

$$K_f\,(10^{-5} m/s) = 165150 \rho_0^{-1.78} (ohm\text{-}m) \tag{8}$$

The inverse relationship between K_f and ρ_0 usually occurs in lithologically homogenous sediments and indicates K_f is primarily determined by variations in aquifer porosity [1]. Once we know the thickness b (m) of the aquifer (Fig. 1) and its resistivity ρ_0 (Fig. 2), and hence K_f, the transmissivity $T\ (m^2/s) = bK_f$ can be easily calculated.

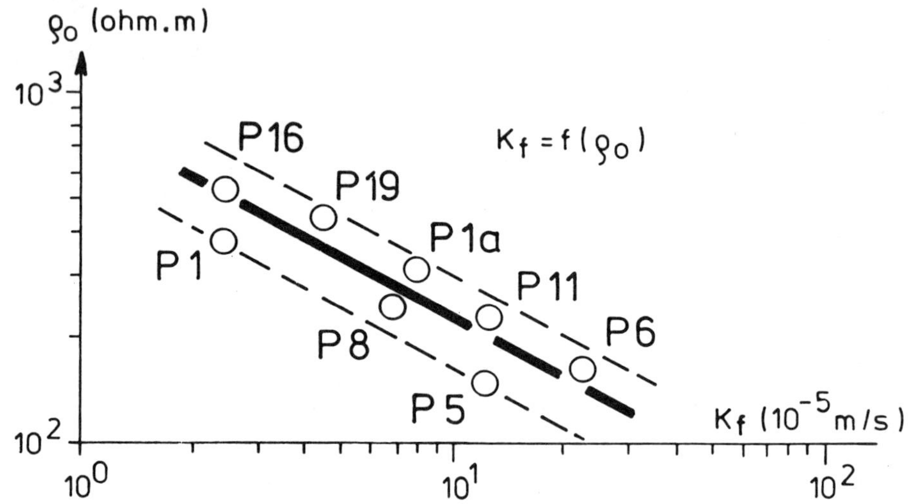

FIG. 6 -- Dependence of $K_f = f(\rho_0)$ based on borehole parametric measurements.

Leakage Coefficient

The leakage coefficient $K_n\ (1/s) = K_f\ /m = f(1/t)$ where K_f is the vertical hydraulic conductivity of an overburden and m is its thickness. It expresses the potential of water to infiltrate through the layers above an aquifer. It is inversely proportional to the protective capacity of the overburden [3] which is considered to be directly proportional to travel time. For a clayey overburden, as in the study area, the protective capacity could be considered as being proportional to the longitudinal unit conductance $S = \sum_{i=1}^{n} \left(\frac{m_i}{p_i} \right)$ where $m_i (m)$ and ρ_i (ohm-m) are the thicknesses and resistivities of the overburden sediments (Table 4).

TABLE 4 -- Calculation of leakage parameters.

Borehole no.	m_1 (m)	ρ_1 (ohm-m)	m_2 (m)	ρ_2 (ohm-m)	$s_i \sim f(t)$ (1/ohm)	$1/s_i \sim f(K_n)$ (ohm)
P 1	0.8	58	0.6	350	0.01551	64.48
P 6	0.5	85	0.85	200	0.01013	98.69
P 8	0.4	120	2.9	254	0.01476	67.73
P 11	0.3	90	2.9	450	0.00977	102.27
P 19	1.0	292	2.4	480	0.00842	118.70

For ground water pollution, this parameter can indicate attenuation of some contaminants as they move through the overburden layers to the water table [3].

CONCLUSIONS

Geoelectric methods can be used in all phases of ground-water contamination investigations. Their information content can be maximized by using them in complementary ways. Supplementary borehole information is required to calibrate the electrical methods. Borehole logging can complement measurements from the surface "mise a la masse" method.

ACKNOWLEDGEMENTS

The authors would like to thank their colleagues from various organizations, who assisted in the field, laboratory and computer work on which this paper is based. Special thanks are due to F. Cahyna, V. Havlikova, P. Matousek, V. Novotna,B. Skuthan for technical assistance in preparing the paper.

REFERENCES

[1] Mazac, O., Kelly, W.E. and Landa, I., "A Hydrogeophysical Model for Relations between Electrical and Hydraulic Properties of Aquifers," Journal of Hydrology, Vol. 79, 1985, pp. 1-19.

[2] Mazac, O., Kelly, W.E., Landa, I., "Surface Geoelectrics for Groundwater Pollution and Protection Studies", Journal of Hydrology, Vol. 93, September 1987, pp. 277-294.

[3] Henriet, I.P., "Direct Applications of the Dar Zarrouk Parameters in Groundwater Surveys," Geophysical Prospecting, Vol. 25, No. 2, 1976, pp. 314-353.

[4] Worthington, P.F., "Hydrogeophysical Equivalence of Water Salinity, Porosity and Matrix Conduction in Arenaceous Aquifers," Ground Water, Vol. 14, No. 4, July-August 1976, pp. 224-232.

[5] Astier, J.L. 1971. Geophysique Appliquee a l' Hydrogeologie, Masson, Paris, 1971.

[6] Mazac, O., Landa, I. and Kolinger, A., "The Feasibility of a Hydrogeological Variant of Mise-a-la-masse Method: An Analysis," Journal of Geological Sciences, Praha, Vol. 14, 1978, pp. 147-177.

[7] Fried, J.J., Groundwater Pollution, Elsevier, Amsterdam, 1975.

[8] Skuthan, B., Mazac, O. and Landa, I., "The Importance of Geophysical Methods for Protecting Groundwaters from Agricultural Pollution," Journal of Geological Sciences, Praha, Vol. 22, 1988, pp. 125-139.

[9] Grinbaum, I.I., Determination of Filtration Properties by Geophysical Methods. Nedra, Moscow, 1965.

[10] Mares, S., Karous, M., Landa, I., Mazac, O., Muller, K. and Mullerova, J., "Geophysical Methods in Hydrogeology and Engineering Geology," SNTL, Praha, 1983 (in Czech).

[11] Bochever, F.M., Bases of Hydrogeological Calculations. Nedra, Moscow, 1969 (in Russian).

[12] Bochever, F.M. and Oradovskaja, A.E., Hydrogeological Bases of Groundwater and Water Basins Protection Against Pollution. Nedra, Moscow, 1972 (in Russian).

[13] Halevy, E., Moser, H. and Zuber, A., "Borehole Dilution Techniques: A Critical Review," in Isotopes in Hydrology, Vienna, I.A.E.A., 1967.

[14] Moser,, H. and Neumaier, F., Die Bestimmung der Sickerwasserstromung unter dem Standamm Rosshaupten mittels radioaktiver Isotope. Wasserwirtschaft, 1958.

[15] Lukes, J., "Filtration Properties of Rocks Determined from Digital Well-logging Measurements," Acta Universitatis Carolinae, Geologia, Praha, Vol. 4, 1973, pp. 289-304.

Vadose Zone Investigation Methods

Jeffrey A. Havlena[1] and Daniel B. Stephens[2]

VADOSE ZONE CHARACTERIZATION USING FIELD PERMEAMETERS AND INSTRUMENTATION

REFERENCE: Havlena, J. A. and Stephens, D. B., "Vadose Zone Characterization Using Field Permeameters and Instrumentation," Current Practices in Ground Water and Vadose Zone Investigations, ASTM STP 1118, David M. Nielsen and Martin N. Sara, Eds., American Society for Testing and Materials, Philadelphia, 1992.

ABSTRACT: In situ characterization of the hydraulic properties of materials within the vadose zone has become increasingly important over the past several years, and is often the only viable means for testing certain geologic and engineered materials. A variety of field methods exist for the hydraulic characterization of unsaturated materials, including ring-type infiltrometers, air and gas permeameters, air-entry permeameters, borehole permeameters, disc permeameters, sorptivity tubes, infiltration ponds, crust tests, and drainage-type instantaneous profile tests. These methods provide information on a number of hydraulic properties, including soil infiltration rate, permeability to air and soil gas, saturated and unsaturated hydraulic conductivity, sorptivity, mean pore size, and matric flux potential. A wide range of materials may be tested in the field, including sandy alluvium, natural clay and engineered clay liners and barriers, fractured and unfractured rock, and soils with macropores. Each of the available tests is appropriate for different geologic materials and testing requirements. The favorable results of a number of these applications demonstrate that most of these methods are reliable, useful, and economical for characterization of unsaturated materials within the vadose zone.

KEYWORDS: vadose zone, permeameters, hydraulic conductivity, in situ testing,

---

[1] Mr. Havlena is a Senior Hydrologist with GRAM, Inc., 8500 Menaul NE, Albuquerque, NM 87112

[2] Dr. Stephens is President of Daniel B. Stephens & Associates, Inc., 6020 Academy, NE, Albuquerque, NM 87109

INTRODUCTION

Over the past several years, in situ characterization of the hydraulic properties of materials within the vadose zone has become increasingly important in site characterization studies. This is due in part to recent advances in the development of new methods and equipment, and to the increased use of analytical and numerical models to predict the fate of contaminants within the vadose zone. Additional incentives for vadose zone characterization arise in response to the RCRA regulations contained in 40 CFR 264 and subsequent Subtitle C amendments which require vadose zone monitoring programs at hazardous waste land treatment facilities and allow exemptions from ground-water monitoring at RCRA sites in certain cases where the characteristics of the vadose zone materials are sufficiently well known and it can be demonstrated that impairment to ground water drinking supplies is unlikely. State and local regulatory agencies also commonly require that vadose zone characterization and monitoring be included as part of the overall monitoring program at many facilities.

An in situ, vadose zone characterization program should be designed to provide the information necessary for the needs of each particular application. Such needs might include: the determination of input parameters for analytical and numerical models to predict seepage rates, travel times, and the source term for ground-water flow and transport models; design criteria for the development of a vadose zone monitoring program; design criteria for the development of vapor extraction systems; design and performance assessment criteria for clay liners and engineered barriers; design criteria for infiltration trenches, galleries, and soil flushing facilities, and; as an aid in the interpretation of soil-gas surveys.

Input parameters for analytical and numerical models that are used to predict aquifer impact, seepage rates, and travel times of contaminants typically include the unsaturated hydraulic conductivity - pressure head (or moisture content) relationship, moisture retention data, and other unsaturated hydraulic characteristics. Currently, these parameters are most often estimated using empirical relationships, or they may be determined in the laboratory using relatively small soil cores. However, frequently a more rigorous determination of these parameters is required, using information obtained from field-scale testing and characterization.

Vadose zone monitoring programs should be designed to utilize the most appropriate monitoring equipment and techniques for the particular materials within the vadose zone and for the conceptual model of unsaturated flow and contaminant seepage. Each type of vadose zone monitoring equipment is suitable for a given range of conditions and frequently can not operate reliably outside of that range. Proper selection and placement of vadose zone monitoring equipment, based on site-specific information and a thorough understanding of the characteristics of the materials within the unsaturated zone, is crucial to the design and operation of the overall system.

Soil gas surveys and vapor extraction systems are highly dependent on the characteristics of the materials within the vadose zone, including moisture content, material heterogeneity, and the permeability of the soil materials to air. Accurate interpretation of soil gas surveys should always consider these factors, as should the design and placement of vapor extraction points.

The vadose zone characterization equipment and methods which can be used to obtain much of the information necessary to achieve the above objectives include ring-type infiltrometers, air and gas permeameters, air-entry permeameters, borehole permeameters, disc permeameters, sorptivity tubes, infiltration ponds, crust tests, and instantaneous-profile tests. These methods can determine permeability at field saturation, field-saturated and unsaturated hydraulic conductivity in three dimensions and the vertical direction, soil sorptivity, infiltration rate, matric flux potential, air-entry pressure, and pore geometry. Most of the equipment and methods require little expense and operational time, are relatively portable and easy to set up, and are capable of producing accurate information at a scale in accord with the size and scope of most vadose zone monitoring systems and investigations. Recent studies have shown that the derived values for hydraulic conductivity obtained from larger-scale soil samples (e.g., in situ testing) have less variability than the values obtained from smaller-scale testing (e.g., laboratory analysis) [1]. This provides an added incentive for using in situ tests in lieu of the more common laboratory analysis of core samples.

One potential drawback to field characterization is the presence of entrapped air within the soil pore space during infiltration and testing. Most field permeameters are influenced by the presence of entrapped air. The values for saturated hydraulic conductivity measured in the field are typically a factor of two (or more) less than actual, fully-saturated hydraulic conductivity values[see ASTM Standard Practice for Comparison of Field Methods for Determining Hydraulic Conductivity in the Vadose Zone, ASTM D 5126, and 2]. For this reason, all field methods determine what is known as "field-saturated" hydraulic conductivity, except where proper precautions (e.g., flooding with CO_2) have been taken to ensure essentially complete saturation during testing. The success of field permeameters is also greatly influenced by the chemistry of the water used during testing. The chemistry of the water used during testing should closely resemble the chemistry of the anticipated infiltrating water (e.g., rainwater or leachate).

Air-Entry Permeameter

The air-entry permeameter (Figure 1) was first proposed by Bouwer [3] as a means to determine the vertical, field-saturated hydraulic conductivity of geologic materials in the field above the water table. Application of the air-entry permeameter has generally been limited to fairly homogeneous sands and silty sands. Recent work [4,5] has extended the use of the air-entry permeameter to lower-permeable clays and engineered clay liners. The air-entry permeameter consists of a single ring, typically 30 cm in diameter and 25 cm deep, which is driven vertically into the material to be tested. A water supply reservoir is

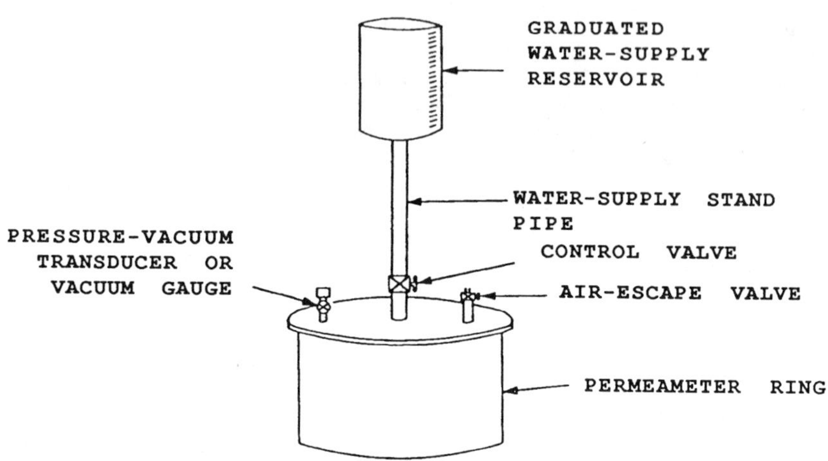

FIG. 1--Air-Entry Permeameter

mounted on top of an adjustable standpipe. The total head applied to the soil surface can be preset according to the characteristics of the materials to be tested (30 to 60 cm is generally sufficient for sands and silts, while 180 cm, or more, may be required for clays).

The operation of the air-entry permeameter requires that water be allowed to infiltrate into the ring, and the rate of decline of head applied at the soil surface is recorded, either through the use of a transducer set within the ring at the soil surface, or by observing the decline of water in the graduated supply reservoir [see ASTM D 5126]. Infiltration is allowed to continue until the wetting front approaches the bottom of the ring, as indicated by a change in soil tension monitored by a tensiometer, the change in electrical conductivity at buried resistance probes, or as inferred by a mass balance approach using an assumed fillable porosity of the soil. The soil in then allowed to drain. The negative pressure created by the draining soil is monitored by a vacuum gauge or pressure transducer mounted at the top of the ring, until a minimum pressure is reached and air begins to enter into the saturated soil. The minimum pressure achieved during this portion of the test is used to calculate the air-entry pressure of the soil.

The determination of vertical, field-saturated hydraulic conductivity is based on the infiltration rate and the assumption that the air-entry pressure approximates the critical pressure, or wetting front pressure head, along the wetting front. The wetting front pressure head has been shown to be the effective capillary drive during infiltration [6], and as such, is a major component of the overall applied hydraulic gradient. The infiltration rate and hydraulic gradient (as given by the total head applied to the soil surface, the depth to the wetting front, and the critical pressure along the wetting front) are used to solve a Green-Ampt type formulation of the infiltration equation.

The air-entry permeameter works best in relatively dry, non-stony soils without an abundance of rootlets or macropores.

Air and Gas Permeameters

Vertical air and gas permeameters--Vertical air and gas permeameters utilize a ring emplaced 5 to 30 cm into the soil similar to an air-entry permeameter installation. Air, or gas (typically nitrogen or carbon dioxide) is allowed to flow at low pressure into the permeameter ring and through the soil profile. The pressure applied to the soil surface is monitored by a pressure gauge or transducer ported into a manifold on the permeameter ring. Falling-pressure and constant-pressure type tests are both possible.

Falling-pressure type tests require that the permeameter and fixed supply reservoir geometry be known, and that the decline of pressure within the permeameter be recorded; constant-pressure tests require that the permeameter geometry, flow rate, and pressure be known. The temperature of the infiltrating air (or gas) and the moisture content of the soil must be known for both type tests (air permeability is highly sensitive to the degree of water saturation). Several different solutions for permeability are applicable [7,8,9,10,11,12] according to the depth of burial of the ring and the type of test (falling or constant pressure).

Although vertical air and gas permeameters may be used in a variety of materials and conditions, these permeameters have the greatest utility in lower-permeability applications where the vertical permeability of the tested material is required. Such applications could include an approximation of the permeability of an engineered cap overlying hazardous tailings which has been emplaced to prevent the upward release of unwanted gasses. It is also possible to estimate the permeability of a material to water using the results of air or gas permeability testing. However, it should be noted that the application of water to soils containing clay minerals may result in swelling, slaking, dispersion of clays, and a resulting decrease in the hydraulic conductivity in most situations. Because of this, it is possible to use the ratio of permeability-to-air to permeability-to-water as an indicator of soil stability and the combined effects of swelling, slaking, and dispersion[12].

The success of vertical air and gas permeability permeameter tests depends largely of the ability to properly emplace the permeameter ring into the soil. The presence of macropores, channels, and soil disturbance can greatly effect the flow rate and the ability to measure soil permeability.

Borehole-type air and gas permeameters--Borehole-type air and gas permeameters provide a more horizontal determination of permeability. Both single-hole and cross-hole type tests may be conducted. The testing equipment for a single-hole type test is set up over a borehole of known geometry, and a pressure or vacuum is applied at the surface over the borehole at a constant head or flow rate. The applied pressure (or vacuum) is monitored by a pressure gauge or transducer, and the flow rate is monitored by a variable area or electronic flowmeter. The temperature, density, and viscosity of the gas and the moisture content of the soil need also be known. The cross-hole type test is similar, except that the pressure (or vacuum) is monitored in a separate borehole at a known radial distance from the source.

The solutions for permeability for borehole-type air and gas permeameter tests most often assume radial, horizontal flow to (or from) the tested borehole. This simplifying assumption allows the use of a one-dimensional, cylindrical solution for compressible gas flow [13]. A more rigorous, multi-dimensional solution may be used where required. However, for most applications where horizontal permeability is greater than vertical permeability, the one-dimensional, radial flow solution is frequently considered appropriate, since most of the airflow is essentially horizontal.

Borehole-type air and gas permeameters may be used to help in the design of vapor extraction systems and as an aid in the interpretation of soil gas data by providing information on the permeability of the tested material to air at the applied vacuum. These permeameters can operate in a wide range of soil types and conditions, including coarse and bouldery alluvium, and clays. Because of the tendency of an applied vacuum to improve the surface seal around the top of the borehole, it is best to use a vacuum-type test in applications where higher pressures or flow rates are required.

Borehole Permeameters

Borehole permeameter tests have long been used to determine field-saturated hydraulic conductivity in the vadose zone [14, ASTM 5126]. Borehole permeameters measure a more three-dimensional hydraulic conductivity than the ring-type permeameters. Both constant head and falling head tests, such as slug tests, are used. Unfortunately, falling head tests are unsupported by a rigorous mathematical foundation at this time. Constant head tests are the most widely used. Constant head-type borehole tests require that water be maintained at a constant level within a borehole by use of a constant head device, and the constant, steady-state flow rate is measured. The testing equipment typically consists of a borehole 5 to 15 cm diameter, a water supply reservoir, and a means of controlling the head of water within the borehole. The equipment used to maintain a constant level of water in the borehole most commonly consists of a float-type stock-tank valve installed within the borehole and connected to a remote water supply reservoir, or the use of a Mariotte device as in the Guelph Permeameter [15] (Figure 2). Borehole permeameters have the advantage over other

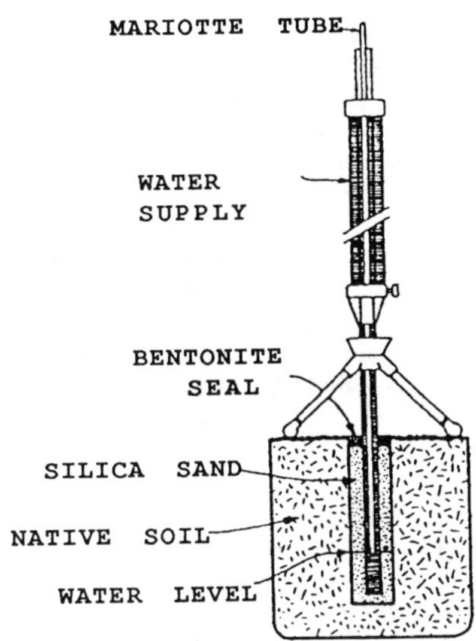

FIG. 2--Guelph Permeameter

test methods in that the equipment can easily be installed at depth. This allows borehole permeameter tests to be used to measure the variability of field-saturated hydraulic conductivity with depth by conducting a series of tests at different depths (or selected horizons in stratified, heterogeneous material) while drilling a borehole.

The solution for field-saturated hydraulic conductivity requires that the steady-state flow rate into the borehole be measured, along with the borehole diameter and depth of ponding. Several different solutions are available, including Glover [16], U.S. Bureau of Reclamation [17], Nasberg [18], Reynolds and Elrick [15], Philip [19], and Stephens, et al. [20] [see ASTM D 5126]. The Glover, U.S. Bureau of Reclamation, and Nasberg solutions do not account for the effects of unsaturated flow away from the borehole. This may lead to erroneously large values for hydraulic conductivity, especially in fine-textured materials where capillary draw is great. The solutions of Reynolds and Elrick, Philip, and Stephens incorporate the effects of capillarity, and are therefore most appropriate for the above conditions and in lithified rock. The Reynolds and Elrick solution used for the Guelph Permeameter requires that two separate tests be conducted within the same borehole; each test at a different height of ponding. This method is appropriate only for unstratified, homogeneous materials. A one-headed solution has recently been developed by Elrick and others [21] which incorporates that an estimate of the ratio of hydraulic to capillary effects be assumed in order to avoid the problems associated with the two-headed test. The Philip solution and the Stephens solution also require that certain capillary parameters be known *a priori*, based on soil texture or measured in the laboratory.

Borehole permeameters may be used for a wide range of applications and materials, including lithified rock. The greatest advantage of the borehole permeameter method over most other methods is their ability to be used at great depths. This is especially useful for investigating the vertical, spatial variability of hydraulic conductivity in areas where the depth to water is great. The test methods which incorporate the effects of capillarity are most exact, especially in fine textured materials. However, the methods which ignore capillarity may be used to approximate field-saturated, hydraulic conductivity in coarser materials, where a less rigorous determination is acceptable. Because of the requirement that steady state be attained, however, borehole permeameter tests may take a considerable time to complete, especially in low-permeability materials.

Disc Permeameter

The disc permeameter has recently been developed [22] as a rapid means to determine field-saturated hydraulic conductivity, sorptivity, macroscopic capillary length, characteristic pore size, and unsaturated hydraulic conductivity at low tensions in a variety of materials. The disc permeameter consists of a circular disc, 20 cm in diameter to which is attached a water supply reservoir and a Mariotte siphon head-control device (Figure 3). Several new varieties of the disc permeameter, called tension infiltrometers, are becoming available at this time. For the determination of field-saturated hydraulic conductivity, the bottom of the disc is set on top of a circular ring a small distance above the soil surface. For determination of the unsaturated parameters, a permeable membrane is attached to the bottom of the disc in contact with the soil. All tests are conducted by measuring the flow rate of water from the ring into the soil surface by observing the decline in water level within the supply reservoir. Disc permeameters induce a three-dimensional flow field. Accordingly, all measurements reflect this three-dimensional orientation.

The solution for field-saturated hydraulic conductivity requires that the steady state flow rate of water into the soil material be measured, along with the geometry of the permeameter. The solutions for the unsaturated parameters also require that the initial and final volumetric moisture content of the tested material be determined. This is usually accomplished by collecting small diameter ring samples of known volume for laboratory analysis after completion of the tests. The unsaturated solutions are based upon the simplifying assumption that the hydraulic conductivity of the soil at the initial soil water potential is negligible in comparison to the hydraulic conductivity at the supply potential. This places some restrictions of the use of the disc permeameter for determination of the various unsaturated parameters in initially moist or wet soils.

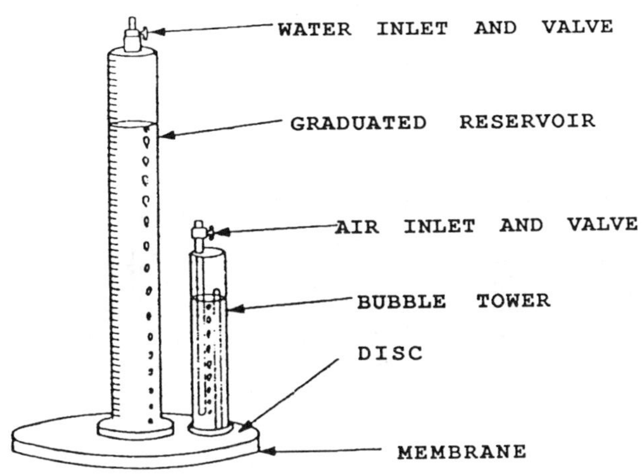

FIG. 3--Disc Permeameter

Disc permeameters are useful in a wide variety of applications, including heterogeneous, gravelly sands, and lithified rock. However, their use is typically restricted to relatively dry soils where the initial soil water tension is below the air-entry pressure. One distinct advantage of disc permeameters is that it can operate in soils containing macropores. Since the hydraulic potential along the bottom of the membrane during unsaturated testing is maintained at a slight tension with respect to atmospheric pressure, water does not flow into macropores. Accordingly, soil matrix properties are measured during these tests, whereas the bulk permeability of the soil-macropore composite is measured during tests for field-saturated hydraulic conductivity conducted at a positive head. Comparison of the results of the two different tests run on the same soil volume allows for an assessment of the contribution of macropores to the overall, bulk saturated hydraulic conductivity.

Sealed Double Ring Infiltrometer

Open-top infiltrometers have long been used to measure the infiltration rate of natural soils [2, ASTM D 5126] . Their range has recently been extended to allow measurement of lower permeable materials and clay liners by the addition of a top cap over the water reservoir to prevent evaporation and disturbance to the water supply [23, and ASTM Standard Test Method for Field Measurement of Infiltration Rate Using a

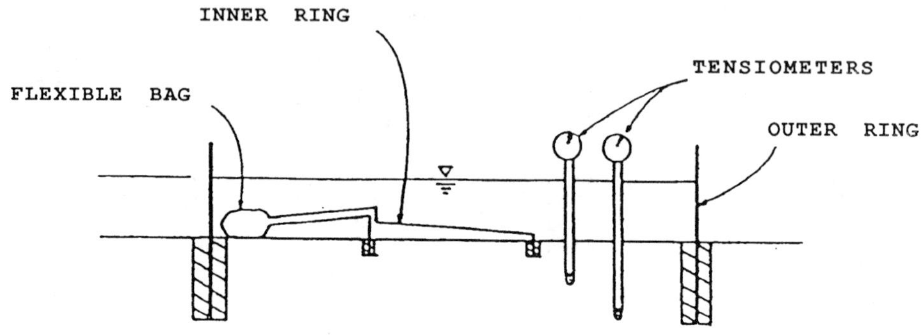

FIG. 4--Sealed Double Ring Infiltrometer

Double Ring Infiltrometer with a Sealed Inner Ring, ASTM Standard D 3385-88]. For this reason, closed-top, or sealed infiltrometers are usually preferred to the open variety. The sealed double ring infiltrometer (SDRI) is currently very popular as a means of approximating field-saturated hydraulic conductivity of low permeable materials. In the strictest sense, SDRI's measure the steady infiltration rate, not field-saturated hydraulic conductivity. However, with some simplifying assumptions, the steady infiltration rate may serve as a conservative approximation of vertical, field-saturated hydraulic conductivity. The SDRI consists of two concentric rings, either circular or approximately square in plan view, set a shallow distance into the material to be tested (Figure 4). The inner ring is commonly two meters across; the outer ring may be one meter larger. The rings are set into a bentonite slurry within two concentric trenches approximately 10 to 20 cm deep each. The bentonite provides a bedding for the rings, and reduces the potential for preferential flow along the sidewalls. Water is allowed to infiltrate between the two rings and within the inner ring. The water infiltrating between the two rings provides a barrier which constrains the water from the inner ring to infiltrate vertically downwards. This assumption of one-dimensional, vertical infiltration is the main advantage of the double ring over the single ring infiltrometer. The inner ring is sealed with an integral cap to limit evaporation during the often long-term tests. The flow rate through the inner ring is periodically measured by weighing the sealed water reservoir used to supply the inner ring. The steady-state flow rate through the inner ring is used to determine the infiltration rate of the material being tested.

Approximation of the vertical, field-saturated hydraulic conductivity using the steady infiltration rate requires several assumptions: 1) the wetting front pressure head is zero, and 2) the flow within the inner ring is essentially one-dimensional downward. Recent modifications have introduced the use of several tensiometers within the soil profile to allow a better estimation of the wetting front pressure head, and consequently a better estimation of hydraulic conductivity. The solution for vertical, field-saturated hydraulic conductivity follows directly from Darcy's relationship.

Because of the typically complicated and lengthy setup and lack of portability of the equipment, use of the SDRI is generally limited to lower permeability materials with field-saturated hydraulic conductivities in the 10^{-9} to 10^{-5} cm/sec range. (For situations where the soil is sufficiently permeable to allow rapid testing, and portability and ease of use are of particular importance, it may be possible to successfully use an open-top infiltrometer.) With proper excavating and rock-cutting equipment, the SDRI may be effectively used in cobbley soils and lithified rock. A further advantage is given by the large volume of soil tested which provides a more representative sampling of actual field conditions than smaller scale tests.

Other In Situ Test Methods

Several other methods of determining field-saturated and unsaturated hydraulic conductivity, sorptivity, and other unsaturated parameters are available. These include the instantaneous profile (or internal drainage) method [24] for determining unsaturated hydraulic conductivity in situ; variants of the sorptivity tube [25,26,27] and other methods for determining sorptivity [27]; the gypsum crust method for determination of unsaturated hydraulic conductivity [28,29]; BAT probe permeameters for measuring field-saturated hydraulic conductivity [30]; and variants of the borehole permeameter method [2] and ring-type infiltrometers [2,31,32, ASTM D 5126]. Most of these testing methods which are used to determine field-saturated hydraulic conductivity may be easily modified to provide actual values for saturated hydraulic conductivity instead of field-saturated hydraulic conductivity by pre-flooding the test material with carbon dioxide gas [33].

CASE STUDIES

The use of several of the above methods by the authors and others over the past few years has allowed a direct comparison of the applications and advantages of each particular test. Fortunately, during several of the tests it was possible to use two or more types of permeameters in the same material to allow more specific comparisons to be made. The following discussion considers the use of a sampling of these permeameters in different soil conditions.

Case 1: Moist Sand

A testing program was recently completed at the site of a chemical spill to determine the field-saturated and unsaturated hydraulic properties of the soil beneath the spill source. This information was to be used for reconstructive and predictive modeling of the history and fate of contaminated water infiltrating from the spill source. The unsaturated zone at the site consisted of poorly-stratified, damp to very moist, fine-to-medium sand which had been subjected to a several-cm rainfall several days before testing began. The equipment used at the site included air-entry permeameters, a Guelph permeameter, and a disc permeameter. Two tests of each permeameter were conducted at each of several test locations established within the property in order to assess the spatial variability of the measured parameters and repeatability of the test results. Owing to the relatively high permeability of the sand, each individual test was completed within one hour or less, and the entire field testing program was finished within two days by the two-member field team. Furthermore, because of the small size and portability of the test equipment, it was possible to ship the permeameters through the mail, allowing the field team to travel to the site by commercial airlines. Table 1 shows the results of the tests for the different permeameters and parameters measured.

TABLE 1--Moist Sand

Test	Hydraulic Conductivity (cm/sec)	Sorptivity (cm/sec$^{1/2}$)
2-1 Guelph (two-head)	-9.13×10^{-4}	1.43
2-1 Guelph (Glover solution)	6.84×10^{-3}	-
2-1 Disc (negative head)	5.70×10^{-4}	4.3×10^{-2}
2-1 Disc (positive head)	1.93×10^{-3}	-
2-1 Air-Entry Permeameter	4.70×10^{-3}	-
2-2 Guelph (two-head)	-4.07×10^{-3}	1.74
2-2 Guelph (Glover solution)	8.03×10^{-3}	-
2-2 Disc (negative head)	2.73×10^{-4}	3.8×10^{-3}
2-2 Disc (positive head)	2.41×10^{-3}	-
2-2 Air-Entry Permeameter	7.60×10^{-3}	-

In general, all of the field-saturated tests at this location are in reasonable agreement with each other and with the expected values for the field-saturated hydraulic conductivity of the sand. The results for field-saturated hydraulic conductivity given by the Guelph two-head approach and by the disc permeameter at negative head do not appear to be valid, most likely because of: (1) the very high initial moisture content of the soil, which violates the simplifying assumptions in the disc permeameter solution, leading to possible erroneous values for hydraulic conductivity, and (2) the possible presence of inhomogeneities within the sand which may have impaired the ability of the Guelph permeameter to determine conductivity by the two-head method. Use of the one-headed method [21] may have avoided this problem. The values for sorptivity determined by the Guelph permeameter are far larger than expected for such soil conditions, and are probably in error. Sorptivity is a measure of the capillary component during infiltration, and are generally less than unity. The apparent differences could be due to the presence of inhomogeneities in the soil. The values for sorptivity given by the disc permeameter appear to be correct for the soil and high initial moisture conditions. It should be noted that the determination of the depth of the wetting front at the conclusion of the air-entry permeameter tests was hampered by the initially high moisture content of the sand. Under these circumstances it is advisable to collect several soil samples at different depths for laboratory determination of the moisture content profile, or to determine the presence of the wetting by using other means, such as tensiometers. The approximate depth to the wetting front can then be determined visually from a plot of moisture content versus depth, or by looking at the matric potential profile given by the tensiometers.

Case 2: Sand, Silty Sand, and Structured Clayey Sand

Several disc permeameter and air-entry permeameter tests were conducted at a UST site in conjunction with a proposed in situ bioremediation program. The results were used in the design of an infiltration gallery system to distribute nutriated water through the vadose zone. Most of the soil at the test sites was silty-sand with considerable organic matter and much surface vegetation. The soil appeared to be weakly structured at the surface, becoming more pronounced at depth. Table 2 shows the results of the testing.

TABLE 2--Sand and structured fill

Test	Material	Hydraulic Conductivity (cm/sec)
AEP-1	gravelly sand	1.5×10^{-3}
AEP-2	structured clay	6.6×10^{-3}
AEP-3	structured clay	5.2×10^{-3}
AEP-4	silty sand	1.7×10^{-3}
Disc-1	igneous rock	1.2×10^{-4}
Disc-2	structured clay	1.4×10^{-3}
Disc-3	silty sand	5.0×10^{-4}

The results appear to be in reasonable agreement and indicate that the hydraulic conductivity of the structured clayey fill may have been controlled by preferential pathways within the fill. Several of the results warrant further discussion:

- Comparison of the results of AEP-4 with Disc-3, conducted in the same weakly-structured silty-sand fill, shows that the disc permeameter operating at a slightly negative head produced a lower value for hydraulic conductivity than the air-entry permeameter did. This is probably due to the presence of macropores within the fill. The results of the air-entry permeameter, operating at a positive head may have been influenced by the presence of macropores. The resulting value for hydraulic conductivity would in effect be an equivalent conductivity for the soil matrix-macropore system. The disc permeameter, operating at a slightly negative head, may not have been influenced by the presence of macropores, and the resulting value for hydraulic conductivity would reflect the soil matrix properties alone.

- Disc-1 was conducted in slightly-weathered, igneous bedrock. The test was conducted in the bottom of a trench where the bedrock was exposed. The value for hydraulic conductivity given by this test (1.2×10^{-4} cm/sec) compares favorably with that from laboratory analysis of a core sample collected from a less weathered horizon of the bedrock (about 5×10^{-5} cm/sec). The disc permeameter was effective in determining the hydraulic conductivity of igneous rock in situ.

Case 3: Engineered Clay Liner Test Fill

Air-Entry permeameter and Guelph permeameter tests were conducted on a clay liner test fill in order to assess their operation in very low-permeable materials, and to compare their results with previous tests conducted at the same test fill with SDRI, BAT probe, and laboratory tests [34]. Because of the very low permeability of the clay fill, special precautions were taken during testing:

- All permeameters were tethered to prevent movement in the wind. A wind break was constructed alongside the permeameters.

- All permeameters were insulated and shielded from the sun.

- All water reservoirs were covered to prevent evaporation.

- The water supply/ head-control reservoirs on the air-entry permeameters were set at about 180 cm in order to apply greater-than-normal hydraulic head to expedite completion of the tests. This required that the permeameter rings be weighted to prevent the hydrostatic head from lifting the permeameters out of the ground.

The air-entry permeameter and Guelph permeameter tests took just under three weeks to complete. The results of the air-entry permeameter tests conducted during this phase of the testing are compared to the results of previous tests in Table 3.

TABLE 3--Test fill

Test Method	Saturated Hydraulic Conductivity (cm/sec)
SDRI	2.4×10^{-8}
BAT Probe:	
Test 1	8.8×10^{-9}
Test 2	7.4×10^{-9}
Air-Entry Permeameter	2.6×10^{-8}
Laboratory Analysis of Shelby Tube Samples:	
Test 1	7.3×10^{-9}
Test 2	2.7×10^{-9}

Although the results of the different test methods are in reasonable agreement, several items should be noted:

- Results of the laboratory testing and the BAT probe tests are nearly one order of magnitude lower than the in situ tests. This may be the result of the different scales of the tests [5]. Both the laboratory core samples and the BAT probes test a very small volume of soil compared to the other tests.

- The SDRI test took more than 40 days to complete; the air-entry permeameter tests took slightly less than three weeks.

- All of these numbers compare favorably with previously published values [35].

CONCLUSIONS

The results of several different programs using field permeameters and instrumentation indicate that these methods can provide an accurate and efficient means of measuring field-saturated and unsaturated hydraulic parameters under a variety of field conditions. The different case studies show that there is reasonable overall agreement between the results obtained from different test methods at one location. The degree of agreement between each of the methods is especially evident if consideration is given to the natural spatial variability of hydraulic properties and the presence of anisotropy, macropores, and structured soils at some test locations. Although it is may be possible to successfully operate one of several different permeameters at a single location, each of the permeameters and test methods is best suited for a given set of soil conditions and testing requirements. The selection and use of the appropriate permeameters for the conditions at a site requires an understanding of the site conditions and of the eventual use of the parameters being measured.

REFERENCES

[1] Sisson, J.B., and P.J. Wierenga, 1981, "Spatial Variability of Steady-State Infiltration Rates as a Stochastic Process," Soil Science Society of America Journal, 45:699-704.

[2] Daniel, D.E., 1989, "In Situ Hydraulic Conductivity Tests for Compacted Clay," Journal of Geotechnical Engineering, ASCE 115(9):1205-1226.

[3] Bouwer, H., 1966, "Rapid Field Measurement of Air-Entry Value and Hydraulic Conductivity of Soils as Significant Parameters in Flow Systems Analysis," Water Resources Research, 2(4):729-738.

[4] Knight, R.B., and J.P. Haile, 1984, "Construction of the Key Lake Tailings Facility," Proceedings, International Conference on Case Histories in Geotechnical Engineering, St. Louis, MO.

[5] Stephens, D.B., Unruh, M., Havlena, J., Knowlton, R.G., Mattson, E., and W. Cox, 1988, "Vadose Zone Characterization of Low-Permeability Sediments Using Field Permeameters," Ground Water Monitoring Review, 8(2):59-66.

[6] Morel-Seytoux, H.J., and J. Khanji, 1974, "Derivation of an Equation of Infiltration," Water Resources Research, 10:795-800.

[7] Corey, A.T., 1959, "Measurement of Water and Air Permeability in Unsaturated Soil," Soil Science Society of America Proceedings, 19:7-10.

[8] Evans, D.D., and D. Kirkham, 1949, "Measurement of Air Permeability of Soil In Situ," Soil Science Society of America Proceedings, 14:65.

[9] Kirkham, D., 1946, "Field Methods for Determination of Air Permeability of Soil in it's Undisturbed State," Soil Science Society of America Proceedings, 11:93-99.

[10] Grover, B.L., 1955, "Simplified Air Permeameters for Soil in Place," Soil Science Society of America Proceedings, 19:414-418.

[11] Groenewoud, H.V., 1968, "Methods and Apparatus for Measuring Air Permeability of the Soil," Soil Science, 106(4).

[12] Reeve, R.C., 1953, "A Method for Determining the Stability of Soil Structure Based Upon Air and Water Permeability Measurements," Soil Science Society of America Proceedings, 17:324-329.

[13] Keller, C., Engstrom, D., and F. West, "In Situ Permeability

Measurements for the Event in U1-C," Consultant's Report.

[14] Stephens, D.B., and S.P. Neuman, 1982, "Vadose Zone Permeability Tests: Summary," ASCE Journal of Hydraulic Engineering, 108(HY5):623-639.

[15] Reynolds, W.D., and D.E. Elrick, 1985, "In Situ Measurement of Field-Saturated Hydraulic Conductivity, Sorptivity, and the a-Parameter Using the Guelph Permeameter," Soil Science, 140(4):292-302.

[16] Glover, R.E., 1953, "Flow for a Test Hole Located Above Groundwater Level," in Theory and Problems of Water Percolation, by C.N. Zangar, U.S. Department of Interior, Bureau of Reclamation, Engineering Monograph No. 8:69-71.

[17] U.S. Bureau of Reclamation, 1978, Drainage Manual, U.S. Government Printing Office, Washington, D.C. pp 74-97.

[18] Nasberg, V.M., 1951, "The Problem of Flow in an Unsaturated Soil for Injection Under Pressure," Izvestja Akademia Nauk, SSSR Odt Tekh Nauk, No. 9.

[19] Philip, J.R., 1985, "Approximate Analysis of the Borehole Permeameter in Unsaturated Soil," Water Resources Research, 21(7):1025-1033.

[20] Stephens, D.B., Lambert, K., and D. Watson, 1987, "Regression Models for Hydraulic Conductivity and Field Test of the Borehole Permeameter," Water Resources Research, 23(12):2207-2214.

[21] Elrick, D.E., Reynolds, W.D., and K.A. Tan, 1988, "Hydraulic Conductivity Measurements in the Unsaturated Zone Using an Improved Single Head Well Permeameter Technique," Ground Water Monitoring Review, Fall, 1988.

[22] Perroux, K.M., and I. White, 1988, "Designs for Disc Permeameters," Soil Science Society of America Journal, 52:1205-1215.

[23] Daniel, D.E., and S.J. Trautwein, 1986, "Field Permeability Test for Earthen Liners," Proceedings, In Situ '86, Blacksburg, VA, 146-160.

[24] Watson, K.K., 1966, "An Instantaneous Profile Method for Determining the Hydraulic Conductivity of Unsaturated Porous Materials," Water Resources Research, 2:709-715.

[25] Dirksen, C., 1975, "Determination of Soil Water Diffusivity by Sorptivity Measurements," Soil Science Society of America Proceedings, 39:22-27.

[26] Clothier, B.E., and I. White, 1981, "Measurement of Sorptivity and Soil Water Diffusivity in the Field," Soil Science Society of

America Journal, 45:241-245.

[27] Chong, S.K., and R.E. Green, 1983, "Sorptivity Measurement and it's Application," p.82-91. in Advances in Infiltration, Proceedings, National Conference on Advances in Infiltration, Chicago, IL, 12-13 December, ASAE, St. Joseph, MI.

[28] Bouma, J., Hillel, D.I., Hale, F.D., and C.R. Amerman, 1971, "Field Measurement of Hydraulic Conductivity by Infiltration Through Artificial Crusts," Soil Science Society of America Proceedings, 33.

[29] Bouma, J., and J.C. Denning, 1972, "Field Measurement of Unsaturated Hydraulic Conductivity by Infiltration Through Gypsum Crusts," Soil Science Society of America Proceedings, 36:846-847.

[30] Torstensson, B.A., 1984, "A New System for Ground Water Monitoring," Ground Water Monitoring Review, 4(4):131-138.

[31] Bouwer, H., 1961, "A Double Tube Method for Measuring Hydraulic Conductivity of Soil in Sites Above a Water Table," Soil Science Society of America Proceedings, 25:334-342.

[32] Daniel, D.E., 1984, "Predicting the Hydraulic Conductivity of Compacted Clay Liners," Journal of Geotechnical Engineering, ASCE, 110(2):285-300.

[33] Stephens, D.B., Lambert, K., and D. Watson, 1983, "Influence of Entrapped Air on Field Determinations of Hydraulic Properties in the Vadose Zone," Paper Presented at Conference on Characterization and Monitoring in the Vadose Zone, National Water Well Association, Las Vegas, Nevada, December 8-10.

[34] Chen, H.W., and L.O. Yamamoto, 1987, "Supplementary Hydraulic Conductivity Testing Clay Test Fill," IT Corporation Internal Report, Project No. XEAD01.

[35] Chen, H.W., and L.O. Yamamoto, 1987, "Permeability Tests for Hazardous Waste Management Unit Clay Liners," Geotechnical and Geohydrological Aspects of Waste Management, D.J.A. van Zyl et al., eds., Lewis Publishing, Inc., Chelsea, MI, 229-243.

Robert G. Knowlton, Jr., Alva M. Parsons and
Kathy N. Gaither

TECHNIQUES FOR QUANTIFYING THE RECHARGE RATE THROUGH UNSATURATED SOILS

REFERENCE: Knowlton, R. G., Jr., Parsons, A. M., and Gaithers, K. N., "Techniques for Quantifying the Recharge Rate Through Unsaturated Soils," Current Practices in Ground Water and Vadose Zone Investigations, ASTM STP 1118, David M. Nielsen and Martin N. Sara, Eds., American Society for Testing and Materials, Philadelphia, 1992.

ABSTRACT: Sandia National Laboratories recently designed a sampling and analysis strategy to implement several techniques for quantifying the recharge rate through thick unsaturated, unconsolidated alluvium. Several methods are employed to estimate the recharge rate, including a soil-physics based (i.e., Darcy's law) method, a chloride mass balance technique, a stable-isotope method, bomb-pulse tritium method, and ^{36}Cl techniques. Each of these techniques yields an independent estimate of the recharge rate which may be used to define background percolation rates of conservative solutes, to approximate the upper boundary flux condition for numerical flow and transport modeling of the site, or as calibration criteria in flow and transport modeling.

The soil-physics based approach is an indirect means of estimating the recharge rate through the application of Darcy's law with measured hydraulic characteristic data. The chloride mass balance method assumes piston displacement to estimate the recharge rate utilizing cumulative chloride versus water content data. The stable-isotope method is a tracer technique used to estimate the transport of the isotopic species deuterium (D or ^2H) and ^{18}O in the water molecule ($H_2{}^{18}O$ and $HD^{16}O$), which in turn allows for the determination of the recharge rate. The tritium and ^{36}Cl versus depth profiles can be used to estimate soil-water movement rates and soil-water age.

KEYWORDS: recharge rate, Darcy's law, chloride, stable isotopes, tracers

Dr. Knowlton and A. Parsons are hydrologists at Sandia National Laboratories, Environmental Impact and Restoration Division, P.O. Box 5800, Albuquerque, NM 87185; K. Gaither is a geologist with Applied Research Technical Services, 4300 San Mateo Blvd., Albuquerque, NM 87110.

INTRODUCTION

Engineers, researchers, and planners have historically been interested in quantifying the recharge rate to an aquifer for water resources planning. The rate and areal extent of water infiltrating below the root zone, coupled with the amount of localized recharge (e.g., mountain front recharge and/or stream infiltration) and underflow, have been used to estimate the amount of reusable ground-water resources in a hydrogeologic basin. If the ground water was pumped at a greater rate than the aquifer was being recharged, then the ground water would be mined. Under this scenario, the aquifer might eventually become dewatered and the resource lost. This is the case in such aquifers as the Ogallala in west Texas.

The technique that was typically employed in water resources planning to estimate the rate and amount of recharge to an aquifer was the water balance approach. In this method, all inputs and outputs to the aquifer are estimated (e.g., precipitation, evapotranspiration, surface runoff, pumping, discharge, and interflow), and the recharge rate is determined as the differential in the water balance equation. The main drawback to this approach is that some of the parameters, such as evapotranspiration, are very difficult to measure or estimate, and therefore uncertain. Probably the largest uncertainty exists in estimating evapotranspiration. This uncertainty in the input parameters then translates into a large uncertainty in the recharge estimate. For this reason, independent methods for estimating the recharge rate were developed.

More recently, engineers and researchers have become interested in estimating the recharge rate, or soil-water movement rate, to assess the potential for contaminant migration through the unsaturated zone. This paper refers to soil-water moving through the vadose zone, below the influence of evapotranspiration, as the recharge rate. Other terms used in this context include seepage velocity and travel time. The recharge rate is a specific flux, the seepage velocity would be estimated by dividing the specific flux by the effective moisture content. The travel time through the vadose zone is estimated by dividing the travel distance by the seepage velocity. Therefore, the key factor in estimating seepage velocities and/or travel times is the recharge rate. The regulatory framework that exists today dictates that these estimates of the recharge rate must be made. The Environmental Protection Agency's (EPA) Resource Conservation and Recovery Act (RCRA) has two requirements for this data. The first requires definition of the rate and direction of ground-water flow. If we choose to interpret ground water as any water existing beneath the land surface, not just the saturated sediments within the aquifer, then this would require the quantification of the recharge rate. More important though, is the second requirement, a risk assessment analysis that is required in the RCRA Facility Investigation/ Corrective Measures Study (RFI/CMS) process prior to selection of a remedial alternative for a treatment, storage, or disposal facility. A similar risk assessment requirement exists in the EPA's Comprehensive Environmental Response, Compensation, and Liability Act (CERCLA) (i.e., Superfund) regulations in the Remedial Investigation/Feasibility Study (RI/FS) process.

In performing a risk assessment, all potential contaminant pathways must be evaluated for their impact on human health and the environment. One of the most obvious pathways for contaminants to impact human health and the environment is migration from a treatment, storage, or disposal facility to the underlying aquifer and subsequent withdrawal by water supply wells. The recharge rate is therefore an important parameter in assessing background percolation rates of solutes, in approximating the upper boundary flux condition for numerical flow and transport modeling of the site, or as calibration criteria in flow and transport modeling.

Low-level and high-level waste repository regulations require a performance assessment of a potential site prior to licensing. Under the Nuclear Regulatory Commission (NRC) regulations for low-level radioactive waste repositories, the potential rate of contaminant movement from the facility needs to be assessed in order to do a performance assessment. The estimation of the recharge rate is therefore in order. Under the NRC and EPA regulations for siting a high-level waste repository, the travel time from the facility to the accessible environment (i.e., a 5 km radius from the site) must be determined. Since the time for contamination to reach the water table is frequently the major component of the overall travel time to the accessible environment, the independent techniques discussed in this paper for estimating the recharge rate should be employed.

Pursuant to the RCRA regulations, Sandia National Laboratories (SNL) recently designed a sampling and analysis strategy to implement several techniques for quantifying the recharge rate through thick unsaturated, unconsolidated alluvium in an arid region of New Mexico. This strategy employs several methods to estimate the recharge rate, including a soil-physics based (i.e., Darcy's law) method, a chloride mass balance technique, a stable-isotope method, bomb-pulse tritium method, and ^{36}Cl techniques. Each of these techniques yields an independent estimate of the recharge rate. By utilizing multiple techniques to quantify the recharge rate (and hopefully realizing agreement between the results) it should add confidence, and possibly reduce the uncertainties, in the results.

RELATED RESEARCH

There have been a number of authors [1,2] that believe there is no direct areal recharge of ground water by infiltration of precipitation in arid regions. This may be true in some areas, but several researchers have shown that in some arid (and even hyperarid) environments recharge to underlying aquifers does occur [3,4,5].

Recent investigations in New Mexico [6,7,8,9,10,11] have shown that direct areal recharge does occur in arid portions of this state, and have shed light on the nature of the recharge process. Stephens and Knowlton [7] utilized a soil physics based approach to quantify the recharge rate and soil-water movement processes at a site in central New Mexico over a two year period of monitoring. Phillips and others [9] sampled subsurface profiles of bomb-pulse chlorine-36 (^{36}Cl) and tritium (^{3}H) at two arid sites in New Mexico, including the site used by Stephens and Knowlton [7]. Bomb-pulse ^{36}Cl and tritium fallout were incorporated in atmospheric moisture and precipitation as a result of atmospheric nuclear testing in the late 1950's and early 1960's. Therefore the bomb-pulse ^{36}Cl and tritium distributions in the vadose zone are the result of soil-water movement over the past 30 years. Phillips and others [9] also utilized the chloride mass balance technique to define the recharge rate. In this technique the rate of chloride deposition to the land surface is considered constant, and the accumulation of chloride with depth in the soil profile is therefore directly related to the recharge rate. Knowlton [11] developed a method to determine the liquid and vapor flux of water in the unsaturated zone using profiles of the stable isotopes oxygen-18 (^{18}O) and deuterium (^{2}H). Knowlton [11] employed the technique at the same site that Stephens and Knowlton [7] used to quantify the recharge rate and help define soil-water movement processes. The recharge estimates from each of these studies (i.e., the soil-physics based work, the ^{36}Cl and tritium profiles, the chloride mass balance results, and the stable-isotope technique) are all in very good agreement with each other (see the subsequent section on Case Study Results).

METHODS

Sandia National Laboratories is instituting a strategy to utilize the soil-physics based approach and the geochemical/isotopic tracer techniques mentioned above in determining soil-water movement rates and recharge throughout its complex. Two related tasks have been formulated. The first task is sampling and analysis of core samples from the coring activities associated with installing deep monitoring wells. The second task involves sampling and analysis of multiple shallow coreholes specifically for determining the spatial and temporal variability of the recharge processes. To date, only the first task has begun. Samples have been collected from the drilling/coring operations related to the installation of one deep monitoring well at SNL.

Core samples were collected using continuous wire-line coring techniques during the drilling of a monitoring well at the Sandia Chemical Waste Landfill site. Both hollow-stem augering (to a depth of about 80 meters) and mud-rotary drilling techniques (to a depth of about 153 meters) were employed. Special care was taken in handling the core to minimize potential disturbance of the samples. Samples intended for hydraulic characterization were obtained in polycarbonate sleeves from discrete intervals during the continuous wire-line coring operation for direct "undisturbed" analyses in the laboratory. Samples collected for geochemical and isotopic analyses were destructively taken from the split-tube core barrel and immediately stored in sealed Mason (i.e., canning) jars. Exposure to air and sunlight is minimized to decrease the potential for evaporation of the soil water, and hence cause fractionation. The Mason jars are filled with soil to nearly full to minimize volatilization within the jar. This rapid transfer of the soil sample into nearly full Mason jars has proven sufficient in obtaining representative geochemical/isotopic analytical results [11].

The "undisturbed" core samples collected for hydraulic characterization are being analyzed for initial moisture content, the soil-moisture characteristic curve (pressure head versus water content relationship), bulk density, porosity, saturated hydraulic conductivity, and grain-size distribution. Samples collected for the geochemical/isotopic analyses are to be performed in the following order on each sample. First, the soil water is extracted from the samples with a vacuum distillation technique [11]. Great care is taken during the distillation process to assure that the soil water is completely extracted from the sample, and that minimal fractionation takes place. This is crucial to achieving representative stable isotope results. Having extracted all the soil water allows us to calculate the gravimetric water content. Next, the soil water is analyzed for the stable isotope compositions of deuterium and oxygen-18, and for bomb-pulse tritium concentrations. The stable isotope data is then utilized to calculate the recharge rate with Knowlton's [11] technique, and the tritium data with Phillips' method [9]. The extracted soil water is analyzed for bomb-pulse tritium at this time also. The third step is to take the dry soil sample and leach the chloride from it, followed by an analysis of the chloride concentration. The chloride concentrations and water content values are then used in the chloride mass balance approach. The final step in processing the soil sample is to prepare a sample of AgCl [8] from the leached chloride for subsequent analyses of the ^{36}Cl concentrations. The ^{36}Cl distribution with depth will be utilized for predicting the bomb-pulse ^{36}Cl movement over the last 30 years, and for age dating the soil water from deep in the soil profile [12]. The following sections give details on how each of these methods is applied.

Soil-Physics Based Approach

A soil-physics based approach will be used to quantify the recharge rate from measured or estimated hydraulic characteristic data. The mathematical basis for this analysis is the unsaturated form of Darcy's law:

$$R = q = -(K(\theta,\psi))\, i = -(K(\theta,\psi))\, (dh/dz) = -(K(\theta,\psi))\, \frac{d(\psi+z)}{dz} \qquad (1)$$

where

R = recharge rate [m s^{-1}]

q = darcy flux [m s^{-1}]

$K(\theta,\psi)$ = unsaturated hydraulic conductivity [m s^{-1}], as a function of volumetric water content, θ, or pressure head, ψ [m]

i = hydraulic gradient

h = total head [m]

z = depth below land surface [m]

Several assumptions must be met in order to apply Darcy's law to estimating the recharge flux. First, the data must be from below the root zone in order to preclude any evapotranspiration uptake from affecting the results. Second, the profile must be under steady state conditions (i.e., the water content and pressure head distributions at depth are slowly varying in time). Third, the soil water must be moving essentially in one dimension, vertically downward.

The hydraulic characterization of the "undisturbed" core samples includes the determination of the initial moisture content, the soil-moisture characteristic curve (pressure head versus water content relationship), bulk density, porosity, saturated hydraulic conductivity, and grain-size distribution. The unsaturated hydraulic conductivity function is being determined from the soil-moisture characteristic curve with van Genuchten's [13] curve-fitting model, using the mathematical formulation of Mualem [14]. A better approximation of the unsaturated hydraulic conductivity function could be obtained through laboratory one-step outflow tests or an in-situ instantaneous profile test. However, these techniques are fairly time consuming and beyond the scope of this program. In applying the Darcy's law formulation in a deterministic fashion it is recommended that the effective unsaturated hydraulic conductivity across any depth interval be the harmonic mean of the individual K estimates [7].

An important point that is quite often overlooked in modeling soil-water movement is that the uncertainty in the unsaturated hydraulic conductivity function is generally quite large. Therefore it is recommended that a stochastic approach be taken to adequately capture the range of possible outcomes for the Darcy's law recharge estimates. More specifically, a Monte Carlo approach, coupled with Latin Hypercube Sampling (to optimally minimize the input parameter combinations), should yield a distribution of recharge estimates, and thereby translate the uncertainty in the model input parameters into the uncertainty in the results. If the objectives of a particular study does not require quantification of this uncertainty because it is a scoping project, then simple bounding calculations may be appropriate.

Stable Isotope Method

The stable isotope method of Knowlton's [11] is based on theoretical work presented by Barnes and Allison [15]. Knowlton's method is a tracer technique used to estimate the transport of the isotopic species deuterium (D or ^2H) and ^{18}O in the water molecule ($H_2{}^{18}O$ and $HD^{16}O$), which in turn allows for the determination of the recharge rate. The main assumption in the development of this mathematical model is that quasi-steady state conditions exist with regard to soil-water movement. Practically speaking, this means that the field sampling should be done during a lengthy period of little or no major precipitation events which could cause significant infiltration. Therefore, the isotope profile should be slowly changing in time. In New Mexico, this requirement is generally met between late spring and mid summer, prior to the 'rainy season' in July and August. An additional assumption must also be met, that the soil water is moving essentially in one dimension, vertically. Significant lateral soil-water movement would invalidate this model. The mathematical model used to describe the shape of the deuterium profile is:

$$R\,(\delta_D - \delta_D^{Rec}) = ((h\,N_{sat}\,D^v)\,/\,\rho)\,(\varepsilon_D + \eta_D)\left[\frac{d\,[\ln(h\,N_{sat}(\varepsilon_D + \eta_D)\,]}{dz}\right] - (D + ((h\,N_{sat}\,D^v)\,/\,\rho))\,\frac{d\,(\delta_D)}{dz} \qquad (2)$$

where

R = recharge rate [m s^{-1}]

δ_D = standardized isotope ratio at any depth [permil]

δ_D^{Rec} = standardized isotope ratio of the recharge water at depth [permil]

h = relative humidity

N_{sat} = saturated water vapor density in air [kg m^{-3}]

D^v = effective diffusivity of water vapor in air [m^{-2} s^{-1}]

D = effective self-diffusion coefficient of water [m^{-2} s^{-1}]

ρ = density of liquid water [kg m^{-3}]

ε_D = equilibrium enrichment factor

η_D = diffusion ratio excess

z = depth below land surface [m]

A discussion on how to estimate or determine these parameters is provided by Knowlton [11], along with a sensitivity analysis.

In this method, the distribution of deuterium, or oxygen-18, and water content with depth are required to estimate the recharge rate. A non-linear least squares analysis is applied to solve equation 2 for the recharge rate. Data required as input to the model includes the stable isotope ratios and water content distributions with depth. As mentioned above, the soil samples collected during the coring operation and stored in Mason jars are subjected to a vacuum distillation procedure [11] to extract the soil water. The soil water is then processed with a CO_2/H_2O equilibration technique or hydrogen reduction method prior to analyzing the stable isotope ratios of $^{18}O/^{16}O$ (referred to as oxygen-18) and $^{2}H/H$ (referred to as deuterium), respectively. The stable isotope ratios are determined on a ratio mass spectrometer.

In addition to estimating the recharge rate, the stable isotope data may be used to quantify or identify other soil-water movement processes. This other information may be useful, especially if there are discrepancies in the results of multiple techniques employed for estimating the recharge rate. As will be discussed below in the case study, the differences in the bomb-pulse tritium and chlorine-36 profiles were attributed to vapor phase movement of tritium, borne out by the results of the stable isotope work. Therefore, the stable isotope data may be used to quantify the distribution of the liquid and vapor flux with depth in the profile [11]. The potential for vapor-phase movement of volatiles might therefore be assessed in this manner. The stable isotope data may also be used to quantify the evaporation rate according to techniques developed by Allison and others [15,16,17]. The evaporation rate may be important in performing a risk assessment, or maybe even in utilizing the water balance method. A third alternative application of the stable isotope data is to help define whether the soil water at depth has been isotopically altered due to infiltration of water from a treatment, storage, or disposal facility. For a case where water of unknown composition (i.e., possibly contaminated) is disposed of in a facility and infiltrates, the extent of vertical migration of the water might be assessed with stable isotopes. Water from a treatment, storage, or disposal facility will probably have a different isotopic signature than that of the soil water that has been affected by natural infiltration and evapotranspiration processes. In a deep vadose zone where the fluid has migrated a finite distance below the facility, this type of characterization might offer a means of convincing a regulator that the imminent threat of contaminants reaching the underlying aquifer is small. Therefore, the likelihood that the regulator would allow vadose zone monitoring in lieu of drilling expensive monitoring wells might be enhanced.

Chloride Mass Balance Approach

The chloride mass balance method has been widely used throughout the world, due to its low cost and ease of use [9,17,18,19,20,21,22]. Two major assumptions are made in applying this technique. The first assumption is that the average rate of chloride deposition from precipitation to the soil is constant. The second assumption is that the chloride moves vertically by piston displacement (i.e., no preferential flow paths, diffusion effects are negligible) below the root zone. Under steady state conditions, the flux of chloride deposited at the land surface is equivalent to the flux of chloride beneath the root zone. Therefore:

$$P\,C_p = R\,C_s \tag{3}$$

where

P = average precipitation rate
C_p = average chloride concentration in precipitation and dry fallout
R = the average recharge rate below the root zone
C_s = the average chloride concentration in the soil

It is interesting to note that in a deep alluvial vadose zone the soil water at depth may be quite old. Therefore the chloride concentrations at depth may not be constant, but may be variable due to past climate change. The chloride mass balance technique may therefore provide us with a means of assessing the change in recharge rates over geologic time scales due to climate change. Therefore the data from this project may be useful as paleoclimatic indicators for integration to global climate modeling scenarios.

Bomb-Pulse Tritium and Chlorine-36 Techniques

The distribution of bomb-pulse tritium and chlorine-36 with depth in the soil profile can be analyzed to determine the soil-water movement processes that have occurred over the past 30 years. The atmospheric nuclear testing performed in the late 1950's and early 1960's has left us with a unique signature of the soil water in the vadose zone that is less than about 30 years old. Therefore, determining the maximum depth of infiltration of the bomb-pulse tritium and chlorine-36 allows us to calculate the modern day recharge rate. Chlorine-36 has a half life of 300,000 years, and tritium has a half life of 12.4 years.

The soil water extracted from the core samples is analyzed for its tritium concentration with a liquid scintillation counting technique. The chloride from the soil samples is processed into AgCl for subsequent analysis of the ^{36}Cl on a Tandem Accelerator Mass Spectrometer.

Chlorine-36 Age Dating

Phillips and others [12] have pioneered the use of cosmogenic ^{36}Cl in age dating very old ground water. In attempting to apply their methodology to date the soil water in a deep vadose zone several assumptions must be met. First, the age of the sediments within the unsaturated zone must be much older than the half life of ^{36}Cl (i.e., 300,000 years) such that the chloride in the profile is a result of infiltration, and not the deposition of the sediments. Second, the soil water must be sufficiently old that there is noticeable decay in the ^{36}Cl with depth. In the present case, it is expected that the soil water at depth in the profile may be quite old, and the ^{36}Cl distribution will exhibit an exponential decay with depth. The decrease in ^{36}Cl concentration with depth is assumed to be attributable to the decay of the isotope species, and therefore the soil-water age can be determined by a simple decay equation. It is interesting to note that these results may be used as paleoclimatic indicator data needed by global climate modelers and those persons involved in the performance assessment activities for the high-level waste program. In the high-level waste arena, the numerical modelers must simulate site conditions over the next 10,000 years, and the data does not exist today to baseline the historical change in recharge rates. Knowing the historical change in recharge conditions due to climate change will allow the Bayesian global climate modelers and performance assessment modelers to condition their results with an established data base rather than by inference.

CASE STUDY RESULTS

As yet, SNL has not received the analyses back from sampling the first core from the deep monitoring well at the Chemical Waste Landfill. However, comparison of results from another site (i.e., a case study) may be used to demonstrate the viability of this approach. As mentioned above, the soil-physics based approach and geochemical/isotopic techniques for quantifying the recharge rate have all been used at a site in central New Mexico by Knowlton and others [6,7,8,9,10,11]. For a complete description of the site and the details on applying the various methodologies consult these references. In brief, the site is located in semi-arid lowland, receives about 20 cm of precipitation annually, and the soils are relatively homogeneous, fine-grained sands to silt loam.

Stephens and Knowlton [7] performed a soil-physics based approach to calculating the recharge rate using two years worth of monitoring data and a wealth of hydraulic characterization data. They deterministically evaluated the average recharge rate over that period of record to be about 0.70 cm/yr.

At the same site, Phillips and others [9] used the chloride mass balance method, and the bomb-pulse tritium and chlorine-36 techniques to quantify the recharge rate. Figure 1 depicts the distribution of ^3H and ^{36}Cl with depth in the profile, along with the results of the chloride mass balance calculations showing chloride age with depth. Figure 1b depicts the chlorine-36 data as a ratio of the chlorine-36 to total chloride concentrations. Figure 1c depicts the tritium data in Tritium Units (T.U.), which is equivalent to one tritium atom in 10^{18} hydrogen atoms. Alternatively, one T.U. is equivalent to 3.24 picocuries per liter (pCi/l) The calculated recharge rates from these three approaches are 0.64, 0.80, and 0.25 cm/yr, respectively. It is interesting to note that the distributions of tritium and chlorine-36 with depth in the profile are markedly different. The current explanation for this discrepancy is that the tritium, which is an isotope of hydrogen and is found in the water molecule, travels both in a liquid and a vapor phase. The chlorine-36 is essentially non-volatile and therefore is not transported in the vapor phase. The bulk of the tritium, therefore, appears deeper in the profile than most of the chlorine-36, due to the enhanced downward vapor-phase movement of the soil water over time [9, 11].

Knowlton [11] sampled this same site for the distribution of deuterium and oxygen-18 with depth in the profile. Figure 2 shows the distribution of deuterium for a profile collected during the summer of 1988. Deuterium (D) is reported as a standardized ratio, δD, signifying the ratio of the deuterium (hydrogen-2) concentration to hydrogen-1 concentration on a permil basis (i.e., one in 10^6) [11]. The profile exhibited in Figure 2 shows that the top meter of the profile has decreasing deuterium concentrations with depth. This is indicative of a downward vapor-phase movement of soil water which is condensing into the liquid phase with increasing depth in the profile Fitting the observed isotope profile data to his model (i.e., equation 2 above), Knowlton calculated the recharge flux at 0.68 cm/yr. The stable isotope method also predicted a net downward movement of soil-water in the vapor-phase within the top meter of the profile, as suggested above by the radioisotope data [9].

Each of these techniques yields an independent estimate of the recharge rate. It is interesting to note how similar the results are, considering the uncertainty in parameters and the differences in assumptions for applying each technique. Table 1 summarizes the results of these recharge estimates for each technique used at this site.

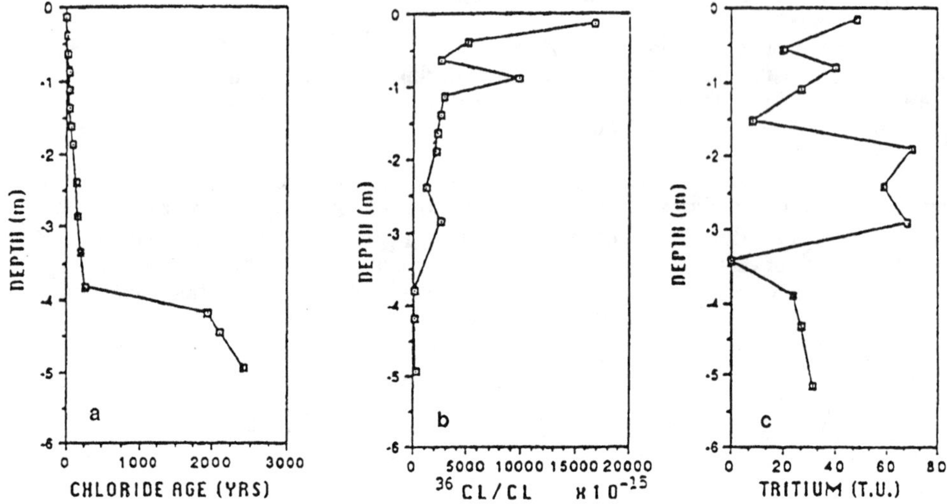

FIGURE 1 -- Chloride Mass Balance Results (a), Bomb-pulse Chlorine-36 Profile (b), and Bomb-pulse Tritium Profile (after Phillips et al., 1988)

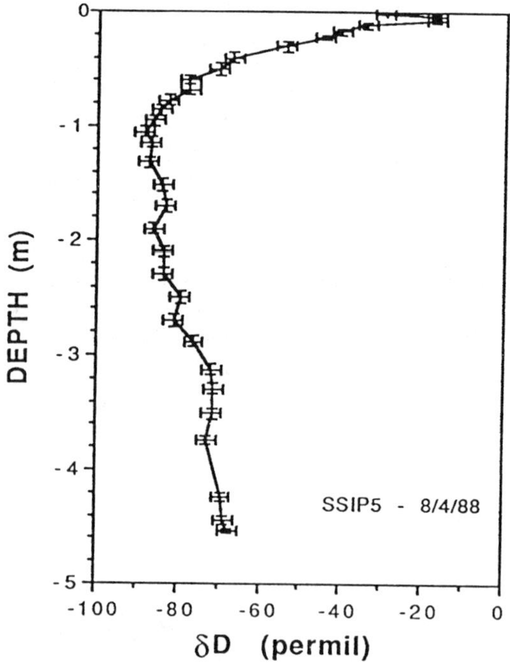

FIGURE 2 -- Depth Versus Deuterium, Y Error Bars Indicate Sampling Interval, X Error Bars Indicate 95% Confidence Interval

TABLE 1 -- Summary of Recharge Estimates

Method	Recharge Rate [cm/yr]
Darcy's law	0.70
Tritium	0.64
Chlorine-36	0.80
Chloride mass balance	0.25
Stable isotope	0.68

SUMMARY

The need for performing independent analyses to quantify the recharge rate has become ever more important with the development of EPA, NRC, and state regulations, as well as DOE orders, pertaining to hazardous, low-level radioactive, and high-level radioactive wastes. These regulations require the determination of potential contaminant travel times to the water table in order to execute a performance or risk assessment. The techniques presented in this paper provide the means of quantifying the recharge rate. These methods include a soil-physics based (i.e., Darcy's law) method, a chloride mass balance technique, a stable-isotope method, bomb-pulse tritium method, and ^{36}Cl techniques. The case study presented here illustrates the viability of performing each method at a given site and potentially reducing uncertainties in the recharge rate by achieving comparable results. Sandia National Laboratories is instituting a program under which they will use these methods to quantify the recharge rate, both in a deep drilling program and a shallow characterization strategy. The recharge rate data is useful in assessing background percolation rates of solutes, in approximating the upper boundary flux condition for numerical flow and transport modeling of the site, or as calibration criteria in flow and transport modeling.

REFERENCES

[1] Falconer, K.L., L.C.Hull, and S.A.Mizell, "A mathematical model for evaluating the suitability of a low-level radioactive waste site", In Waste Management '82 (Post, R.G., ed.), Vol. 2, Proc. Symp., Tucson, Arizona, 8-11 March 1982, p.245-258.

[2] Mann, J.F., Jr, "Waste waters in the vadose zone of arid regions: Hydrologic interactions", Groundwater. 14:367-372, 1976.

[3] Verhagen, B.Th., P.E.Smith, I.McGeorge, and Z.Dziembowski, "Tritium profiles in Kalahari sands as a measure of rain-water recharge", Isotope Hydrology, Proceedings of the International Symposium. 1978, IAEA, 2:733-751, 1979.

[4] Dincer, T., A.Al-Mugrin, and U.Zimmerman, "Study of the infiltration and recharge through the sand dunes in arid zones, with special reference to the stable isotopes and thermonuclear tritium", Journal of Hydrology. 23:79-109, 1974.

[5] Bergstrom, R.E., and R.E.Aten, "Natural recharge and localization of fresh water in Kuwait", Journal of Hydrology. 2:213-231, 1965.

[6] Stephens, D.B., R.G.Knowlton, Jr., J.McCord, and W.Cox, Field study of natural ground water recharge in a semi arid lowland, New Mexico Water Resources Research Institute Technical Completion Report No. 177, New Mexico State University, Las Cruces, New Mexico, 1985.

[7] Stephens, D.B. and R.G.Knowlton, Jr, "Soil-water movement and recharge through sand at a semi-arid site in New Mexico", Water Resources Research. 22:881-889,. 1986.

[8] Mattick, J.L., F.M.Phillips, and T.A.Duval, Quantification of ground-water recharge rates in New Mexico using bomb-36Cl and bomb-3H as soil-water tracers, New Mexico Water Resources Research Institute Technical Completion Report No. 220, New Mexico State University, Las Cruces, New Mexico, 1986.

[9] Phillips, F.M., J.L.Mattick, and T.A.Duval, "Chlorine-36 and tritium from nuclear-weapons fallout as tracers for long-term liquid and vapor movement in desert soils", Water Resources Research. 24:1877-1891, 1988.

[10] Knowlton, R.G.,Jr., F.M.Phillips, and A.R.Campbell, A stable-isotope investigation of water vapor during ground-water recharge in New Mexico. Technical Completion Report No. 237, New Mexico Water Resources Research Institute, New Mexico State University, Las Cruces, NM, 88p, 1989.

[11] Knowlton, R.G., Jr., A stable isotope study of water and chloride movement in natural desert soils, Dissertation, New Mexico Institute of Mining and Technology, Socorro, New Mexico, 241p., 1990

[12] Phillips, F. M., H. W. Bentley, S. N. Davis, D. Elmore, and G. B. Swannick, "Chlorine-36 dating of very old ground water, II, Milk River aquifer, Alberta", Water Resour. Research, 22: 2003-2016, 1986.

[13] van Genuchten, M.T., Calculating the Unsaturated Hydraulic Conductivity with a New Closed-Form Analytical Model, Res. Rep. 78-WR-08, 65 pp., Princeton University, Princeton, New Jersey,. 1978.

[14] Mualem, Y, "A New Model for Predicating the Hydraulic Conductivity of Unsaturated Porous Media", Water Resources Research, 12(3):513-522,. 1976.

[15] Barnes, C.J., and G.B.Allison, "The distribution of deuterium and ^{18}O in dry soils: 3. Theory for non-isothermal water movement", Journal of Hydrology. 74:119-136, 1984.

[16] Allison, G.B., and M.W.Hughes, "The use of natural tracers as indicators of soil-water movement in a temperate semi-arid region", Journal of Hydrology. 60:157-173, 1983.

[17] Sharma, M.L. and M.W.Hughes, "Groundwater recharge estimation using chloride, deuterium and oxygen–18 profiles in the deep coastal sands of western Australia", Journal of Hydrology. 81:93–109, 1985.

[18] Gardner, W.R., "Water uptake and salt distribution patterns in saline soils", Isotope and Radiation Techniques in Soil Physics and Irrigation Studies, Proc. Symp. Istanbul, IAEA, Vienna, pp.335-340, 1967.

[19] Eriksson, E. and Khunakasem, V., "Chloride concentrations in groundwater, recharge rate and rate of deposition of chloride in the Israel coastal plain", Journal of Hydrology, 7:178-197, 1969.

[20] Sharma, M.L., "Recharge estimation from the depth-distribution of environmental chloride in the unsaturated zone - Western Australian examples", Estimation of Natural Groundwater Recharge, I. Simmers, editor, pp. 159-174, 1987.

[21] Edmunds, W.M. and Walton, N.R.G., "A geochemical and isotopic approach to recharge evaluation in semi-arid zones - past and present", Application of Isotope Techniques in Arid Zone Hydrology, Proc. Advisory Group Meeting, V'ienna, 1978, IAEA, pp.47-68, 1980.

[22] Allison, G.B. and Hughes, M.W., "The use of environmental chloride and tritium to estimate total local recharge to an unconfined aquifer", Australian Journal of Soil Research, 16:181-195, 1978.

David E. Daniel, Peter M. Burton, and Su-Dar Hwang

EVALUATION OF FOUR VADOSE ZONE PROBES USED FOR LEAK DETECTION AND MONITORING

REFERENCE: Daniel, D. E., Burton, P. M., and Hwang, S.-D., "Evaluation of Four Vadose Zone Probes Used for Leak Detection and Monitoring," Current Practices in Ground Water and Vadose Zone Investigations, ASTM STP 1118, David M. Nielsen and Martin N. Sara, Eds., American Society for Testing and Materials, Philadelphia, 1992.

ABSTRACT: Gypsum and fiberglass resistance devices, a heat dissipation sensor, and a miniature resistivity probe were evaluated for vadose-zone monitoring. Tests were performed to determine how well the probes would detect changes in the moisture content of soil or in the chemistry of soil water. The probes were imbedded in a clayey soil. The electrical resistivity of gypsum blocks changed with water content over a broad range of water content in calibration cells. Gypsum blocks successfully indicated changes in water content in columns of soil that were wetted. Wet blocks tended to disintegrate when they were repeatedly removed from and inserted into the soil, but no disintegration of blocks left in unsaturated soil for up to 3 months was observed. The electrical resistance of fiberglass devices decreased with increasing water content and salinity of the soil water in calibration cells. The devices did not work well in tests in which the soil was wetted to near saturation -- the readings of the fiberglass devices sometimes changed in the opposite sense to changes in the moisture content of the soil. Heat dissipation sensors provided useful information over a relatively wide range in moisture content, yielded stable readings over periods of several months, and did not appear to be particularly vulnerable to degradation in soil. Miniature electrical resistivity probes yielded stable readings over periods of many months. The probes were found to undergo particularly large changes in readings when the soil was wetted with a liquid having a high electrolyte concentration.

KEYWORDS: vadose zone, monitoring, probes, sensors, water content, suction, soil water potential, contaminantion

David Daniel is an associate professor of Civil Engineering at The University of Texas at Austin, Austin, Texas 78712; Peter Burton is a Geotechnical Engineer with Burns and McDonnell, 4800 East 63rd St., Kansas City, MO 64141; Su-Dar Hwang is a former Graduate Research Assistant at The University of Texas at Austin, 78712.

INTRODUCTION

A variety of facilities that are constructed for the treatment, storage, or disposal of waste materials have the potential to contaminate ground water if unexpected releases occur. Ground-water monitoring serves as an effective means for determining whether ground-water contamination is occurring. However, devices that are buried in the vadose zone could potentially provide much earlier warning of an unexpected release and afford the opportunity for clean-up of the contaminated materials before ground-water contamination has occurred.

One of the most commonly-used techniques for vadoze-zone monitoring is the pressure-vacuum lysimeter [1-3]. The pressure-vacuum lysimeter is a device for extracting soil water samples from unsaturated soils. Such sampling can be extremely effective in confirming that contamination has occurred. However, pressure-vacuum lysimeters can become plugged over time, cannot extract samples from relatively dry soils, may be difficult to operate if located beneath a land disposal unit, may alter the chemistry of the soil water during sampling, can be expensive to use (due to sampling and analytical costs), and can expose personnel to contaminant sources.

The purpose of this research was to evaluate relatively inexpensive probes that might be used as indirect indicators that unexpected releases are occurring into the vadose zone. In principle, relatively large numbers of the probes would be buried in the vadose zone and monitored electronically over time. If any of the probes indicated the development of an anomaly, the soils near that probe would be investigated further, e.g., by installing or operating an already-installed pressure-vacuum lysimeter.

Four probes were evaluated in this research: gypsum resistance block, fiberglass resistance block, heat dissipation sensor, and a miniature resistivity probe. The purpose of the research was to evaluate the physical stability and the stability of readings of the probes over time, and to evaluate the capability of the probes to indicate a change in water content and/or electrolyte concentration. Another appropriate evaluation criterion for sensors would be the zone of influence of a probe, but, in part because the zone of influence would be site-dependent, this criterion was not included in the scope of this project.

MATERIALS AND METHODS

Monitoring Probes

Gypsum Blocks. -- Gypsum blocks are prismatic or cylindrical blocks of gypsum that change electrical resistance when they change water content [4]. The gypsum block is placed in the unsaturated soil and the gypsum either takes in water from or gives up water to the surrounding soil until thermodynamic equilibrium is established. The electrical resistance of gypsum varies with water content: the higher the water content, the higher the electrical conductivity and, hence, the lower the electrical resistance. Two electrodes are typically embedded in the gypsum, and the resistance is usually measured with high-frequency alternating current.

The original gypsum block built by Bouyoucos and Mick [4] is shown in Fig. 1a. The problem with this design is that electric current can short circuit the gypsum by going from an electrode, out of the block through soil, and into the other electrode. A cylindrical block (Fig. 1b) prevents short circuiting [5,6].

Because gypsum is partly soluble in water, gypsum blocks placed in wet soils tend to disintegrate. Resins may be added to gypsum to improve their longevity [7]. Bouyoucos and Mick [5] report that gypsum blocks may function for more than 5 years in dry soils but as little 3 months in wet soils.

Gypsum blocks may be calibrated with suction by placing the blocks in a pressure plate or pressure membrane apparatus [5]. The calibration is hysteretic for wetting-drying cycles[6,8]; however, the degree of hysteresis is not well defined. Blocks may also be imbedded in representative soil samples moistened to a range of water contents.

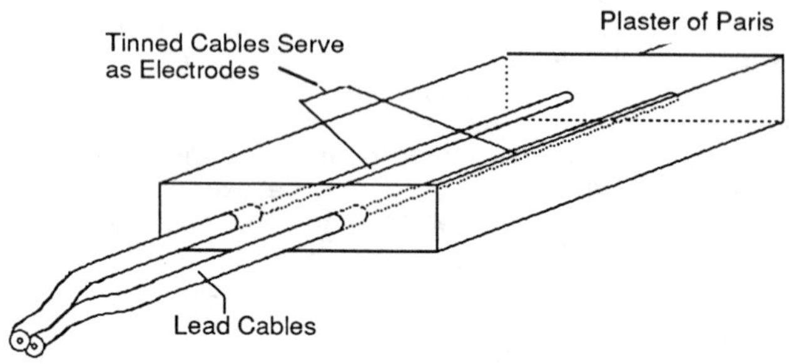

(A) Bouyoucos Block [4]

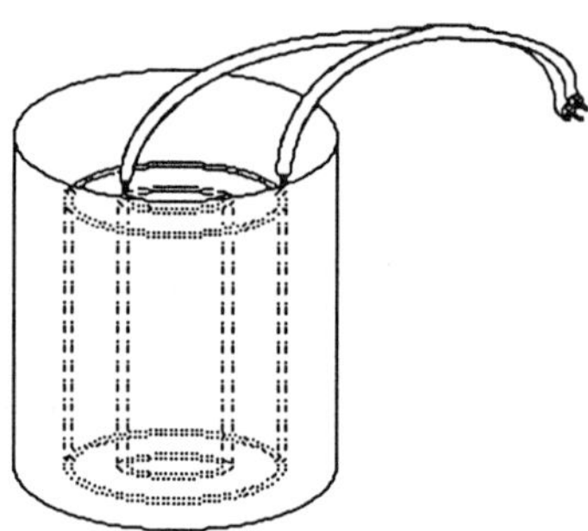

(B) Cylindrical Block

Figure 1. Gypsum Blocks.

The electrolyte concentration of water in the gypsum block remains more-or-less constant at 300 ppm regardless of the salinity of the surrounding soil. However, for salt concentrations greater than 5,000 ppm in the surrounding soil, the electrolyte concentration in and electrical resistance of gypsum blocks can be affected [9]. Thus, gypsum blocks are best suited for soils salt levels < 5,000 ppm.

The gypsum blocks used in this study were Model 5201 blocks manufactured by Soilmoisture Equipment Corporation. The resistances were measured with a Model 5910. A soilmoisture meter manufactured by the same company. The readings from this meter were checked and calibrated against a conductivity bridge (Model 31 bridge, Yellow Springs Instrument Co.).

Fiberglass Resistance Cells. -- Fiberglass resistance cells (Fig. 2), which have also been called Coleman blocks after the work of Coleman and Hendrix [10], work in essentially the same way as gypsum blocks. A porous material [fiberglass cloth] is placed in soil; the fiberglass gains or loses water until thermodynamic equilibrium is reached. Calibrations enable one to translate electrical resistance to either soil suction or water content [10]. The fiberglass cloth is contained in a monel case with holes to permit moisture movement between the fiberglass and soil (Fig. 2). A temperature-measuring probe may be a part of the unit. Other characteristics are similar to gypsum blocks, except for the relative insolubility of fiberglass compared to gypsum. No information concerning hysteresis of response could be found in the literature for fiberglass cells.

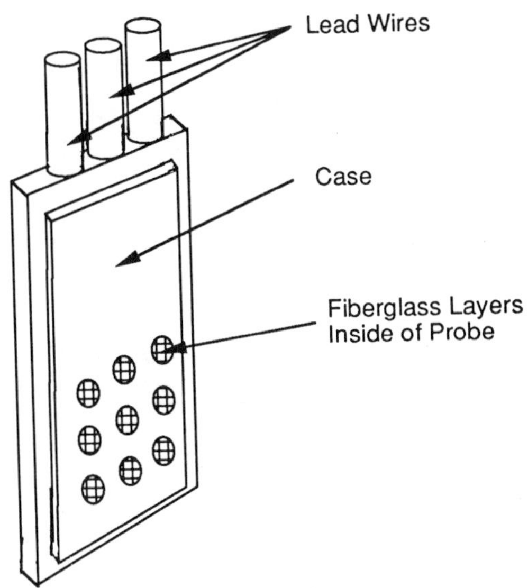

Figure 2. Fiberglass Resistance Cells

The fiberglass cells used in this research were Model MC-314 cells obtained from Soiltest. The same readout equipment described earlier for the gypsum blocks was used for the fiberglass cells.

Heat Dissipation Sensors. -- Heat dissipation sensors [11-13] measure the rate of dissipation of heat in a porous body. Heat dissipation sensors consist of a porous matrix, heating coils, and a temperature-sensing diode (Fig. 3). The probe is inserted into unsaturated soil, and the porous material wets or dries until equilibrium is reached with the soil. Temperature of the probe is measured. A current is supplied to the heating coil for a period of time. Temperature is measured again a fixed time after the supply of heat is discontinued. The temperature rise is related to the amount of heat dissipated which in turn is a function of the water content of the porous material. The size of the sensor and magnitude of heating are such that the pulse of heat is contained entirely within the sensor during the period of measurement. The rate of heat dissipation is calibrated under conditions of known soil suction or water content as described earlier for gypsum blocks.

Model A1001C heat dissipation sensors manufactured by Agwatronics were used in this study. The readout device was the Model A1011 AGWA meter.

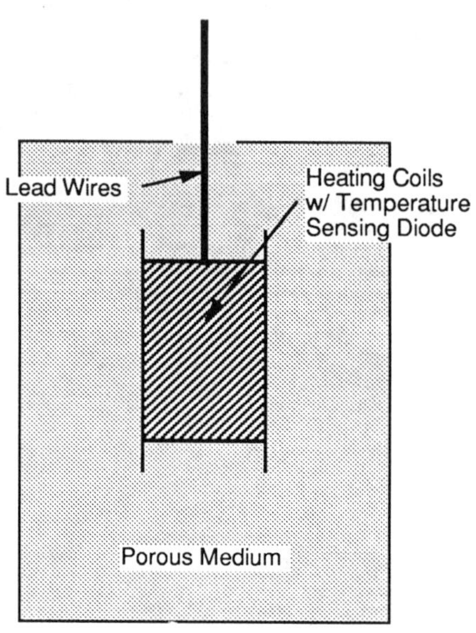

Figure 3. Heat Dissipation Sensor.

Resistivity Probe. -- Soil moisture sensing with earth resistivity probes has been described by several researchers [14-17]. Four-electrode probes measure the electrical resistivity of the soil, which is dependent primarily upon soil salinity and water content. The probes constructed for this research (Fig. 4) were constructed by wrapping four equally-spaced copper wires around a 12-mm-diameter acrylic rod approximately 6 mm apart from one another. The probe was operated as a Wenner array by passing a fixed current between the outer two electrodes and measuring the voltage drop between the inner two electrodes using a Bison 2350B resistivity meter. Further details on operation of the probe are given by Burton [18].

Tensiometers. -- Soil suctions were measured with two types of tensiometers. Small tensiometers for laboratory measurements were constructed from 6-mm-diameter, 1-bar, porous ceramic cups glued to acrylic bodies. A larger tensiometer (about 25 mm in diameter) manufactured by Irrometer Company was also used in the laboratory to represent a field-scale tensiometer. All tensiometers were monitored with an electrical pressure transducer, digital multimeter, and strip-chart recorder.

Thermocouple Psychrometers. -- Thermocouple psychrometers were used to measure suction in soils that were too dry to permit measurement of suctions with tensiometers. The psychrometers and their use are described by Hwang [19].

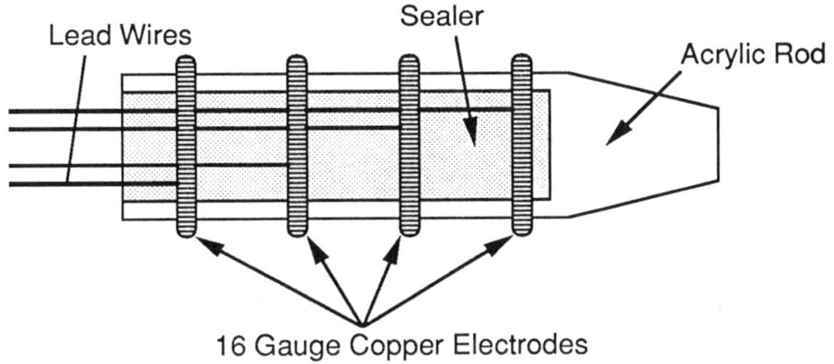

Figure 4. Electrical Resistivity Probe.

Soil. -- The soil used in this research is kaolin Hydrite R purchased from Georgia Kaolin Co. The soil has liquid and plastic limits of 58% and 34%, respectively. This soil was selected because it is commercially available, because it has been studied extensively at University of Texas laboratories, because experience has shown that its properties are consistent over time, and because it has a large water-holding capacity.

Procedures

Two types of tests were performed: tests in sealed containers and tests in columns that were wetted at one end. Sealed containers measured 100 mm in diameter and 143 mm in height. The columns had the same diameter but a height of 300 mm. For both cells, the soil was moistened with either distilled water, tap water, or various concentrations of NaCl to produce a volumetric water content in the soil of about 20%. The soil was compacted using compaction procedures similar to ASTM Method D698. For columns, the soil was compacted in 75-mm-thick lifts using the hammer and height of drop in ASTM Method D698. The probes were placed near the middle of the 143-mm-high sealed containers and near the quarter or fifth points of the columns. All probes for columns tests were installed through horizontal holes drilled through the wall of the tubes containing the soil and into the soil. With the columns, time was allowed for the probes to equilibrate, and then water was introduced by ponding water on the surface of the column. Further details are given in Refs. [18,19].

The moisture characteristic curve for compacted kaolinite was developed as a part of this investigation to provide a mechanism for relating moisture content and soil water potential. The moisture characteristic curve is shown in Fig. 5.

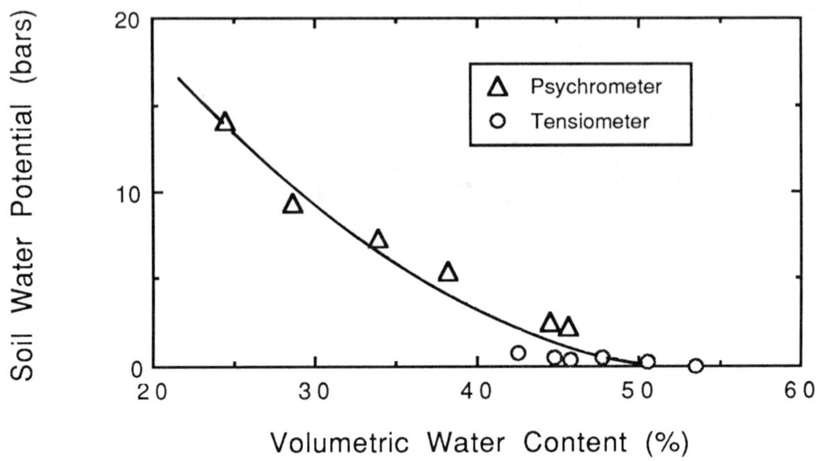

Figure 5. Moisture Characteristic Curve for Compacted Kaolinite.

RESULTS

Gypsum Blocks

It was found that when gypsum blocks were inserted in a pool of distilled water, steady readings were achieved in 2 minutes. For a wet block installed in the relatively dry kaolinite, 48 hours were needed for the readings to stabilize. Similar response times were observed when (as recommended) dry gypsum blocks were inserted in the soil.

The calibration curves for several gypsum blocks are shown in Fig. 6. To develop these calibration curves, the soil was wetted with distilled water. The "meter reading" represents percent relative moisture for agricultural soils. Because kaolin is not an agricultural soil, the value of the meter reading has little significance. Calibrations with actual resistance are reported by Hwang [19]. For this soil, gypsum blocks were sensitive to changes in water content over a range in volumetric water content of approximately 25% to 50%, which corresponds to soil suctions between 0 and 15 bars. However, the sensitivity to changes in water content was not nearly as good over a range in water content of 25% to 45% as it was at higher water contents, e.g., 45% to 50%.

The salinity of the soil water had a significant impact upon calibration curves for an extremely saline pore fluid (Fig. 7). The calibrations for distilled water and 0.01 M NaCl were approximately the same, and increasing the salinity to 0.1 M NaCl did not cause a large change in calibration. However, introduction of an extremely saline pore fluid (2.8 M NaCl) significantly shifted the calibration curve. Fortunately, increased salinity increased the meter reading (decreased soil resistance), which is the same effect as increasing the soil water content. Thus, if the soil were wetted by a saline liquid (such as a leachate from a landfill), the resistance of the gypsum would decreased both as a result of wetting of the block and as a result of increased salinity. It is fortunate that the two effects (water content and salinity) are not offsetting.

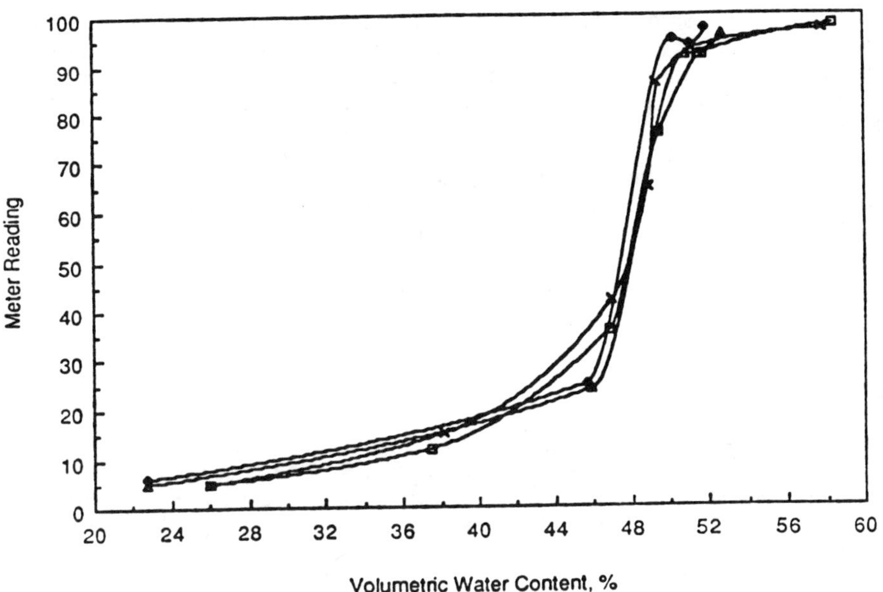

Figure 6. Calibration Curves for Four Gypsum Blocks.

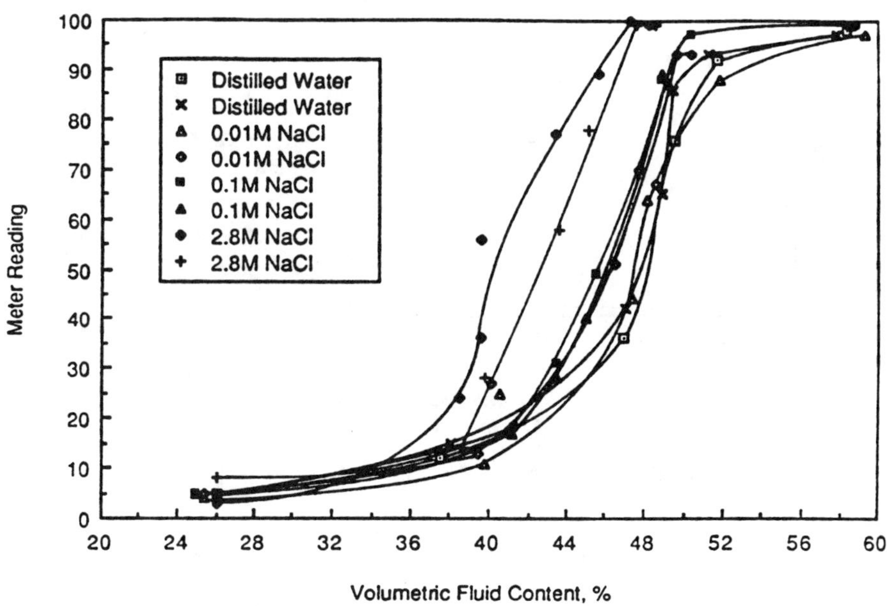

Figure 7. Effect of Soil Salinity upon Calibration of Gypsum Blocks.

Column tests were performed by introducing an occasional slug of distilled water or 0.1 M NaCl to the top of a column and then observing the response of gypsum blocks located at various depths. The uppermost gypsum block responded with an increased meter reading (indicating that the soil had been wetted) and later, after the supply of water can been discontinued, with a decreased reading. Data are given in Ref. [19].

Only a few problems were experienced with disintegration of the gypsum blocks during repeated use over several months. Several probes that were repeatedly inserted into soil and removed from soil cracked or deteriorated after several months of use. No disintegration of gypsum blocks was observed for blocks left in the soil columns or calibration chambers for periods of up to 3 months. Definitive information on longer-term performance on the probes is lacking in the literature and is an important area for further investigation.

Fiberglass Resistance Probes

The fiberglass resistance probes were calibrated in the same manner as the gypsum blocks. Calibration data are shown in Fig. 8. The fiberglass resistance probes were effective sensors of soil water content over a similar range of water content as the gypsum block, although the fiberglass devices functioned better at slightly higher water contents than the gypsum blocks.

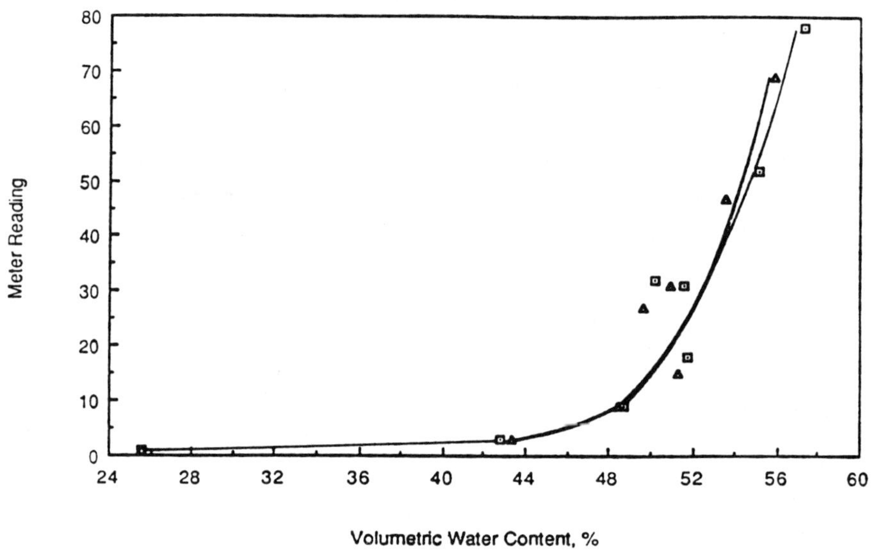

Figure 8. Calibration of Fiberglass Resistance Devices.

The fiberglass resistance probes were also sensitive to salinity of the pore fluid in the soil (Fig. 9). Again, fortunately, increasing either salinity or water content tends to decrease the resistance of the probe, which is favorable for sensing movement of electrolyte-rich fluids in the vadose zone.

In column experiments, where the soil was wetted at one end, the fiberglass devices gave somewhat unsteady results. Over a 2-week period, when the water content near the top of the column was constant at 52%, some readings were slowly increasing and some were decreasing when all should have been steady. The cause for instability could not be identified, but the data leads to questions about long-term stability of the probes.

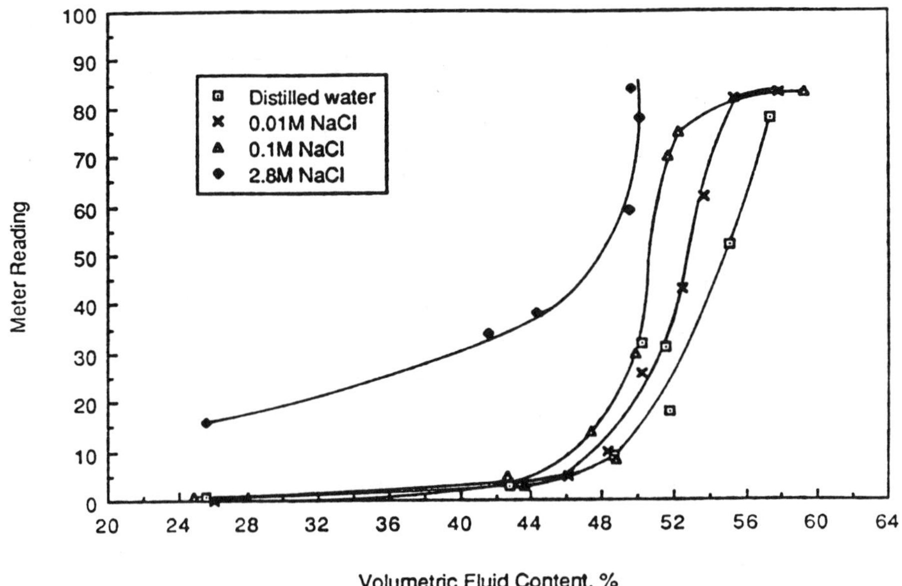

Figure 9. Effect of Soil Water Salinity on Fiberglass Resistance Probes.

Heat Dissipation Sensors

The heat dissipation sensors were calibrated with tap water in a manner similar to the calibration procedures used for the gypsum blocks and fiberglass resistance probes. Results are shown in Fig. 10 for 3 probes. Each probe had a significantly different calibration curve, pointing to the need to calibrate each probe individually. The meter reading from heat dissipation sensors responded most significant y to changes in water content over a range in water content of about 30% to 50%, which corresponds to a suction of 0 to 10 bars. Heat dissipation sensors were not calibrated in soils with different electrolyte concentration. However, the response of the probes was not expected to be highly sensitive to electrolyte concentration.

A column test was performed by ponding tap water on top of the column and monitoring the response of heat dissipation sensors embedded within the soil. The column was allowed to equilibrate for 797 hrs (33 days) and then tap water was introduced. The response of the probe embedded 70 mm below the top of the soil column is shown in Fig. 11. Prior to introduction of water in the column, it took about 400 hours for the reading of the heat dissipation sensor to stabilize. At 800 hours into the test (3 hours after the water was introduced), the meter reading suddenly decreased significantly, indicating wetting of the soil. A similar response was seen at later times in other sensors imbedded deeper in the soil column. The heat dissipation sensors clearly indicated an increase in water content of the soil.

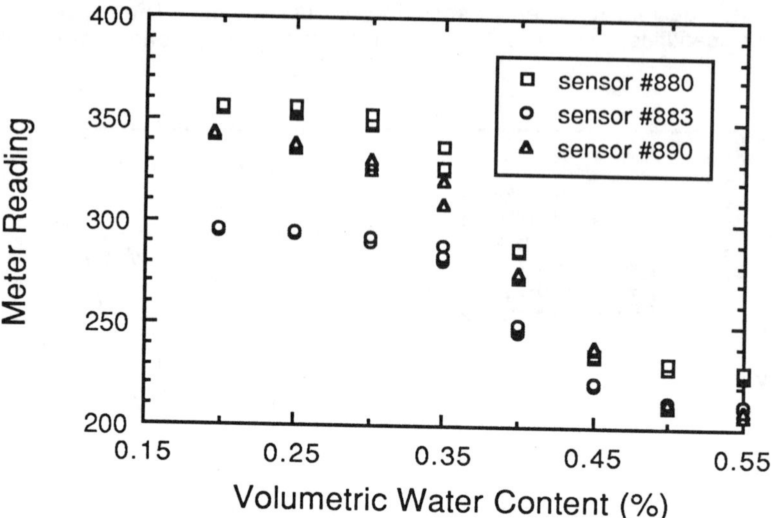

Figure 10. Calibration of Heat Dissipation Sensors.

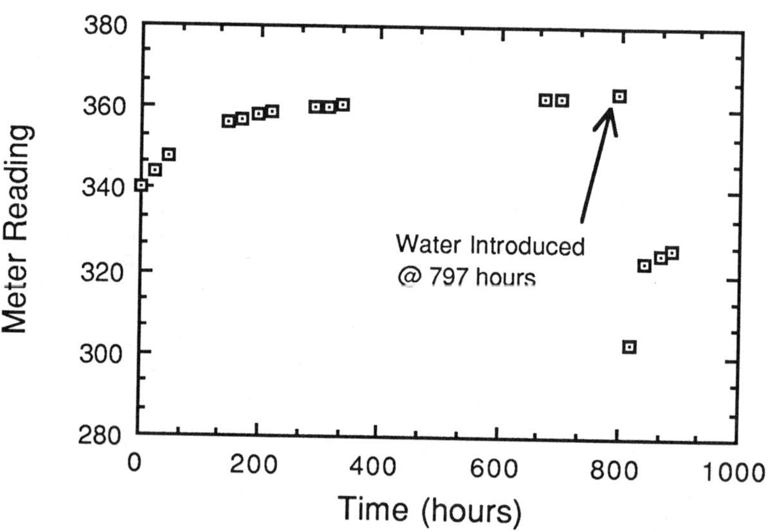

Figure 11. Response of Uppermost Heat Dissipation Sensor in Column Test.

The porous element that forms the body of the heat dissipation sensor underwent no significant disintegration or decomposition during several months of usage. The readings were stable and tended to undergo little fluctuation during the calibration procedures or column tests.

Electrical Resistivity Probes

The electrical resistivity probes were calibrated in the same manner as the other probes. Results are shown in Fig. 12 for 6 probes embedded in soil that was wetted with tap water. Each probe had a significantly different calibration curve, which is to be expected since the actual resistivity depends on many factors such as quality of contact of electrodes to the soil. Indeed, the readout from the resistivity probe is itself of no significance; however, shifts in the resistivity can indicate a change in water content. The meter reading from resistivity probes responded most significantly to changes in water content over a range in water content of about 20% to 35%, which corresponds to a suction of 5 to 15 bars. The resistivity probes did not indicate significant changes in readout when the volumetric water content was above about 35%.

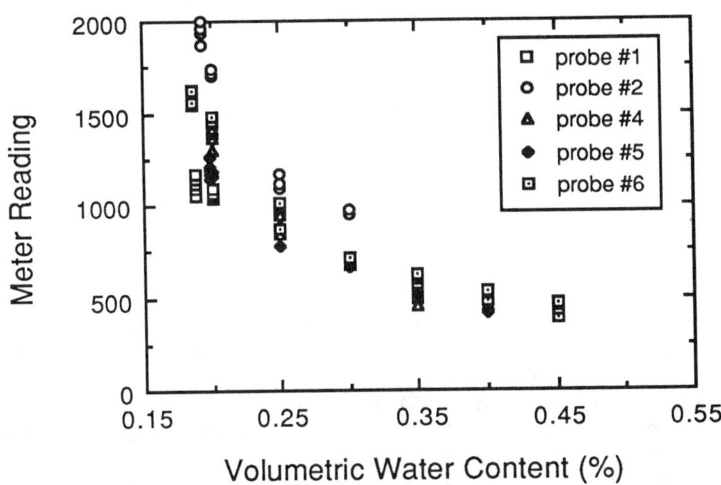

Figure 12. Calibration of Electrical Resistivity Probes for Soil Wetted with Tap Water.

The electrical resistivity of soil located between probes depends not only upon the water content but also upon the salinity of the soil water. In general, increasing either the water content or the salt content of the soil water will reduce the electrical resistivity of the soil. Calibrations were also performed for soil wetted with 0.2 M NaCl; results are shown in Fig. 13. Calibrations for the more saline solution are similar to those for tap water except that, for a given probe, the resistivity decreases much more quickly with an increase in moisture content when the soil was wetted with the more saline solution. Fortunately, resistivity decreases with the soil is wetted or when the electrolyte concentration increases -- this means that if the vadose zone is wetted with a leachate, the resistivity will tend to increase for two reasons (increased water content and increased salinity) and that the effects of water content and salinity do not tend to offset one another.

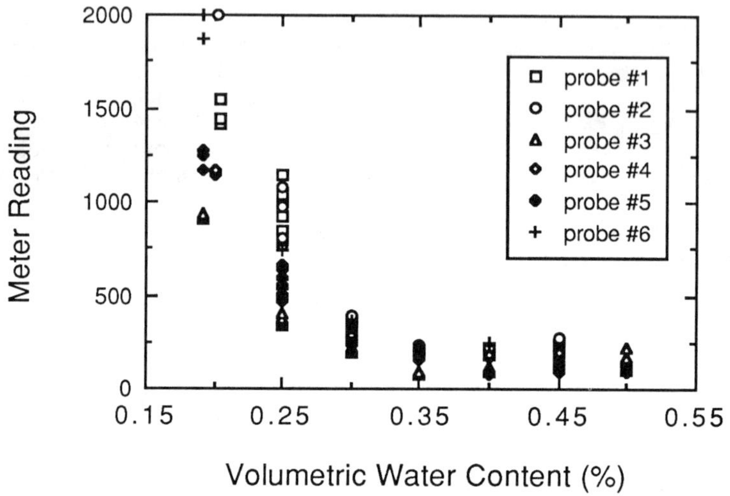

Figure 13. Calibration of Electrical Resistivity Probes for Soil Wetted with 0.2 M NaCl.

A column test was performed by ponding water on top of the column and monitoring the response of resistivity sensors embedded within the soil. The column was allowed to equilibrate for 797 hrs (33 days) and then tap water was introduced. The response of the probe embedded 150 mm below the top of the soil column is shown in Fig. 14. The readings on the resistivity probe stabilized almost immediately and remained steady throughout the period that preceded introduction of water at the top of the column. The resistivity decreased significantly several hours after water was introduced at the top of the column.

SUMMARY AND CONCLUSIONS

The purpose of this research was to evaluate the usefulness of four small, inexpensive probes that might be used for leak detection and vadose zone monitoring. Although the probes have been used for decades [20], no single investigation has been reported in which the performance of the probes was compared for the same soil. The probes that were investigated were gypsum blocks, fiberglass resistance devices, heat dissipation sensors, and miniature resistivity probes. The probes were inserted into a compacted kaolinite soil at an initial volumetric water content of 20% and initial suction of 15 bars. Calibration curves were developed between the reading of the probe and water content by incrementally wetting the soil with water or a solution of NaCl and water. The response of the probes inserted in columns of soil that were wetted from the top was also observed.

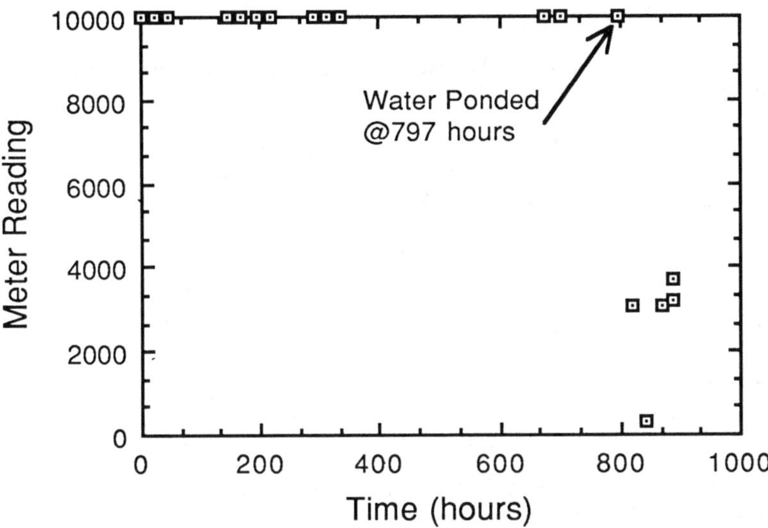

Figure 14. Response of Electrical Resistivity Probe Embedded in Soil Column.

All four probes enjoyed some degree of success in detecting changes in water content of the unsaturated soil. Each type of probe had a different optimum range of water content over which its sensitivity was greatest. For this soil, these were the optimum ranges:

Type of Monitoring Probe	Range Over Which Probe Exhibited Maximum Sensitivity of Response	
	Vol. Water Content (%)	Soil Suction (bars)
Gypsum Block	20 - 50	0 - 15
Fiberglass Res. Device	40 - 55	0 - 4
Heat Dissipation Sensor	30 - 50	0 - 10
Resistivity Probe	20 - 35	5 - 15

Except for a few minor problems all of the probes worked reasonably well as indicators of changes in water content and/or increases in soil water salinity. It is concluded that all the probes investigated can be useful indicators of changes in soil water conditions in the vadose zone and might be used to advantage for monitoring of the vadose zone.

REFERENCES

[1] Everett, L. G., and L. G. McMillion, "Operational Ranges for Suction Lysimeters," Groundwater Monitoring Review, Summer, 1985, pp. 51-59.

[2] Peters, C. A., and R. W. Healey, "The Representativeness of Pore Water Samples Collected from the Unsaturated Zone Using Pressure-Vacuum Lysimeters," Groundwater Monitoring Review, Spring, 1988, pp. 96-101.

[3] Everett, L. G., McMillion, L. G., and L. A. Eccles "Suction Lysimeter Operation at Hazardous Waste Sites," in Ground-Water Contamination: Field Methods, ASTM STP 963, A. G. Collins and A. I. Johnson, Eds., American Society for Testing and Materials, Philadelphia, 1988, pp. 304-307.

[4] Bouyoucos, G.J., and A.H. Mick, "An Electrical Resistance Method for the Continuous Measurement of Soil Moisture under Field Conditions," Michigan Agricultural Experiment Station, Technical Bulletin 172, 1940, 38 p.

[5] Bouyoucos, G.J., and A.H. Mick, "Improvements in the Plaster-of-Paris Absorption Block Electrical Resistance Method for Measuring Soil Moisture under Field Conditions," Soil Science, V. 63, 1947, pp. 455-465.

[6] Bourget, S.J., Elrick, D.E., and C.B. Tanner, "Electrical Resistance Measurements: Their Moisture Hysteresis, Uniformity, and Sensitivity," Soil Science, V. 86, 1958, pp. 298-304.

[7] Bouyoucos, G.J., "More Durable Plaster-of-Paris Moisture Blocks," Soil Science, V. 76, 1953, pp. 447-451.

[8] Tanner, C.B., and R.J. Hanks, "Moisture Hysteresis in Gypsum Moisture Blocks," Soil Science Society of America Proceedings, V. 16, 1952, pp. 48-51.

[9] Hancox, N., and J. Walker, "The Influence of Liquid Resistivity Changes on Plaster-of-Paris Resistance and Capacitance Moisture Gauges," British Journal of Applied Physics, Vol. 17, 1966, pp. 827-833.

[10] Colman, E.A., and T.M. Hendrix, "The Fiberglass Electrical Soil-Moisture Instrument," Soil Science, V. 67, 1949, pp. 425-438.

[11] Phene, C.J., Hoffman, G.J., and S.L. Rawlins, "Measuring Soil Matric Potential In Situ by Sensing Heat Dissipation within a Porous Body: I. Theory and Sensor Calibration," Soil Science Society of America Proceedings, V. 35, 1971, pp. 27-33.

[12] Phene, Rawlins, S.L., and G.J. Hoffman, "Measuring Soil Matric Potential In Situ by Sensing Heat Dissipation within a Porous Body: II. Experimental Results," Soil Science Society of America Proceedings, V. 35, 1971, pp. 225-229.

[13] Lee, R.K.C., and D.G. Fredlund, "Measurement of Soil Suction Using the MCS 6000 Sensor," Proceedings, Fifth International Expansive Soils Conference, 1984, pp. 50-54.

[14] Kirkham, D., and G.S. Taylor, "Some Tests of a Four-Electrode Probe for Soil Moisture Measurement," Soil Science Society of America Proceedings, V. 14, 1949, pp. 42-46.

[15] Rhoades, J.D., and J. van Schifgaarde , "An Electrical Conductivity Probe for Determining Soil Salinity," Soil Science Society of America Journal, V. 40, 1976, pp. 647-651.

[16] Rhoades, J.D., "Inexpensive Four-Electrode Probe for Monitoring Soil Salinity," Soil Science Society of America Journal, V. 43, 1979, pp. 817-818.

[17] Wilson, L. G., "Monitoring the Vadose Zone: Part III," Ground Water Monitoring Review, V. 3, N. 1, 1883, pp. 155-156.

[18] Burton, P. M., "An Evaluation of Three Vadose Zone Monitoring Techniques," M. S. Thesis, Univ. of Texas, Austin, Texas, 1987, 130 p.

[19] Hwang, S. D., "Four Types of Probes for Monitoring the Vadose Zone," M. S. Thesis, Univ. of Texas, Austin, Texas, 1987, 118 p.

[20] Johnson, A. I., "Methods of Measuring Soil Moisture in the Field," U. S. Geological Survey Water Supply Paper 1619, 25 p.

James Farrell[1] and Martin Reinhard[2]

MEASUREMENT OF ORGANIC VAPOR ISOTHERMS ON WET SOILS AND AQUIFER MATERIALS

REFERENCE: Farrell, J. and Reinhard, M., "Measurement of Organic Vapor Isotherms on Wet Soils and Aquifer Materials," Current Practices in Ground Water and Vadose Zone Investigations, ASTM STP 1118, David M. Nielsen and Martin N. Sara, Eds., American Society for Testing and Materials, Philadelphia, 1992.

ABSTRACT: A technique for measuring organic vapor sorption onto water coated soils and aquifer materials is presented. The method has several advantages over other techniques such as microbalance, extraction, and headspace methods. The solids used are coated with a surface layer of water and are in equilibrium with a gas phase at 100% relative humidity, simulating conditions in the unsaturated zone. Advantages of the method include the ability to measure isotherms over at least 5 orders of magnitude in organic vapor concentration, the ability to measure sorption onto very low sorbing solids, and the ability to allow an indefinite time for equilibrium to be reached.

KEYWORDS: organic vapor adsorption, chlorinated solvents, vadose zone partitioning

While much work has been done in characterizing sorption of halogenated organic solvents onto soils and aquifer materials in water saturated systems, few studies have addressed the sorption of these compounds in unsaturated systems, i.e. conditions similar to those in the vadose zone. The high vapor phase diffusivity of these solvents make it possible for a source of these contaminants to rapidly spread to areas of uncontaminated soil or ground water. A knowledge of the vapor phase sorption behavior of these organics is vital in assessing the potential of these vapors to migrate through the vadose zone.

[1]Graduate Research Assistant, Department of Civil Engineering, Stanford University, Stanford, CA 94305

[2]Professor, Department of Civil Engineering, Stanford University, Stanford, CA 94305

The purpose of this paper is to present a technique for measuring organic vapor sorption onto water coated solids such as soils and aquifer materials, and to present results from measurement of chloroform (CF), trichloroethylene (TCE) and perchloroethylene (PCE) sorption onto a sandy aquifer material. The solids used have a surface layer of adsorbed water and are in equilibrium with a gas phase at 100% relative humidity, simulating conditions in the unsaturated zone.

The technique which has been developed possesses several advantages over other methods of vapor sorption measurement such as microbalance [1], extraction [2] and headspace analysis [3,4]. Advantages of the method to be described include: (1) the ability to measure isotherms over at least 5 orders of magnitude in organic vapor concentration, (2) the ability to measure sorption onto very low sorbing solids, and (3) the ability to allow for an indefinite time for equilibrium to be reached.

One factor which makes measuring organic vapor sorption more difficult than measuring sorption from aqueous solution is that the amount of compound sorbed is often less than the amount of compound in the void space surrounding the solids. For example, the K_d (mg/l in the gas phase per mg/g in the sorbed phase) as measured in our lab for CF on Borden sand (a natural aquifer material of low surface area and low organic carbon content,[5]) predicts that for a 1 g sample contained in a 1 ml sample holder, there will be roughly 14 times more CF in the void space than is sorbed. Thus, in order to measure the sorption of CF onto Borden sand it is necessary to minimize and accurately measure the void space around the sample.

The large amount of adsorbate in the void space limits the ability of each of the three aforementioned techniques for measuring sorption onto low sorbing solids. Microbalance methods [1] are of limited usefulness since they generally have a large volume of void space compared to sample volume. Extraction techniques [2] suffer from the fact that the adsorbate in the void space cannot be differentiated from adsorbate which is adsorbed. Low sorbing solids present additional problems for most headspace techniques [3,4] because of sorption onto septa and system losses.

Additionally, the usefulness of the microbalance technique is limited to measuring adsorption of a single sorbate at a time. The microbalance technique cannot be used for water coated solids since adsorption of the organics could not be distinguished from loss of water by the sample. Due to a reduction in the surface tension in the adsorbed water layer, the capillary force holding some of the water onto the solids is reduced by the presence of an organic vapor which may lead to loss of some of the adsorbed water. We have experimentally observed this effect by weighing our water coated soil columns before and after sorbing an organic vapor. For TCE, PCE and CF the total weight of the sample (water coated solids with adsorbed organic) was reduced after adsorbing one of these vapors.

EXPERIMENTAL

Materials and Methods

The method to be presented has been used to measure partitioning isotherms for TCE, PCE and CF vapors onto Borden sand, nonporous glass beads, and silica gels. The isotherms were measured by sorbing the organic vapor onto a column of the solid followed by eluting the organic while monitoring the effluent with a flame ionization detector (FID). A schematic of the experimental apparatus is shown in Figure 1. To ensure reproducibility, all experiments were performed in duplicate.

The sorbent sample holder consisted of a 25 cm long mirror-finished stainless steel column with an inner diameter of 7 mm. A 95 to 1.6 mm stainless steel reducing union was used to connect the column to a six-port switching valve. To both hold in the solids and reduce dead volume, the void space of the unions were packed with silanized glass wool. All tubing was 1.6 mm o.d. stainless steel.

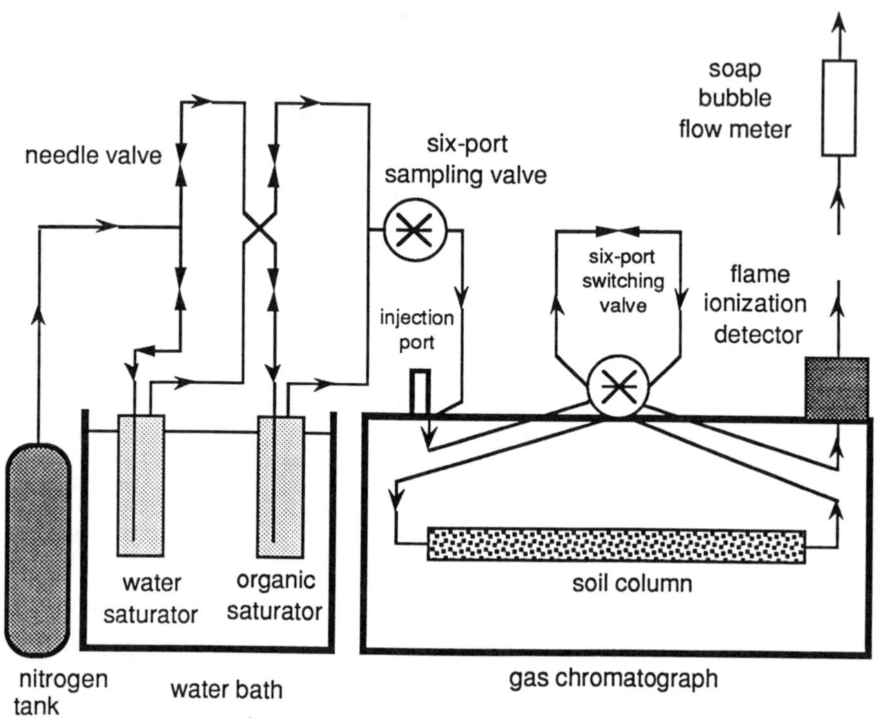

Fig. 1 -- Sorption Apparatus

Zero Grade nitrogen was used as the carrier gas and was further purified by passing through drierite (anhydrous CaSO4, #8 mesh, W.A. Hammond Co.) and a molecular sieve (alumina-silicate, 3Å, Linde). After purification, the nitrogen was passed through a gas bubbler to saturate the carrier stream with water vapor. The flow then entered a tee where it was split into a fraction which passes through a second gas bubbler containing one of the organic liquids, and a fraction which bypasses the organic saturator. The flows are recombined at a mixing-tee before entering a six-port valve. All organic liquids were supplied by Aldrich with purities of: TCE (99% +), PCE (99.98%) and CF (99% +).

When breakthrough profiles were measured, the desired plateau concentration was obtained by adjusting the fraction of the carrier gas which passed through the organic bubbler. After the organic saturated and pure streams were recombined, the flow entered a six-port switching valve mounted on a Hewlett Packard (model 5890) gas chromatograph (GC).

The two positions of this valve allowed the carrier stream to be directed either through the column or through a column bypass loop before going to the detector. The FID signal was recorded with a PE-Nelson (Perkin Elmer) data system on an IBM AT personal computer.

To determine the dry weight of the adsorbent, the packed column was placed in the oven and purged with dry nitrogen at 105° C for at least 24 hours. The column was then removed from the oven and weighed to the nearest 0.0001 g. The solids were then water coated by passing a carrier stream at 100% relative humidity through the column. The solids were allowed to equilibrate with the water saturated carrier stream until the weight of the column was constant to the nearest 0.0001 g. To ensure that the water vapor saturation corresponds to the temperature of the column, an additional gas bubbler not shown in Fig. 1 was placed in the oven, next to the column.

All experiments were performed isothermally at 30° C. This temperature was chosen because it was the lowest temperature which could be maintained steadily in the GC oven. After the solids were equilibrated with the humidified carrier stream, the column was weighed to determine its water content.

Next, the pore volume of the column was measured by injecting 30 ml of a nonretained tracer, 5% methane in argon, through the column and through the bypass loop. Since the injected tracer peaks were symmetrical, the difference in the retention volumes between the column loop and the bypass loop was equal to the pore volume of the column.

Both breakthrough and elution profiles for the organics were measured. An organic vapor breakthrough profile was generated by first equilibrating the carrier stream at the desired organic concentration while passing it through the bypass loop to the detector. Once a stable organic concentration was reached, the six-port valve was switched and the flow was then diverted through the column before going to the detector. The organic vapor stream was allowed to pass through the column until the breakthrough profile appeared complete. The column was then removed from the six-port valve, sealed with stainless steel caps, and allowed to equilibrate for 7 days before an elution profile was measured.

Considerable tailing in the breakthrough and elution profiles was observed. This tailing, combined with slight instabilities in the influent organic concentration, precluded the use of breakthrough curves for measuring equilibrium sorption. The slow rate of sorbate uptake by the solids after the apparent breakthrough was too low to quantitatively measure. This agreed with Karickhoff's observation [6] of a long-term, slow uptake of adsorbate after the initial rapid uptake of the majority of the adsorbate. Because of this slow uptake during frontal breakthrough, elution profiles, which can accurately measure tailing over at least 5 orders of magnitude in concentration, must be used to measure equilibrium sorption.

To determine a sorption isotherm, in this case a desorption isotherm, a series of elution profiles was used. To measure an elution profile, the switching valve which closes off the column was opened allowing pure, water-saturated carrier gas to pass through the column. The flow rate of the purge gas was approximately 3.5 ml/min, which is roughly one column pore volume per minute. At this low flow rate, the pressure drop across the column was less than 1400 Pa (0.2 psi). After several minutes, the switching valve was closed, sealing off the column. The column was then allowed to equilibrate again for 7 days before another elution profile was measured.

Seven days was chosen as the equilibration period since it was found that for equilibration periods of up to 5 days, a deviation from equilibrium could be measured. When identical experiments were performed on a porous aquifer material, elution profiles for columns equilibrated for periods ranging from 5 days to 3 weeks were identical, while those equilibrated for periods of 3 days or less exhibited measurable differences.

The initial eluent plateau concentration after equilibration (see Fig. 2) is taken as the equilibrated vapor phase organic concentration in the column. The total amount sorbed is determined by adding the masses of sorbate eluted during each purging of the column. The purging and equilibration sequence is repeated until the vapor phase organic concentration becomes too low to accurately measure. After the last purging of the column, the sorbent was extracted with methanol to assess the amount of extractable organic remaining on the solid. The extract was analyzed by a gas chromatograph equipped with an electron capture detector. The amount of organic extracted from the solid after the last purging was always so low that it was quantifiable to only one significant figure. However, any errors in this measurement are inconsequential since the mass of sorbate remaining on the solid after the repeated purgings was 4 or 5 orders of magnitude less than that which was measured by purging.

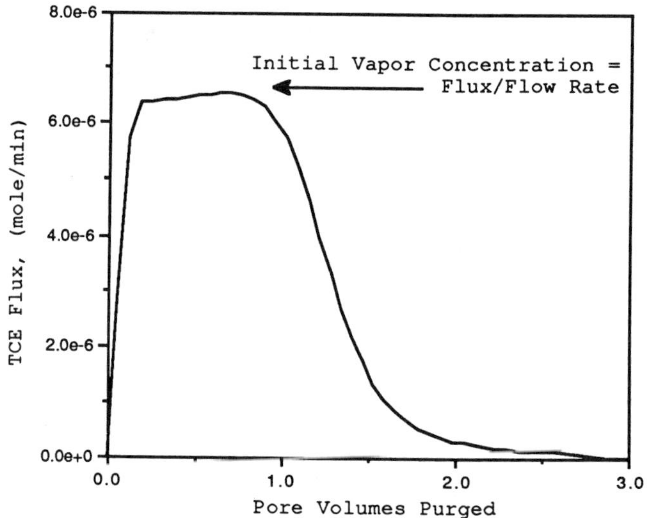

Fig. 2 -- TCE Elution Profile.

Data Evaluation

An isotherm is constructed from a series of elution profiles by adding up the amount of sorbate eluted during each purging of the column. The principle behind the method of calculation is that the amount sorbed at the start of a given purging is the sorbate that is eluted during that and all subsequent purgings less the amount that is in the void space at the start of the given purge, plus the amount extracted at the end. Table 1 contains the data for the column A PCE isotherm in Figures 4a & 4b. The table is constructed from the data measured during the ten purgings of the column. The data taken during each purging was the initial equilibrated vapor phase concentration (C_i), the total PCE eluted during the purging (p_i), and the purge gas flow rate. The mass of PCE in the column void space before each purge (C_i x v) is calculated by multiplying the initial vapor concentration by the void volume of the column. After the tenth purging of the column, the solids were extracted with methanol and it was determined that 3 x 10^{-8} g/g of extractable PCE remained on the solid.

The amount of PCE sorbed at the start of a given purge cycle, i, at a vapor phase concentration C_i, is given by Eq. 1:

$$m_i = \left(\sum_{j=i}^{n} p_j \right) - v \times C_i + x \quad (1)$$

where

i = purge number
n = total number of purging steps
p_i = mass of PCE purged
C_i = initial vapor concentration
v = void volume of the column, and
x = mass of PCE extracted by solvent.

TABLE --1 Purge Data

i purge number	**C_i** initial vapor conc. (g/ml)	**p_i** mass PCE purged (g/g)	**C_i x v** PCE in void space (g/g)	**m_i** PCE sorbed (g/g)
1	1.60×10^{-4}	5.49×10^{-5}	4.49×10^{-5}	2.77×10^{-5}
2	3.47×10^{-5}	1.39×10^{-5}	9.73×10^{-6}	7.99×10^{-6}
3	5.12×10^{-6}	2.07×10^{-6}	1.44×10^{-6}	2.41×10^{-6}
4	1.55×10^{-6}	6.81×10^{-7}	4.35×10^{-7}	1.34×10^{-6}
5	6.02×10^{-7}	2.54×10^{-7}	1.69×10^{-7}	9.25×10^{-7}
6	3.93×10^{-7}	1.83×10^{-7}	1.10×10^{-7}	7.30×10^{-7}
7	2.99×10^{-7}	3.67×10^{-7}	8.38×10^{-8}	5.73×10^{-7}
8	1.39×10^{-7}	6.83×10^{-8}	3.90×10^{-8}	2.51×10^{-7}
9	8.71×10^{-8}	1.24×10^{-7}	2.44×10^{-8}	1.97×10^{-7}
10	4.74×10^{-8}	6.73×10^{-8}	1.33×10^{-8}	8.40×10^{-8}

The accuracy of this method of isotherm measurement was corroborated using a gravimetric technique on dry silica gel (Davisil, Alltech Assoc.). Adsorption of CF onto dry silica, instead of on a water coated soil or sediment, was used to check the method for two reasons. The silica was dry, i.e. not water coated, because wet silica does not adsorb enough CF to accurately measure gravimetrically. Additionally, loss of water from wet solids would interfere with the gravimetric method.

The gravimetric method consisted of first drying a column of silica gel by purging it at 200 °C for at 72 hours with dry nitrogen. Chloroform was then adsorbed onto the dry silica by passing a carrier stream of dry nitrogen saturated with chloroform vapor through the column. The column of CF laden silica gel was then subjected to several purgings with dry nitrogen and weighed both before and after each purging. The mass of CF purged as measured by weighing (to the nearest 0.0001g) was then compared to the mass purged as measured by the FID. Figure 3 compares the cumulative CF purged as measured by both methods. The close agreement between the two methods, as indicated by the proximity of the 5 data points to the line of unit slope, shows that summing the masses purged from the column does not lead to a systematic build-up of errors in measurement and is therefore an adequate method of isotherm determination.

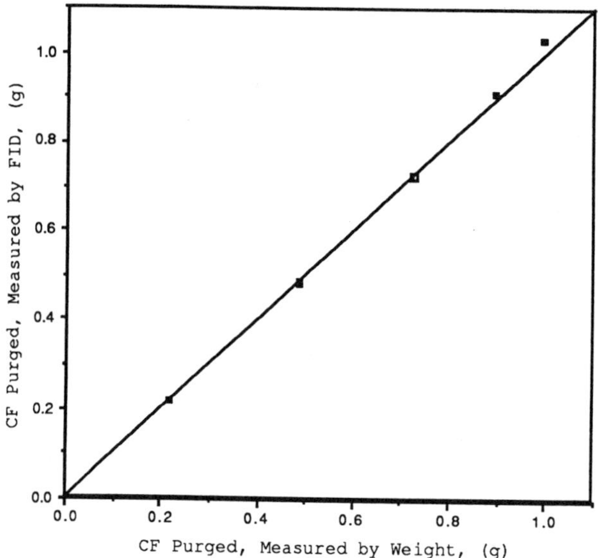

Fig. 3 -- Comparison of Gravimetric and FID Methods.

Figure 4a shows two independently measured isotherms for PCE on Borden sand. The first two data points in Table 1 were omitted from Fig. 4a to allow the isotherm nonlinearity to be viewed more easily on a linear plot. Figure 4b shows all the PCE data from Table 1 in logarithmic form. The data for both columns A and B are in close agreement. Isotherms for TCE and CF on Borden sand are shown in Figures 5 and 6.

As Figure 4a shows, PCE is more strongly adsorbed at decreasing vapor concentrations. The same is true for both CF and TCE onto Borden sand. Although the isotherms are nonlinear, it is useful for comparative purposes to calculate apparent partition coefficients for ranges of the isotherms which can be approximated as linear. For $P/Po > 0.006$, the average PCE K_d value of 0.144 ml/g determined in this study at 30 °C is in reasonable agreement with a K_d value of 0.30 ml/g obtained by Ball and Roberts [5] at 20 °C in a water saturated system of PCE and Borden sand.

The isotherms were found to be insensitive to the water content of the solids. For TCE sorption on Borden sand the water content of Column A was 43% greater than that of Column B, but as Figure 5 shows, the amount sorbed is nearly identical. This is in agreement with Chiou and Shoup [2] who found that at relative humidities above 90%, organic vapor adsorption was independent of water content and similar to that found in saturated systems.

The desorption isotherms measured in this investigation all appear to be of Type II [7] in that there is stronger adsorption at lower vapor concentrations. This type of behavior has not been seen in other organic vapor adsorption studies on water coated solids. Chiou and Shoup [2] found that for sorption of benzene and chlorobenzenes onto Woodburn soil at RH greater than 50%, all isotherms were practically linear for organic $P/Po < 0.5$. Peterson et al. [8] found that adsorption of PCE and TCE vapors onto humic acid coated aluminum oxide particles followed their linear model for the entire range studied.

One explanation for this finding of nonlinear sorption onto the water coated solids may be due to differences in the experimental techniques. The aforementioned studies measured sorption over less than 2 orders of magnitude in concentration and neither allowed more than 12 hours for equilibration. Additionally, the two previous studies measured adsorption while this investigation measured desorption, and it is known that adsorption and desorption processes may be hysteretic.

This Type II sorption behavior would lead to reduced vadose zone vapor transport rates than those which would be predicted from K_d measurements made at high vapor concentrations. The difference in transport rates at high and low concentrations can be significant. Column A data for PCE on Borden sand gives an apparent $K_d = 0.139$ ml/g for $P/Po > 0.006$. However, the slope for the low concentration region (between P/Po=2.1e-4 and 4.2e-4) is 2.54 ml/g. The difference in isotherm slope, and hence the retardation of vapor movement, in this case is approximately a factor of 20. Also, this stronger sorption at lower vapor concentrations could possibly lead to slower removal rates in vapor stripping operations of vadose zone contaminants.

SUMMARY AND CONCLUSION

The method presented here has been found to give consistent results in measuring vapor sorption isotherms spanning 5 orders of magnitude in vapor concentration. This wide range in vapor concentrations allows measurement of sufficiently low concentrations to detect nonlinearity in the partitioning. By accurately measuring the void space of and around the sample, this method is capable of measuring vapor sorption onto low sorbing solids. Additionally, by allowing sufficient time for equilibration, the method is capable of measuring the long term, slow uptake and release of a small fraction of the adsorbate. Although this slow sorbing fraction is small, it may significantly impact soil remediations because of its resistance to mobilization.

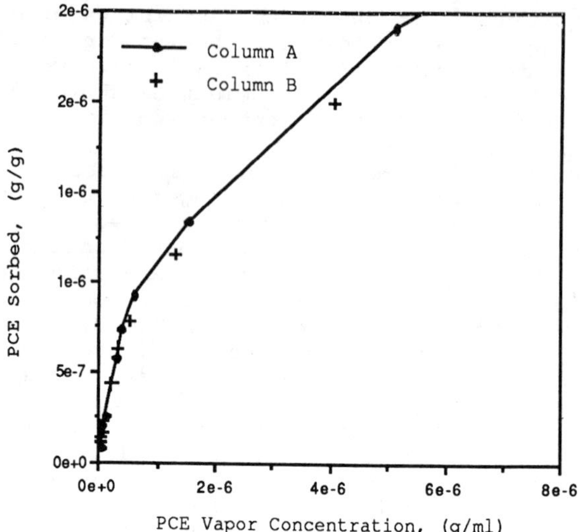

Fig. 4a -- PCE Isotherms on Borden Sand.

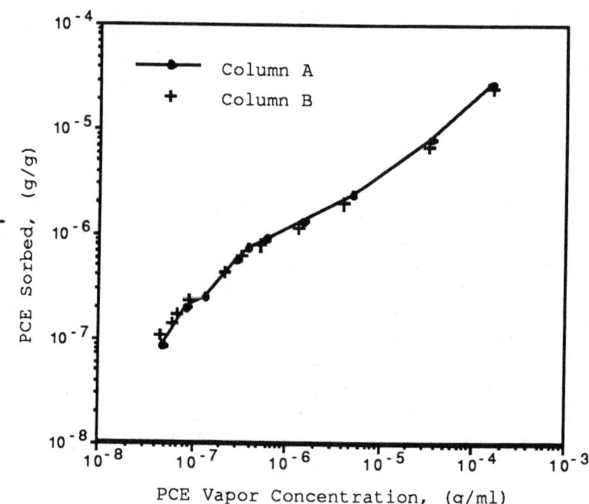

Fig. 4b -- PCE Isotherms on Borden Sand.

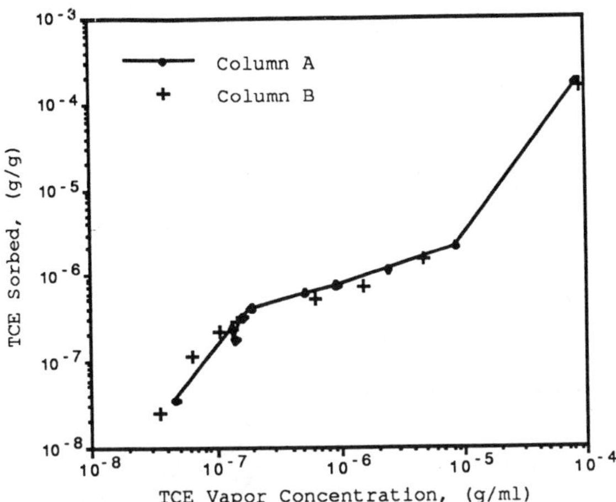

Fig. 5 -- TCE Isotherms on Borden Sand.

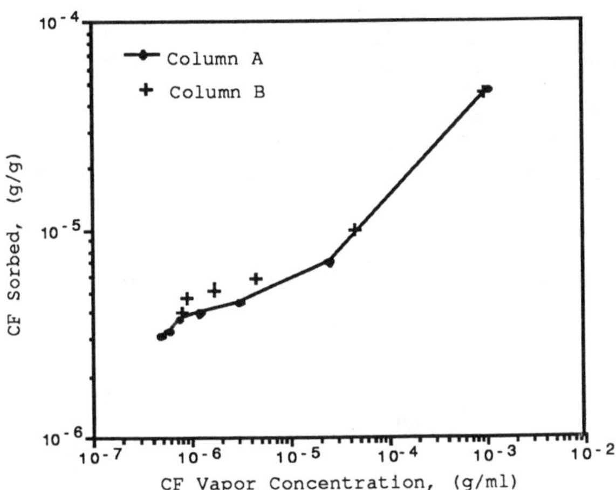

Fig. 6 -- CF Isotherms on Borden Sand.

ACKNOWLEDGEMENTS

This project was supported by the U.S. EPA Office of Research and Development under agreement R-815738-01 through the Western Region Hazardous Substance Research Center. The content of this paper does not necessarily represent the views of the agency.

REFERENCES

[1] Chiou, C. T., D. E. Kile, and R. L. Malcolm, "Sorption of Vapors of Some Organic Liquids on Soil Humic Acid and its Relation to Partitioning of Organic Compounds in Soil Organic Matter," Environmental Science and Technology, Vol.22, No.3, March 1988 pp.298-303.

[2] Chiou, C. T. and T. D. Shoup, "Soil Sorption of Organic Vapors and Effects of Humidity on Sorptive Mechanism and Capacity," Environmental Science and Technology, Vol.19, No. 12, December 1985, pp.1196-1200.

[3] Garbarini, D., and L. Lion, "Evaluation of Sorptive Partitioning of Nonionic Pollutants in Closed Systems by Headspace Analysis," Environmental Science and Technology, Vol. 19, No. 11, November 1985, pp. 1122-28.

[4] Crittenden, J., T. Rigg, D. Perram, S. Tang, and D. Hand, "Predicting Gas-Phase Adsorption Equilibria of Volatile Organics and Humidity," Journal of Environmental Engineering, Vol. 115, No. 3, June 1989, pp.560-73.

[5] Ball, W. and P. Roberts, "Sorption Kinetics of Synthetic Organic Chemicals in Sandy Aquifer Material," Geological Society of America, 1987 Annual Meeting, Phoenix, AZ, Oct. 26-29, 1987.

[6] Karickhoff, S. W. "Sorption Kinetics of Hydrophobic Pollutants in Natural Sediments," in Contaminants and Sediments, R. A. Baker (ed.), Vol. 2., Ann Arbor Science Inc., 1980.

[7] Adamson. A. W., Physical Chemistry of Surfaces, 4th ed., Wiley Interscience, John Wiley & Sons, New York, 1984.

[8] Peterson, M. S., L. W. Lion, and C. A. Shoemaker, "Influence of Vapor-Phase Sorption and Diffusion on the Fate of Trichloroethylene in an Unsaturated Aquifer System," Environmental Science and Technology, Vol.22, No. 5, May 1988, pp 571-78.

Thomas R. Heins, Henry B. Kerfoot, Donald J. Miller, and Donald A. Peterson

QUALITY ASSURANCE IN SOIL GAS SURVEYS: FALSE POSITIVE ACETONE IDENTIFICATIONS AT A CODISPOSAL SITE

REFERENCE: Heins, T. R., Kerfoot, H. B., Miller, D. J., and Peterson, D. A., "Quality Assurance in Soil Gas Surveys: False Positive Acetone Identifications at a Codisposal Site," Current Practices in Ground Water and Vadose Zone Investigations, ASTM STP 1118, David M. Nielsen and Martin N. Sara, Eds., American Society for Testing and Materials, Philadelphia, 1992.

ABSTRACT: This paper points out the importance of quality assurance in soil gas surveys through the implementation of quality control measures. A soil gas survey was undertaken at the Cortese Landfill to assess vapor migration. Soil gas samples were analyzed from 54 locations for 15 volatile organic compounds by gas chromatograph (GC) methods. Quality control measures included GC calibration, replicate, spike and blank samples and analysis of split samples by an off site analytical lab. Acetone was detected at many of the soil gas sample locations at unexplainable concentrations relative to concentrations observed in groundwater. A second GC, equipped with a different column, was used to confirm presence of acetone, but at concentrations lower than those estimated using the first column. This indicated that coeluting compounds may have caused false positive results (misidentifications) in the earlier data. This approach to verification of field analytical data quality assurance proved to be beneficial in preventing costly additional sampling and analysis.

KEYWORDS: soil gas, landfill, codisposal, quality control, quality assurance, acetone.

T.R. Heins, Project Manager, GZA GeoEnvironmental of New York, 364 Nagel Dr, Buffalo, NY, 14225; H.B. Kerfoot; Kerfoot and Associates, Inc.; D.J. Miller, Senior Engineer, Golder Associates, Inc., 3730 Chamblee Tucker Rd, Atlanta, GA 30341; D.A. Peterson, Region Environmental Scientist, Waste Management of North America, Inc., 1387 Fairport Rd, Fairport, NY, 14450.

I. INTRODUCTION

Soil gas surveys have been used in various environmental studies to assess volatile organic contamination. They are frequently employed in preliminary site characterization studies, providing immediate data at less expense than conventional subsurface sampling and chemical analysis. Soil gas surveys have been used to identify sources of volatile organics from leaking underground storage tanks, enhance monitoring well placement strategy, and in remedial work to support the design of vapor phase extraction systems *[1,2,3,4]*.

During this study, a soil gas survey was used to evaluate vapor phase contaminant releases from the Cortese Landfill. This application presents some particular challenges due to the large number of compounds normally present in landfills and the scrubbing effect landfill gases (predominantly methane and carbon dioxide), derived from municipal waste, exert on the migration pattern of volatile organic constituents associated with codisposed industrial waste. Thus, volatile compounds tend to move with the landfill gases.

The usefulness of soil gas surveys is, to a large extent, dependent on adequate quality assurance planning prior to engaging in any data gathering efforts. The objectives of every program should be clearly defined, and site information evaluated to establish appropriate procedures, equipment, and the necessary precision and accuracy of the survey. This paper focuses on how quality assurance planning can improve the usefulness of soil gas surveys and avoid problems and expense associated with interpretation of bad data.

II. BACKGROUND

The Cortese Landfill site, located in Narrowsburg, New York, is bordered to the northeast by a very steep bedrock escarpment, and to the southwest by a railroad track embankment (Figure 1). The Delaware River is approximately 122 meters (400 feet) southwest of the site. The landfill area slopes gently to the south and southwest, toward the railroad embankment. The topography within the site's boundaries is relatively flat, with minor irregularities due to past excavation and filling activities. On-site elevations range from approximately 216 meters (710 feet) near the edge of the escarpment to 210 meters (690 feet) at the site's southern boundary; however, the bedrock escarpment bordering the east edge of the site rises to 325 meters (1068 feet) in elevation, over 107 meters (350 feet) above the site. The lowest elevation in the area is near the Delaware River (approximately 205 meter (672 feet)).

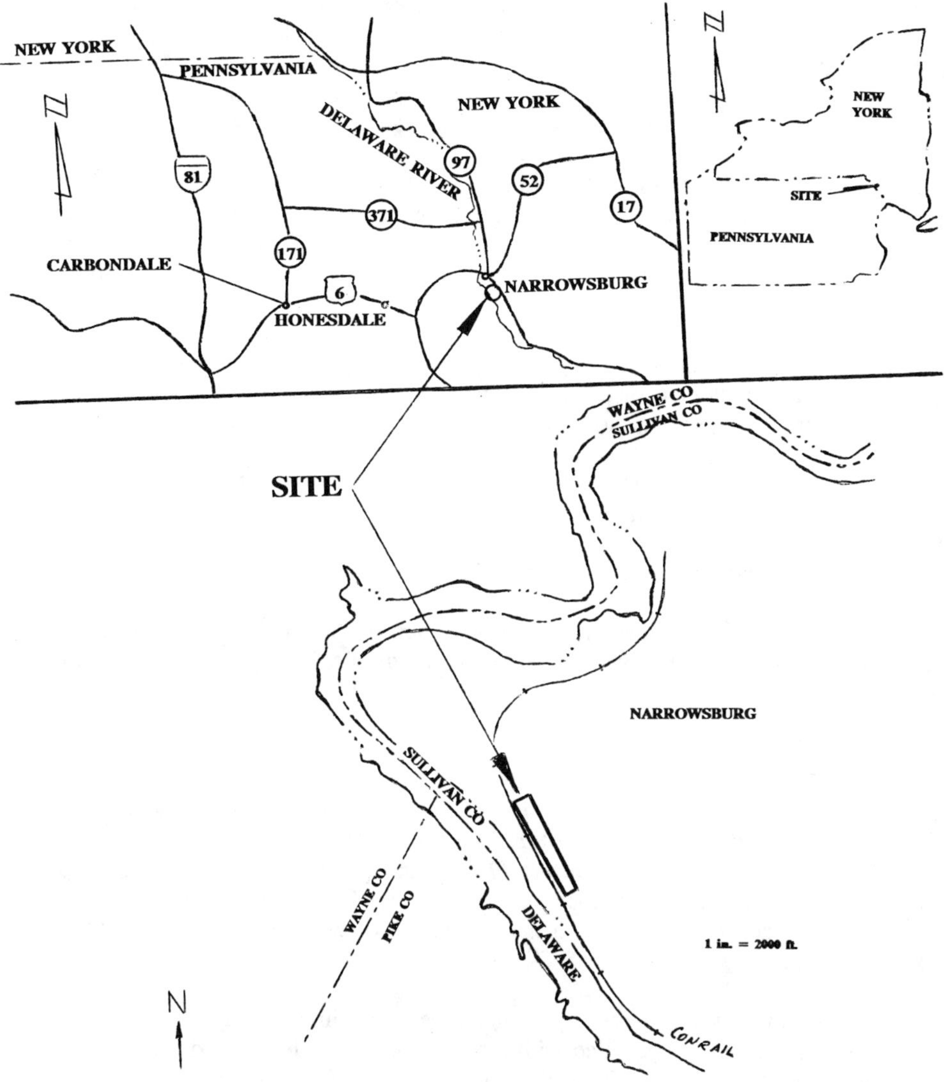

FIGURE 1 - SITE LOCATION

In order to determine the extent of landfill gas migration, methane was used as an indicator compound, since gases in landfills typically contain large amounts of methane relative to VOC concentration. Soil gas samples were analyzed from 54 locations within and near the landfill to identify areas of increased methane concentration indicating horizontal gas migration.

To evaluate migration of other VOCs and possible dissolution into infiltrating water, it was necessary to analyze for additional compounds. The VOCs found in groundwater downgradient of the site were analyzed (Table 1). This table also lists the compounds physical properties. All sample locations and depths were arrived at through negotiation with New York State regulators.

Data quality objectives for the survey were United States Environmental Protection Agency Level II, which call for identification of analyte specific organic compounds by gas chromatography at detection levels in the low part per billion range *[5]*.

A quality assurance plan was prepared prior to doing the work. The plan described the field sampling and analysis procedures as well as QC measures. QC measures included GC calibration requirements, replicate analysis, spike analysis, equipment blanks and off-site laboratory GC/MS analysis.

This paper focuses on how the measures taken during this study helped to detect problems in the original study plan. In particular, we discuss interferences that caused false positive results for acetone identification and quantification.

IV. FIELD METHODS

Soil gas samples were collected at a total of 54 locations (refer to Figure 2); 13 locations on the landfill and 41 off-site. Locations within the landfill limits were sampled at a depth of about 0.6 meters (2 feet) thought to be just below the cap material. Off of the landfill, samples were collected at 0.9 meters (3 foot) intervals from ground surface to a depth of 5.5 meters (18 feet), refusal or until groundwater was encountered.

Each soil gas sample was collected by driving a new 13 millimeter (1/2 inch) outside diameter galvanized steel pipe fitted with a removable tip to the desired depth. The probe was then retracted about 50 millimeters (2 inches), the tip removed and a new length of polyethylene tubing inserted. The use of new tubing precludes the necessity of decontamination. The tubing was extended to the bottom of the probe and sealed at the top to prevent ambient air from entering the system. The steel probe was sealed at ground surface with bentonite and water paste to prevent preferential flow along the annulus.

To collect soil gas samples, an air sampling pump, a three way sampling tee and photoionization detector (PID) were attached in series to the top of the polyethylene tubing. The pump was engaged and the PID was monitored until a stable reading was

Much of the site lies within the historic flood plain of the Delaware River which is dissected by the railroad embankment. The embankment rises about twenty feet above the surrounding ground to an elevation of approximately 218 meters (715 feet) near the site. This structure constrains surface water drainage in the site area creating ponding in surface depressions during wet periods.

The Cortese Landfill is a 2 hectare (5 acre) closed codisposal landfill. The landfill operated from July 1970 to July 1981, and principally received municipal and industrial waste. Interviews indicate drummed industrial waste (approximately 5000 drums) was transported to the site and disposed in unlined trenches.

Essentially no cap exists over the landfill and waste is at or very near the ground surface. We expect that drums are located within 6 meters (20 feet) of ground surface. The depth to water is 4.6 to 6.0 meters (15 to 20 feet). Leachate from the landfill is known to be impacting groundwater. The groundwater flow direction is to the southwest toward the Delaware River. Moderate to low level volatile organic compounds (VOCs) have been found in downgradient wells near and adjacent to the landfill. Downgradient wells near the Delaware River show minor detectable levels of VOCs.

Residences exist to the west between the site and the Delaware River. An inactive municipal well is located north-west of the site, outside of the observed groundwater plume.

The geology of the Narrowsburg area may be generally characterized as shallow bedrock of Upper Devonian age mantled with Pleistocene glacial ice contact and outwash deposits. The Delaware River incised a deep, frequently steep walled channel in the bedrock. The escarpment east of the landfill is an outcrop of this bedrock. The valley is filled with dense packed sands with zones of either silt or gravel. These sediments have been reworked by the river into scours, channels, and bars. A thin mantle of silt, averaging less than ten feet in thickness, overlies the sands. These silts are overbank deposits resulting from post-glacial flooding within the valley.

III. SURVEY DESIGN

The goal of the soil gas survey was to evaluate gas-phase migration of VOCs either directly from the landfill or from the groundwater plume. Landfills can produce gas at rates that support pressure-driven subsurface flow, resulting in vertical and horizontal gas-phase migration of VOCs. The specific concerns were potential migration of landfill gases into residences and the potential horizontal gas-phase migration of VOCs followed by dissolution in infiltrating water and migration into groundwater. The second question arose because of the proximity of the municipal well. Since characterization of the groundwater plume demonstrated contaminant migration away from the municipal well, the question of gas-phase migration of volatile compounds toward the well surfaced.

TABLE 1.--

TARGET COMPOUNDS AND THEIR PHYSICAL CHEMICAL PROPERTIES

Compound	Molecular Weight	Vapor Pressure (atm)	Henry's Law Constant ([PPMv gas]/ [ug/l water])	Ionization Potential (eV)
Vinyl Chloride	62.5	3.2	1.3	12.6
Acetone	58.1	0.28	$4.5x10^{-4}$[b]	9.7
Methylene Chloride	84.9	0.46	$2.3x10^{-4}$	9.3
1,2 trans-dichloroethene	96.9	0.34	$6.8x10^{-2}$	9.6
Methyle Ethyl Ketone	72.1	0.31	$2.0x10^{-4}$[b]	9.5
1,1,1 trichloroethane	133.4	0.16	$1.1x10^{-1}$	11.3
Benzene	78.1	0.13	$7.2x10^{-2}$	9.2
Toluene	92.1	0.037	$6.9x10^{-2}$	8.5
Tetrachloroethene	165.8	0.019	$1.6x10^{-1}$	9.3
Chlorobenzene	112.6	0.016	$3.3x10^{-2}$	9.1
Ethylbenzene	106.2	0.013	$6.1x10^{-2}$	8.8
Xylenes	106.2	0.012	$6.6x10^{-2}$	8.5
1,4 dichloro-benzene	147.0	0.0019	$2.0x10^{-2}$	
Methane	16.0	270	41	9.1

[a] Source: Mackay and Shiu, J. Phys. Chem. Ref Data, 10(4), 1981, 1175-1199.

[b] Henry's Law constants are presented for comparative purposes, solutions of these compounds actually follow Raoult's Law.

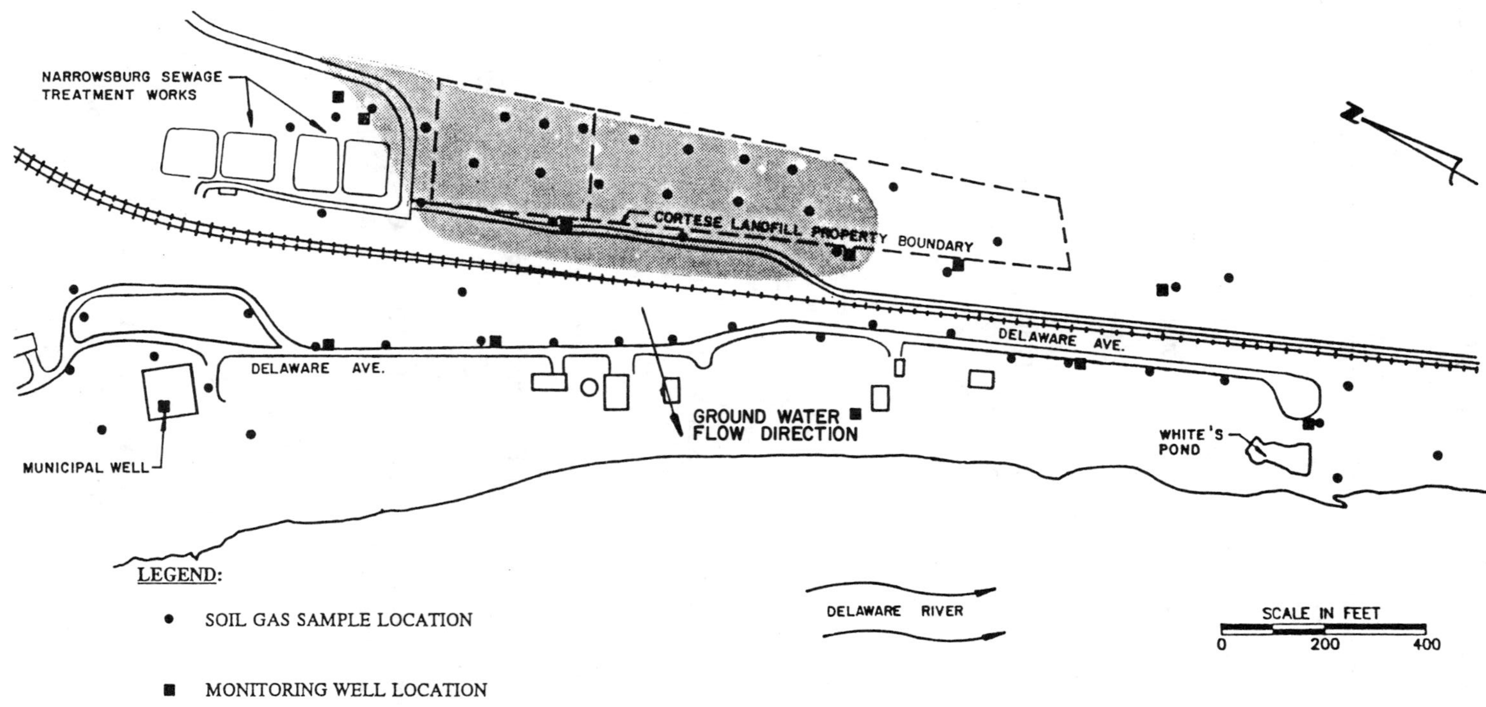

FIG. 2 -- EXTENT OF METHANE DETECTION

obtained. A sample of the soil gas was then taken with a gas tight syringe through the septum port of the sampling tee for analysis.

Each soil gas sample was analyzed in the field for the compounds listed in Table 1. Since the target compounds exhibit different chromatographic properties, it was necessary to use several GCs to do the analysis. The GC's were calibrated at least twice a day.

Replicate samples were analyzed at a rate of 1 per 20 samples to evaluate method precision. System losses were evaluated through the analysis of two spike samples during the work. Blanks were run between each analysis.

Methane was analyzed using a AID model 580 GC equipped with a flame ionization detector and a DC-200 packed column.

The remaining compounds were analyzed using a Photovac GC, equipped with a PID. Due to the variety of compound retention times, it was necessary to use two separate columns for the analysis. Initially, analysis was done using CPsil-5CB and KCl/Alumina columns. This configuration allowed identification of each compound with methylene chloride and 1,2 trans-dichloroethene being detected on both columns.

A preliminary assessment of the data was done as the field work was being completed. This analysis revealed that acetone concentrations being observed were much higher than expected, as discussed below in Section V. Acetone has a relatively short retention time. A review of chromatographic data revealed that several naturally occurring compounds have similar retention times to acetone on the CPsil-5CB column [6]. These compounds include carbon disulfide, dimethyl sulfide, dimethyldisulfide, and others. Thus, these compounds could cause false positive identifications and/or quantifications.

It was decided to use a GC configuration that allowed acetone identification on both columns to gain a greater degree of confidence in the data. Additional analysis was done using a CPsil-5CB and Carbowax B column. This configuration allowed dual column identification of vinyl chloride, acetone, methylene chloride, 1,2 trans-dichloroethene and methyl ethyl ketone.

This new configuration was used for the remainder of the work. Additionally, eight of the locations previously analyzed were retested to provide corroborative data between the two configurations.

V. DISCUSSION

Data collected during the soil gas survey showed it was unlikely that landfill gas was migrating from the site. As shown in Figure 2, methane was detected in soil gas at the landfill, but not at locations away from it. However, acetone was detected at locations away from the landfill, as discussed above. Effort was then concentrated on evaluating the possibility that this was due to migration from the landfill or the associated groundwater plume.

The presence of acetone in soil gases at sampling locations is not proof of significant horizontal migration, because these detections could arise from several other sources. As such, it is useful to consider the potential sources of volatile organic compounds in soil gas and the ways in which each would affect the soil gas data.

Groundwater contamination as a source of soil gas VOCs exhibits gradual horizontal concentration changes and increasing concentrations with depth. Shallow soil contamination exhibits high soil gas concentrations and very steep horizontal concentration gradients. Vertical concentration profiles in such instances show an increase to a maximum followed by a decrease with depth (depending upon the age of the spill) *[1 and 7]*. Landfill gas migration will exhibit gradual horizontal changes, high methane and/or carbon dioxide concentrations, and can show a sudden increase in concentration with sampling depth once barriers to vertical gas migration are penetrated.

Figure 3 is a contour map of soil gas concentrations of acetone as reported by the initial analytical system. Clearly, these data do not support the hypothesis of landfill gas migration from the facility as concentrations of acetone are higher at locations away from the landfill, near the municipal well. Additionally, the vertical concentration profile did not indicate the increase with depth characteristics of a groundwater source.

Figure 4 presents acetone concentrations detected in groundwater. As shown, there is a localized hot spot near the southern end of the landfill. Comparing Figures 3 and 4 shows that the soil gas survey also detected a hot spot near the southern end of the landfill consistent with the groundwater data. However, the elevated levels of acetone detected north-west of the site were not consistent with the groundwater data. Potential sources of this apparent acetone were investigated further.

Attention was focused on the analytical system in consideration of potential sources of error in the data. In an evaluation of gas chromatographs, a tendency to misidentify compounds was exhibited by 'universal' or non-selective gas chromatograph detectors (such as the PID) *[8]*. Because of this and the fact that concentrations of "acetone" reported by the analytical system were not reasonable, the validity of the identifications of 'acetone' was tested in the second round of sampling and analysis through dual-column gas chromatographic analyses and remote gas chromatography/mass spectrometric analysis of replicate samples.

Table 2 presents the results of analysis for acetone on the two columns used in the later part of the work. As shown, the concentration of "acetone" was higher in each case when analyzed on the CPsil-5CB column. This clearly indicates that other

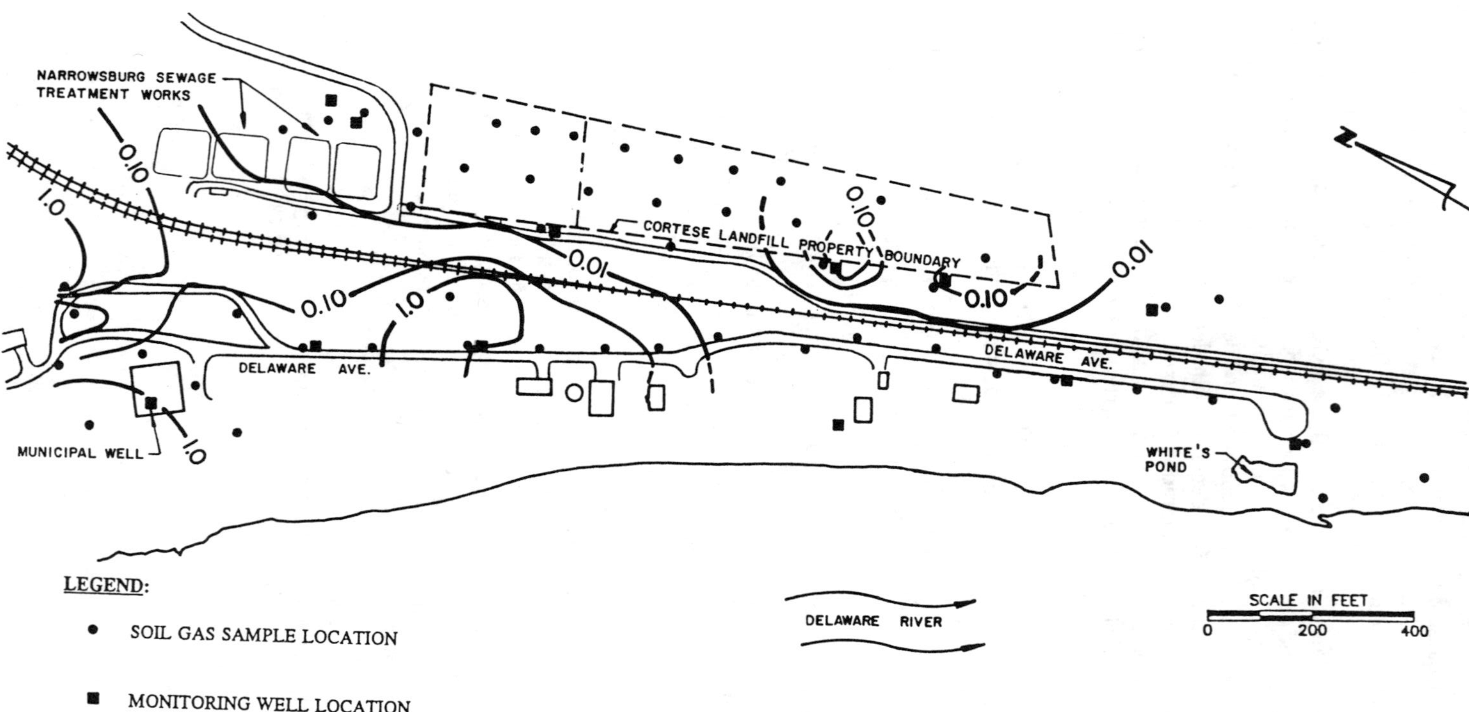

FIG. 3 -- ACETONE DISTRIBUTION IN SOIL GAS

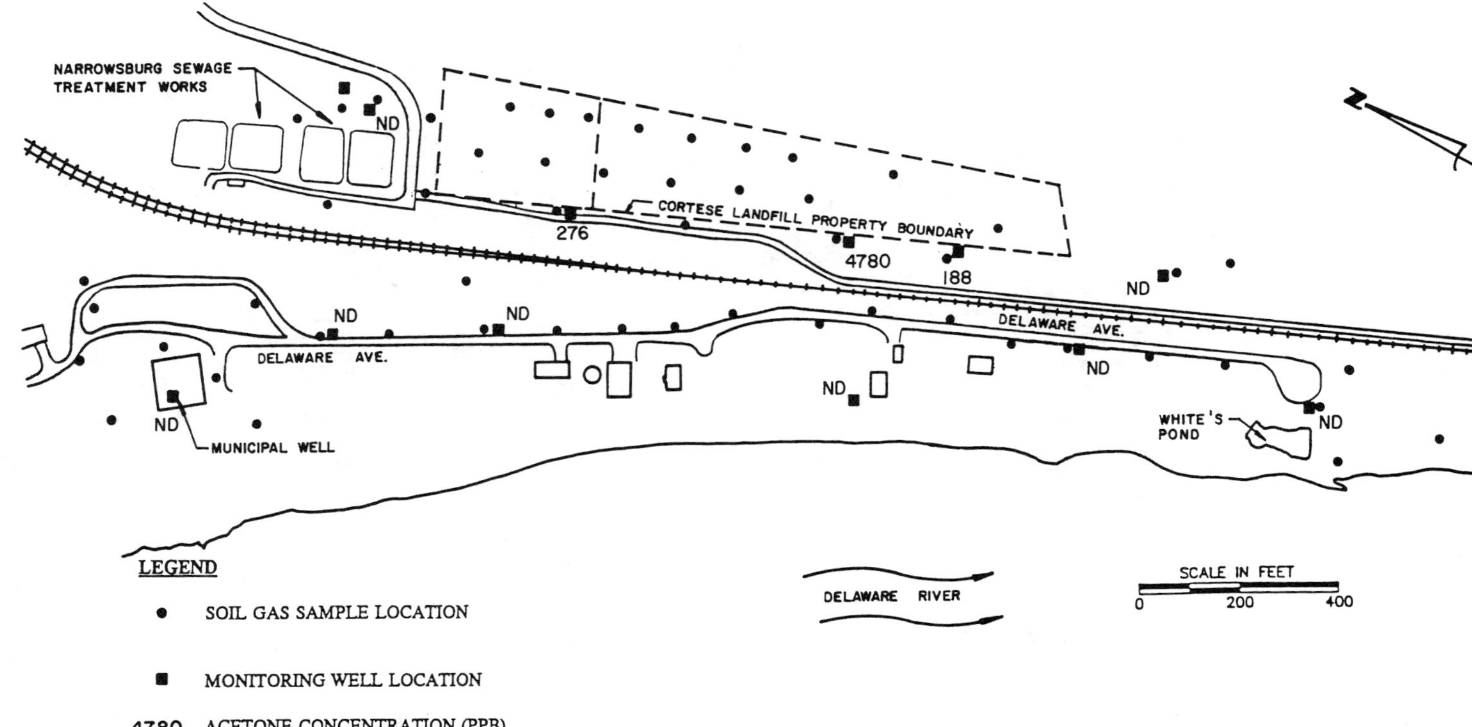

FIG. 4 -- ACETONE DISTRIBUTION IN GROUNDWATER

TABLE 2.--

ACETONE ANALYTICAL TEST RESULTS (PPMv)

Sample	CPsil-5CB Column	Carbopack-B Column
103.5	1.2	0.1
107.3	0.35	0.05
110.2	0.36	0.13
111.3	2.9	0.22
115.2	1.6	0.18
118.4	0.76	0.20
125.1	0.22	0.14
136.3	0.58	0.15

compounds may have been coeluting with acetone when the CPsil-5CB column was used. Independent off-site laboratory analysis by GC/MS methods indicated low or non-detectable levels of acetone.

Coeluting compounds are hypothesized to include sulfur compounds such as carbon disulfide and dimethyl sulfide that are biologically formed under anerobic conditions where methane and sulfur are present (i.e. the gases formed in codisposal facilities). These sulfur compounds would be readily detected by the photoionization detector used in this survey and have significant Henry's Law constants, which would allow them to be present in pore gases.

Results of remote laboratory analysis by gas chromatography with a mass spectrometric detector (GC/MS) of soil gas samples showed significant carbon disulfide in the sample from the location that had the highest "acetone" concentration reported originally, while the dual-column GC analyses showed the presence of coeluting compound(s). Since sulfides have relatively low ionization potentials, low concentrations in samples could result in large signals on the PID and positive errors in the reported acetone concentrations.

These results clearly indicate that there were misidentifications and misquantitations of acetone in the data set initially produced. This situation, if not identified and pointed out, would have resulted in a massive and senseless soil sampling and analysis effort.

VI. CONCLUSIONS

When the potential compounds present in soil gas samples are not well defined, caution must be exercised in assigning identities to the compounds due to the potential for misidentifications. When gas samples from landfills are analyzed misidentfification of compounds, including acetone, is common. The potential for uncertainty in compound identifications is increased by the use of a non-selective gas chromatographic detector, such as a PID with a high-energy lamp and isothermal (non-temperature-programmed) gas chromatograph operation.

Potential errors in identification are most likely for early-eluting compounds. This is because the resolution of early-eluting compounds is usually much less than later-eluting ones. Naturally occurring volatile compounds (which typically are early-eluting) can have the same or almost the same retention time as a target compound. Frequently, in order to optimize sample throughput and maintain the cost/benefit advantage of soil gas surveys, analyses are performed under conditions that are a compromise between chromatographic separation and total analysis time. Naturally formed compounds than can elute at or near the same time as early-eluting target compounds include dimethyl sulfide, dimethyldisulfide, nitrous oxide, carbon dioxide, and others. Coelution of these compounds with target compounds can cause qualitative and quantitative errors and result in overreacting to data.

All of these mundane observations about the science of producing accurate and defensible data are very important when regulators are forced to make decisions on the basis of uncertain or inaccurate data. In this instance, soil gas surveys at locations with a large number of potential components should be conducted with a flexible work plan, with target compounds selected for reasons driven by data needs for risk assessments or design of cleanup systems, and not a laundry list of compounds. Target compounds should be selected based upon a complete analysis of the potential source (i.e. leachate, landfill gas, etc.). The predominant and most reliably detected compounds should be selected for analysis in soil gas. The data reported to regulators should contain an identification of compounds that were frequently encountered but were not target compounds. Such identifications may require confirmatory setups such as dual-column confirmation of identities, off-site laboratory GC/MS analyses of samples taken in valid sample containers, or other steps.

Because of the potential cost of errors in the identities of compounds, it is a good idea to ensure that experienced degreed chemists be responsible for analytical data interpretation. In addition, personnel should be intimately familiar with the principle and operation of the instruments used. A thorough quality assurance plan should have measures to evaluate the validity of uncertain identifications, along with corrective actions to be taken if identification errors above an acceptable level are detected. The plan should clearly specify the contaminants of concern that will not be reliably detected or identified by the analyses planned. In addition, the specific analytical system to be used should have a known degree of reliability; the analytical system including the operator and its performance should be known before field deployment.

VII. REFERENCES

[1] Marrin, D.L.; H.B. Kerfoot, "Soil-Gas Surveying Techniques", Environmental Science Technology, Vol. 22, No. 7, 1988, pp. 740-745.

[2] Schlender, M.H.; Rowe, Journal of Environmental Progress, Vol. 8, No. 4, 1989, pp. 231-234.

[3] Kerfoot, H.B. Proceedings of the U.S. EPA Soil Vapor Extraction Technology Workshop, June 1989 Edison, NJ

[4] Nadeau, R.J., T.S. Stone, G.S. Klinger, Case Study, undated "Sampling Soil Vapors to Detect Subsurface Contamination; A Technique and Case Study", USEPA, Region 8 Environmental Response Team, p. 215-226.

[5] Guidance for Conducting Remedial Investigations and Feasibility Studies Under CERCLA, EPA/540/G-89/004, October, 1988.

[6] Photovac Technical Bulletin No. 5 "Monitoring of Some Widely Used Industrial Solvents in Ambient Air Using the Photovac 10S series Gas Chromatographs", Photovac, Inc., Thornhill, Ontario.

[7] Kerfoot, H.B., 1990. "Soil-Gas Sampling for Detection and Delineation of Groundwater Contamination", GLC '90, HMCRI: Silver Spring, Md.

[8] Kerfoot, H.B., S.J. Pierett, E.N. Amick, D.W. Bottrell, "Analytical Performance of Four Portable Gas Chromatographs Under Field Conditions", Journal of the Air and Waste Management Association, Vol. 40, No. 8, Aug. 1990, pp. 1106-1113.

Direct Push Technology in Ground-Water Investigations

Andrew I. Strutynsky, Raymond E. Sandiford, and Dennis Cavaliere

USE OF PIEZOMETRIC CONE PENETRATION TESTING WITH ELECTRICAL CONDUCTIVITY MEASUREMENTS (CPTU-EC) FOR THE DETECTION OF HYDROCARBON CONTAMINATION IN SATURATED GRANULAR SOILS

REFERENCE: Strutynsky, A. I., Sandiford, R. E., and Cavaliere, D., "Use of Piezometric Cone Penetration Testing With Electrical Conductivity Measurements (CPTU-EC) for the Detection of Hydrocarbon Contamination in Saturated Granular Soils," Current Practices in Ground Water and Vadose Zone Investigations, ASTM STP 1118, David M. Nielsen and Martin N. Sara, Eds., American Society for Testing and Materials, Philadelphia, 1992.

ABSTRACT: Piezometric Cone Penetration Testing with soil Electrical Conductivity measurements (CPTU-EC) was used for the detection of hydrocarbon saturated granular soils at two airport fuel storage areas. Details of the CPTU-EC equipment and site subsurface conditions are provided. Test program phases, including laboratory testing, field insitu testing and CPTU-EC data interpretation are described. Comparisons are made between CPTU-EC and adjacent monitor well data. Limitations to the CPTU-EC method are discussed.

KEY WORDS: Piezometric Cone Penetration Test, Soil Electrical Conductivity, free phase petroleum, hydrocarbon product contamination, airport fuel tank farms, computerized data acquisition, continuous soil profiling.

INTRODUCTION

Programs to remediate ground water accumulations of free phase petroleum hydrocarbon products, consisting primarily of aviation jet fuels, are ongoing at the fuel tank farms at John F. Kennedy (JFKIA), La Guardia, and Newark International Airports, in and around the city of New York. The Port Authority of New York and New Jersey (Port Authority) frequently requires supplemental ground water information in addition to that acquired in monitor wells at the sites.

Andrew Strutynsky, PE, is the Technical Director of STRATIGRAPHICS, 439 Taylor Ave., Glen Ellyn, IL 60137. Raymond Sandiford, PE, is a Senior Geotechnical Engineer and Dennis Cavaliere is a Geotechnical Engineer at the Port Authority of New York and New Jersey, 1 World Trade Center, New York, NY 10048.

STRATIGRAPHICS, a consulting company specializing in penetrometer data acquisition, was retained by Port Authority to evaluate the use of penetrometer soil electrical conductivity measurements (CPTU-EC) to delineate the accumulations of hydrocarbon products in the subsurface at the sites. STRATIGRAPHICS personnel had previously conducted similar studies on using penetrometer conductivity measurements for the detection of hydrocarbon contaminated soils and for the detection of ground water ice crystals in Arctic permafrost soils (Reference 1).

The penetrometer technique provides various advantages during geo-environmental subsurface investigations. These advantages include a relatively non-destructive test procedure; immediate, computerized data reporting and interpretation; continuous profiling; a high degree of exploration personnel safety; and lower exploration costs and higher productivity as compared to borehole techniques.

The experimental program for evaluating the applicability of penetrometer soil electrical conductivity measurements for the detection of free phase petroleum hydrocarbon products in ground water consisted of two phases. The first was a laboratory study using typical site soils, ground water, and jet fuel. A series of 60 tests was performed in order to establish a range of expected field measurements. The initial laboratory study was followed by field studies at the Satellite and Bulk Fuel Farms at JFKIA.

CPTU-EC soundings were performed adjacent to monitor wells for comparisons between penetrometer and monitor well data. CPTU-EC soundings were also performed at intermediate locations for correlation to the areal distribution of hydrocarbon product accumulations. A total of 48 CPTU-EC soundings were performed during the field study (Figures 1 and 2).

SOIL ELECTRICAL CONDUCTIVITY

Soil electrical conductivity is controlled by the conductance of the system of soil particles and fluids occupying the soil pore spaces. Factors affecting soil electrical conductivity, especially for sand aquifers, include:

<u>Mineralogy</u> Siliceous sand grains are essentially non-conductive, so granular soil electrical conductance is dependent on the quantity and conductance of the soil pore fluid. Clay minerals have some electrical conductance due to adsorbed water and ionic charges, thus clay conductance depends on both mineralogy and pore fluid characteristics.

<u>Pore Fluid</u> The electrical conductance of pore fluids plays the major role in granular soil electrical conductivity. Sands saturated with conductive fluids, such as saline water or landfill leachates, have a relatively high conductivity. Sands saturated with petroleum hydrocarbon products typically have low electrical conductivity because most petroleum hydrocarbon products are poor conductors.

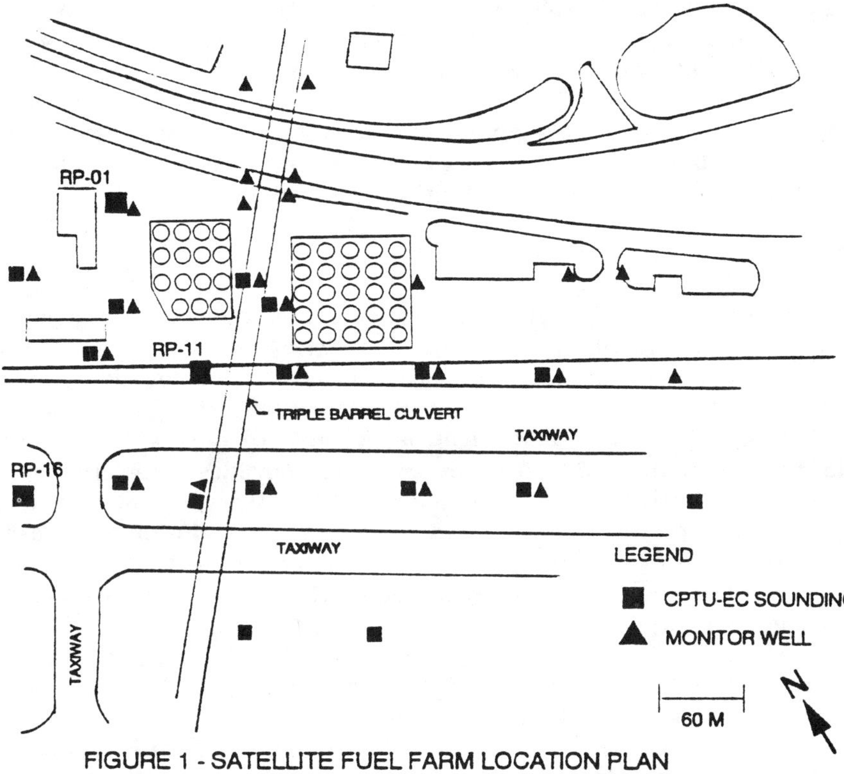

RP-01
RP-11
TRIPLE BARREL CULVERT
TAXIWAY
RP-16
TAXIWAY
LEGEND
CPTU-EC SOUNDING
MONITOR WELL
TAXIWAY
60 M
N

FIGURE 1 - SATELLITE FUEL FARM LOCATION PLAN

TIDAL BASIN
(SALT WATER)
SLURRY CUT-OFF WALL
BBP-12
BBP-35
LEGEND
CPTU-EC SOUNDING
MONITOR WELL
60 M
N

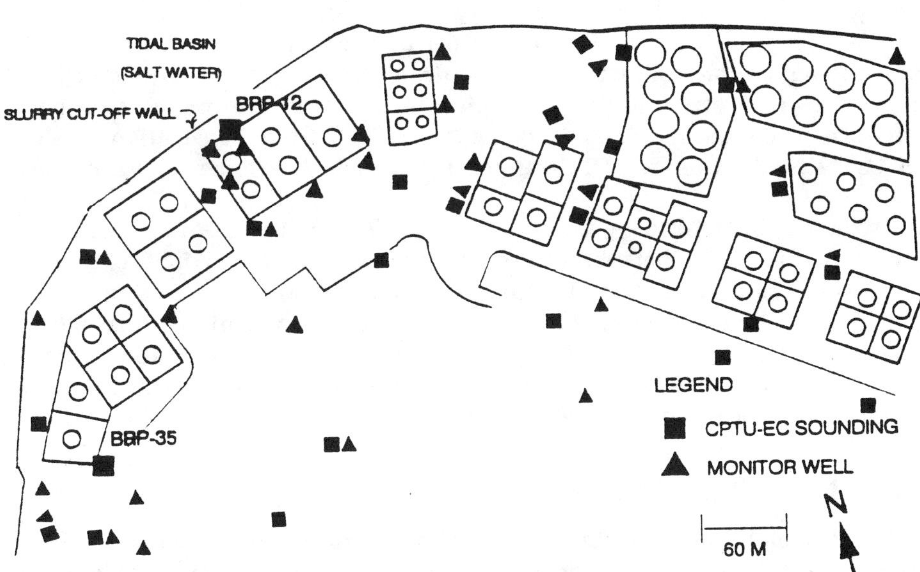

FIGURE 2 - BULK FUEL FARM LOCATION PLAN

Saturation The degree of soil saturation has a pronounced effect on soil electrical conductivity. Conductivity increases with increases in water saturation. Partially saturated sands have low electrical conductivity.

Porosity Soil porosity has an affect on soil electrical conductivity (Reference 2). Less pore fluid is required to fully saturate the pore space of a low porosity (dense) soil, resulting in lower soil electrical conductivity.

LABORATORY PROGRAM

The laboratory program to determine the effects of free phase petroleum hydrocarbon contamination on granular soil electrical conductivity included a total of 60 tests. Samples of soils from JFKIA site excavations, brackish (salty) ground water from site monitor wells, and samples of jet fuel were used to provide a range of variables that might be encountered during field testing. Soil samples were compacted to different porosities (densities) to determine the sensitivity of test results to porosity changes. Over the range of porosities expected to be representative for field conditions, the effects of porosity variation on soil electrical conductivity were considered to be relatively minor when compared to the changes in soil conductivity induced by variation in hydrocarbon content.

Laboratory testing showed that the electrical conductivity of the JFKIA sand samples depended primarily on the amount of water filling the soil pore spaces (degree of water saturation). Soil conductivity decreased with increasing substitution of pore water by jet fuel (Figure 3). A jet fuel saturated sand sample had an electrical conductivity similar to that of a dry sand sample.

The laboratory study indicated that in order to discriminate between dry sands above the water table, and free phase petroleum hydrocarbon product saturated sands below the water table, data on soil saturation was also required. A pore water pressure transducer, used during Piezometric Cone Penetration Testing (CPTU), was added to the CPTU-EC penetrometer to determine soil saturation.

Soil stratigraphy defined by Cone Penetration Test (CPT) measurements can be used to distinguish between the effects of soil type and pore fluid chemistry on measured soil conductivities. Thus, the soil shear resistance measurements of the CPT penetrometer, the piezometric measurement of the CPTU penetrometer, and soil electrical conductivity measurements were all combined in a CPTU-EC penetrometer in order to provide sufficient data to define petroleum hydrocarbon product contamination of saturated granular soils.

PENETROMETER TECHNIQUE

CPTU-EC penetrometer testing consists of smoothly pushing a small diameter (0.044 m - 1.7 inch), instrumented probe (penetrometer) directly into the ground, while a computer data acquisition system displays and records the soil shear resistance, pore water pressure response and soil electrical conductivity during penetration (Figure 4).

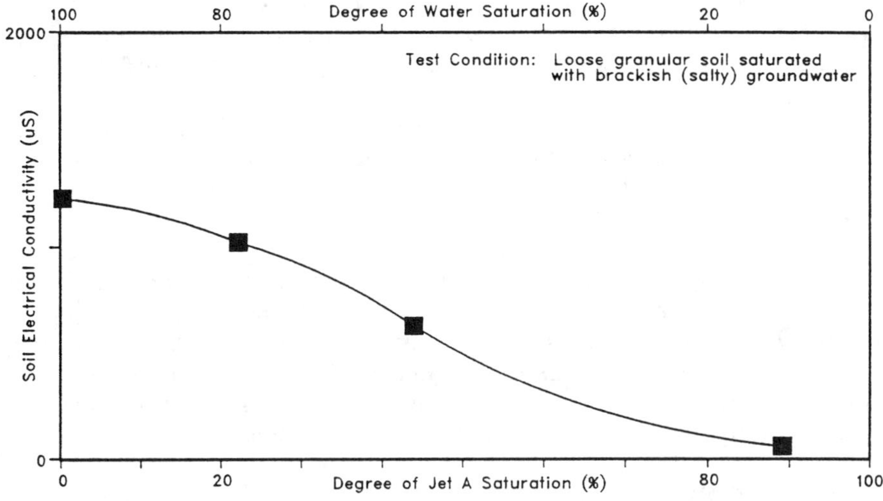

FIGURE 3 — EFFECT OF JET A PRODUCT CONTAMINATION ON GROUNDWATER SATURATED GRANULAR SOIL ELECTRICAL CONDUCTIVITY

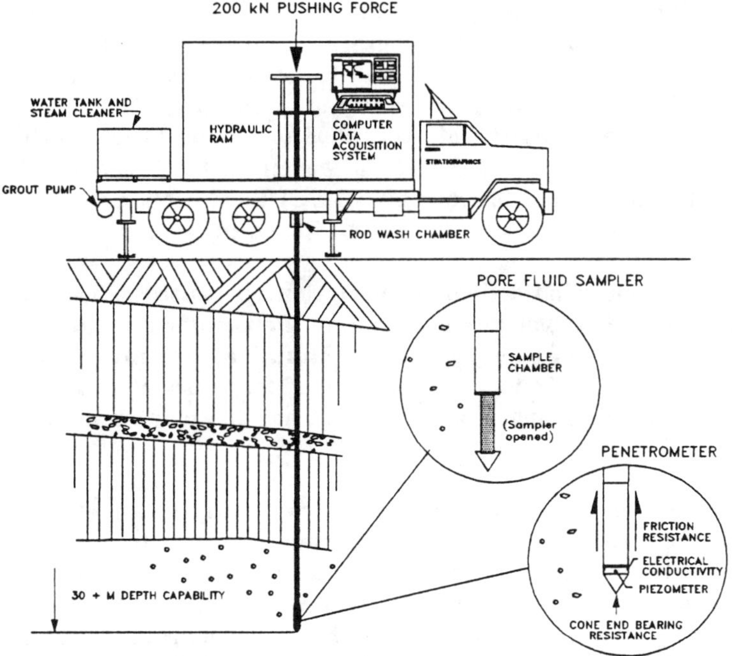

FIGURE 4 — PENETROMETER SUBSURFACE EXPLORATION SYSTEM

The penetrometer is mounted at the downhole end of a string of sounding rods. A hydraulic ram is used to smoothly push the penetrometer and rod string directly into the ground, without drilling a borehole, at a constant rate of 0.02 m/sec (4 ft per minute). Electronic signals from downhole sensors inside the penetrometer are transmitted by a cable, strung through the hollow sounding rods, to a data acquisition and display computer system at the surface.

CPTU-EC data are used to develop continuous profiles of geotechnical, hydrogeological, and gross geochemical soil conditions rapidly, accurately and economically. Penetrometer samplers can be used to obtain ground water or soil samples for laboratory testing (Reference 3).

Site disturbance is minimized since no borehole cuttings or drilling fluids are generated during penetrometer operations. Personnel exposure to contaminated soil is significantly less than exposures during drilling and sampling. Penetrometer downhole equipment is easily decontaminated by steam cleaning during retrieval. The small open hole left in soils above the water table after penetrometer retrieval is readily grouted.

CPTU-EC PENETROMETER MEASUREMENTS

The CPTU-EC penetrometer incorporates cone resistance, friction sleeve resistance, piezometric, thermal and soil electrical conductivity sensors. The resistance of a soil to penetration is measured on the tip and along the sides of the CPTU-EC penetrometer. The soil resistance acting on the cone tip is controlled primarily by soil grain size and porosity. The cone resistance measurement has a resolution of about 0.05 to 0.10 m (2 to 4 inches). The sliding friction between the soil and the penetrometer is measured along a sleeve mounted just behind the cone tip. The CPT-EC friction sleeve resistance measurement has a resolution of about 0.15 m (6 inches).

A pressure transducer in the tip of the penetrometer is used to measure the soil pore water pressure response to penetration. Pore water pressure response is primarily controlled by the degree of saturation, potentiometric surface, compressibility and horizontal permeability of the penetrated soil (Reference 4). The CPTU-EC piezometric measurement has a resolution of about 0.03 m (1 inch).

The soil electrical conductivity is measured between two electrodes also mounted in the tip of the CPTU-EC penetrometer. The electrodes are insulated from the steel body of the penetrometer by plastic insulators. The CPT-EC soil electrical conductivity measurement has a resolution of about 0.04 m (1.5 inches). A thermistor inside the CPTU-EC penetrometer provides data on downhole equipment temperatures. These data can be used to adjust the measured soil conductivity to a corrected conductivity at a reference temperature of 25 degrees C.

CPTU-EC data are acquired as analog signals from the transducers inside the penetrometer. The analog signals are transmitted by cable strung through the sounding rod string to a computerized data acquisition system inside the penetrometer truck. The data acquisition system translates the analog signal to a digital value using a 16-bit, analog to digital (A/D) converter. The 16-bit conversion provides a digital data resolution of 1 part in 32,768.

The CPTU-EC data are logged at a 2 Hz frequency. This logging frequency provides insitu soil data at about 0.01 m (3/8 inch) depth intervals. Data appear on a high resolution, color computer monitor in real time. Real time data display allows for the immediate definition of site conditions. Data are logged on hard disk for permanent storage. A preliminary, hard copy sounding log is generated at the conclusion of each test. Recorded data are computer processed to develop interpretations of site conditions.

GENERAL CPTU-EC DATA INTERPRETATION

Correlations between penetrometer data and soil type classifications have been developed from geotechnical soil bearing capacity theory, and observational criteria from adjacent CPT soundings and drilled and sampled boreholes (Reference 5). The CPT cone resistance increases exponentially with increases in soil grain size. The CPT friction ratio (the friction sleeve resistance divided by the cone resistance) increases with increases in the fines content of a soil. A correlation scheme based on the cone resistance and friction ratio values (Figure 5) has proved most useful in interpreting soil types from CPT measurements.

Soil saturation is evaluated using the CPTU-EC piezometric data. Atmospheric (zero) water pore pressure is measured in unsaturated soils. Hydrostatic pore water pressures are generally recorded in high permeability, granular soils below the water table. High pore water pressures are recorded in saturated, fine grained soils during penetrometer advance.

CPTU-EC FIELD TESTING PROGRAM

A total of 48 CPTU-EC soundings were performed at the JFKIA Satellite and Bulk Fuel Farms. The stratigraphy at the two sites is somewhat similar. The surficial soils at both sites consist of a hydraulically placed, fine to medium sand fill, ranging in thickness from about 1.5 to 4.6 m (5 to 15 ft).

At the Satellite Fuel Farm site, this sand fill overlies a discontinuous tidal flat deposit, which consists of about 0 to 1.5 m (0 to 5 ft) of silty clay and peat. At the Bulk Fuel Farm, heterogeneous deposits of refuse and silt interlayer the hydraulic sand fill and tidal flat deposits. Underlying the tidal flat deposits at both sites is a fine to medium sand stratum in excess of 30.5 m (100 ft) thick.

The tidal flat deposits form discontinuous aquitards across the sites, resulting in both locally perched and water table (unconfined) aquifer ground water conditions. At the Bulk Fuel Farm site, the ground water has been partially contained by a slurry cut-off wall. The deeper ground water at both sites is brackish (somewhat salty) with moderate electrical conductivity. Shallow ground water is typically less salty and less conductive, probably reflecting a recent rainwater origin.

The JFKIA Satellite and Bulk Fuel Farms have significant subsurface accumulations of aviation jet fuel, as determined in monitor wells at the two sites. For the Satellite Fuel Farm, free phase petroleum hydrocarbon product thicknesses interpreted from CPTU-EC data were compared to product thicknesses measured in nearby monitor wells. This comparison showed that the general thickness patterns were very consistent, but that the insitu CPTU-EC data indicated product thicknesses to be generally 25 to 50% less than the monitor well product thicknesses.

These results confirm the hypothesis that monitor wells generally contain a thicker accumulation of free phase petroleum hydrocarbon product than is actually present in the soil. This occurs because most products float on the capillary zone above the water table. Thus, the product fills a monitor well for the thickness of the capillary zone and for a depth below the ground water table required to achieve buoyancy equilibrium between the product and ground water.

An uncontaminated, water table (unconfined) aquifer is indicated by the CPTU-EC sounding log at the Satellite Fuel Farm Location RP-16 (Figure 6). The shallow stratigraphy consists of a homogeneous sand stratum. The piezometric measurements indicate the sand to be of medium to high permeability, and indicate a water table at a depth of 1.65 m (5.4 ft).

The soil electrical conductivity increases just above the water table, reflecting increasing soil water content. Soil conductivities are relatively low and constant below the water table, reflecting low ground water salinity conditions. It was subsequently determined that a nearby water main was leaking, and the fresh water leakage was probably responsible for the low soil electrical conductivity measurements.

An accumulation of free phase petroleum hydrocarbon product is indicated by the CPTU-EC sounding log at the Satellite Fuel Farm Location RP-01 (Figure 7). The piezometric measurements indicate a free fluid surface at 1.83 m (6.0 ft) of depth. The very low soil conductivity between 1.83 and 2.10 m (6.0 and 6.9 ft) depths indicates a thin layer of product. Increasing soil conductivity below the product layer indicates increasing ground water salinity and density with depth.

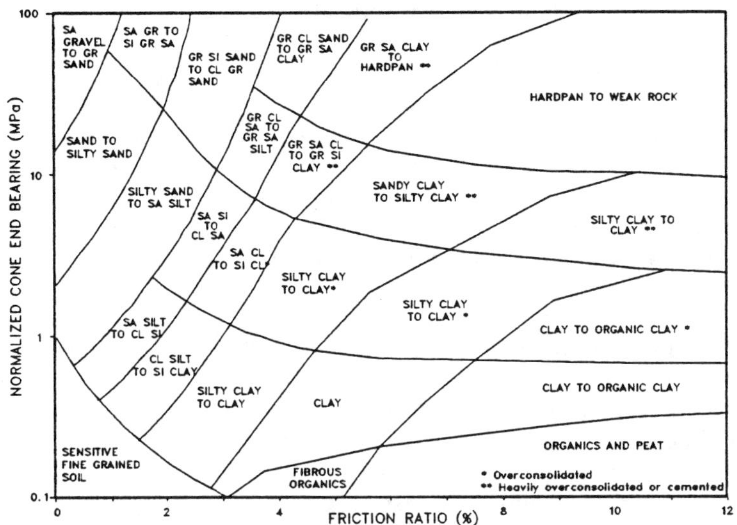

FIGURE 5 – CORRELATION CHART FOR CPT SOIL TYPES

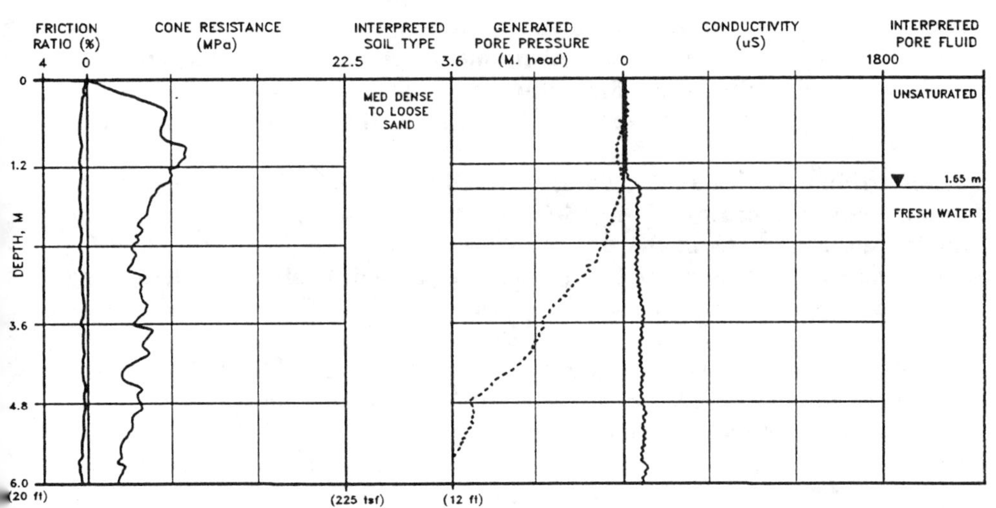

FIGURE 6 – CPTU–EC SOUNDING RP–16

Unusual results were obtained at the Satellite Fuel Farm Location RP-11 (Figure 8) next to a product recovery well. The CPTU-EC data indicate 0.85 m (2.8 ft) of a ground water-petroleum hydrocarbon product mixture, overlying a 0.76 m (2.5 ft) thick layer of product. The product layer overlies another mixed layer, which in turn overlies ground water.

This unexpected sequence is thought to be due to rapidly changing ground water conditions. Record rainfalls during the autumn of 1989 are conjectured to have both raised the locally depressed water table and filled in the cone of depression created by the nearby recovery well. The former surficial product layer became inundated. It is interpreted that due to soil permeability effects, insufficient time had passed prior to the December, 1989 CPTU-EC study for fluid density equilibrium between product and ground water to have been re-established .

This interpretation has been corroborated by CPTU-EC soundings combined with penetrometer ground water sampling at other project sites with similar rapidly changing ground water conditions. A monitor well, typically screened 1.5 m (5 ft) above and 3.0 m (10 ft) below the water table, would provide no hint of this phenomenon, because density equilibrium would occur almost instantaneously in the monitor well riser pipe.

Many of the CPTU-EC sounding logs at the Bulk Fuel Farm were not as definitive as those at the Satellite Fuel Farm. Product thickness trends from the CPTU-EC soundings did generally correspond with monitor well defined trends. However, the presence of perched ground water and product, numerous trapped product lenses, and complex ground water flow conditions caused by the slurry cut-off wall and the discontinuous aquitard, caused CPTU-EC data interpretation to be much more subjective than at the Satellite Fuel Farm.

Ground truthing the CPTU-EC data, acquired at 0.01 m (3/8 inch) intervals to monitor wells screened over 5 m (15 ft) lengths may not be appropriate for the complex site conditions at the Bulk Fuel Farm. The CPTU-EC sounding log at Location BRP-12 (Figure 9) illustrates some of the difficulties in interpreting data at sites with a complex hydrostratigraphy.

The CPTU-EC results were definitive in areas of the Bulk Fuel Farm where uniform conditions existed. The presence of free phase petroleum hydrocarbon product overlying a water table aquifer is indicated by the CPTU-EC sounding log at Location BRP-35 (Figure 10). Soil electrical conductivity is very low between the free fluid surface at a depth of 2.80 m (9.2 ft) and a depth of 3.29 m (10.8 ft), indicating a 0.49 m (1.6 ft) thick layer of product. A 0.15 m (0.5 ft) thick transition zone underlies the product layer and probably consists of soil saturated with both product and ground water.

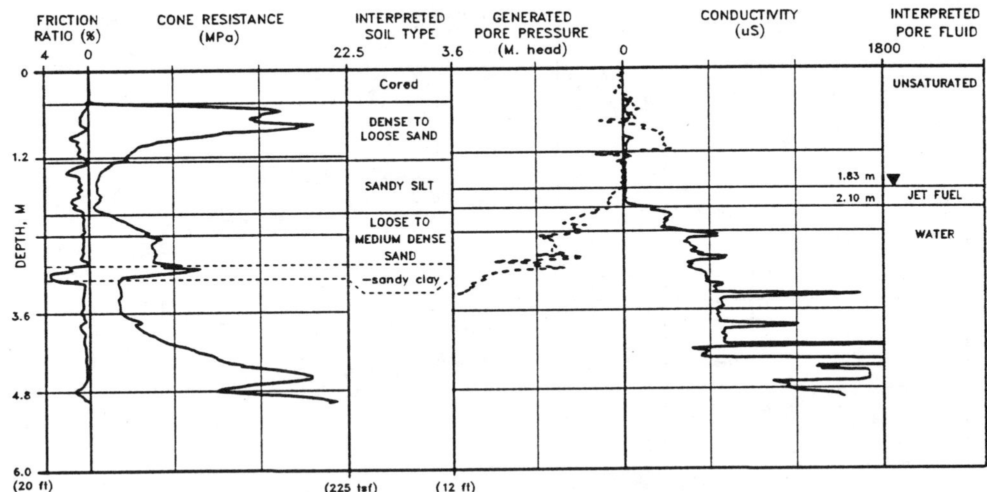

FIGURE 7 - CPTU-EC SOUNDING RP-01

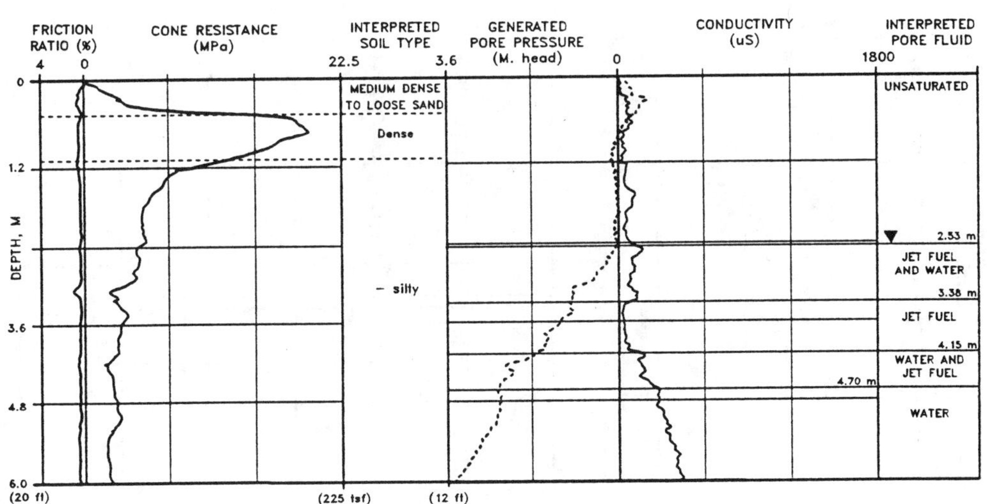

FIGURE 8 - CPTU-EC SOUNDING RP-11

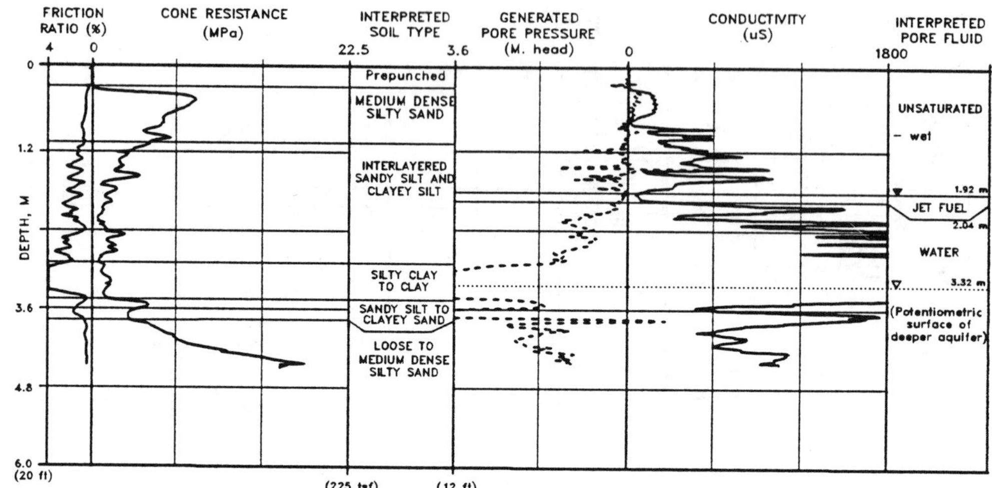

FIGURE 9 — CPTU-EC SOUNDING BRP-12

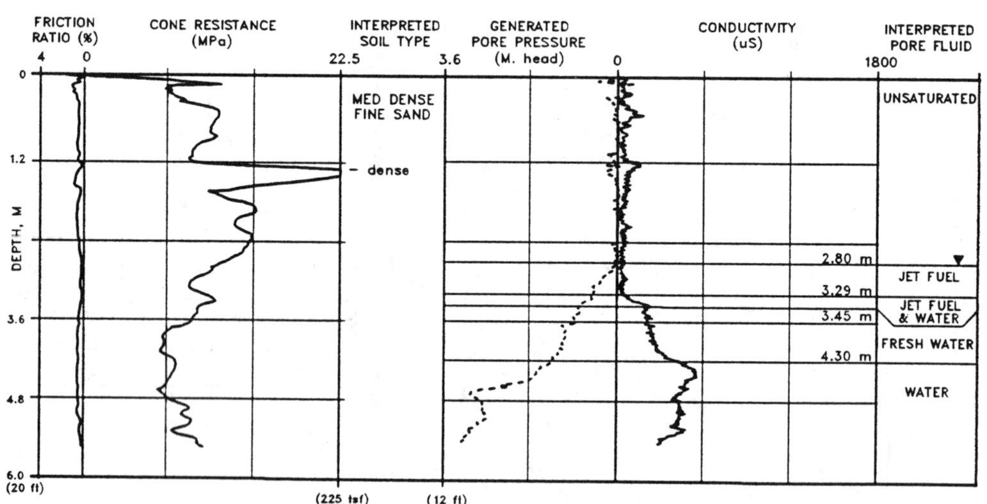

FIGURE 10 — CPTU-EC SOUNDING BRP-35

CPTU-EC COSTS AND PRODUCTIVITY

A comparison of production rates and costs for CPTU-EC and conventional monitor well surveys is as follows:

	CPTU-EC	Monitor Wells
Production Rate	8 to 12/day	1 to 2/day
Unit Cost	$66/m ($20/ft)	$131/m ($40/ft)
Cost per location, 6.1 m (20 ft) depth	$485/ea.*	$1066/ea.*

*** includes data interpretation or inspection.**

The CPTU-EC method provides a more rapid means of surveying an area, and is less than one half the cost of conventional monitor well survey methods on a per location basis.

CONCLUSIONS

The CPTU-EC penetrometer method has been shown to provide a rapid means of surveying sand aquifers for free phase petroleum hydrocarbon product contamination. In areas of more complex stratigraphy, additional testing is necessary to verify the applicability of CPTU-EC methods. Monitor wells with long screened lengths may not provide the best method of ground truthing CPTU-EC measurements at sites with complex hydrostratigraphic conditions.

Penetrometer ground water sampling should be included in CPTU-EC field investigation programs to provide direct samples of CPTU-EC identified anomalous ground water zones. Sensitive CPTU-EC piezometric transducers should be used to provide high accuracy in water table location.

The rapidity and the relative non-destructive nature of the CPTU-EC method especially provides advantages in areas of high priority usage or sensitivity, such as active apron areas of airport terminals, or in residential areas surrounding contamination sources. The CPTU-EC method in many cases, allows for more rapid and better definition of the true thickness of free phase petroleum hydrocarbon products in ground water. Cost savings in initial survey work should translate into better placement of permanent monitor and recovery wells, resulting in decreased overall remediation/investigation program costs.

REFERENCES

[1] Strutynsky, A.I., B.J. Douglas, L.J. Mahar, G.F. Edmonds, and E. Hencey, 1985. Arctic Penetration Tests Systems. Civil Engineering in the Arctic Offshore, ASCE, pp 162-168.

[2] Kutter, K.L., K. Arulanandan, and Y.F. Dafalias, 1979. A Comparison of Electrical and Penetration Methods of Site Investigation. Offshore Technology Conference Proceedings.

[3] Strutynsky, A.I., T.J. Sainey, 1991. Use of Piezometric Cone Penetration Testing and Penetrometer Ground Water Sampling for Volatile Organic Contaminant Plume Detection. Petroleum Hydrocarbons and Organic Chemicals in Ground Water. API/NWWA, Conference Proceedings, pp 71-84.

[4] Saines, M., A.I. Strutynsky, and G. Lytwynyshyn, 1989. Use of Piezometric Cone Penetration Testing in Hydrogeologic Investigations. First USA/USSR Hydrogeology Conference, Moscow, USSR, NWWA.

[5] Douglas, B.J., R.S. Olsen, 1981. Soil Classification using the Electric Cone Penetrometer. Cone Penetration Testing and Experience, Conference Proceedings, ASCE, pp 209-227.

Paul D. Kuhlmeier and Thomas E. Sturdivant

DELINEATION OF LITHOLOGY AND GROUNDWATER QUALITY IN A COMPLEX FLUVIAL ESTUARINE DEPOSITIONAL ZONE

REFERENCE: Kuhlmeier, P. D. and Sturdivant, T. E., "Delineation of Lithology and Groundwater Quality in a Complex Fluvial Estuarine Depositional Zone," Current Practices in Ground Water and Vadose Zone Investigations, ASTM STP 1118, David M. Nielsen and Martin N. Sara, Eds., American Society for Testing and Materials, Philadelphia, 1992.

ABSTRACT: In the development of a shallow subsurface investigation design strategy, a fundamental objective is the quantification of pollutants both vertically and horizontally within the subsurface. The dilemma produced in the design phase is one of a tradeoff between study costs and sufficient data collection to meet the study objectives. A novel application of two devices normally used in geotechnical engineering was tested at Naval Air Station Moffett Field for their ability to provide low cost, and timely hydrogeologic and chemical data.

The first device, a cone penetrometer, is capable of determining variations in soil media physical properties, both in saturated and unsaturated media, by measuring friction and pore water pressure. It provides a continuous log of unconsolidated materials through depth thus allowing for delineation of aquifer and aquitard zones. The second device, a hydropunch, is similar in nature to a well point and is capable of collecting in-situ pore water samples from depths up to 250 feet. A volume of 500 ml may be retrieved through a porous stainless steel tip and removed by closing upper and lower check valves.

Over 130 such CPT/Hydropunch measurements were used to characterize the mixed rivertine and estuarine sediments at the Mountain View, California Naval Air Station. The base which is on the National Priorities List contains numerous sites where pollutants have been identified. The study described in this paper focuses on the methods and results obtained from comprehensive use of the CPT/Hydropunch™ screening methodology in variably saturated media, to a depth of 60 feet.

KEY WORDS: estuarine sediments, water quality, lithology, San Francisco Bay, Cone Penetrometer, Hydropunch

Dr. Kuhlmeier is Manager of Special Projects at Morrison Knudsen Corp., P.O. Box 73, Boise, ID 83729; Mr. Sturdivant is a Senior Hydrogeologist at Automated Sciences Group Inc., Oliver Springs, TN, 83704.

The recent expansion of activity in hazardous waste site investigation has resulted in the development of new technologies and the application of technologies from other fields. The need for a method of groundwater sampling that does not require the expense and commitment of a monitoring well may be met with the Hydropunch™ (Hp). The need for lithology determination and aquifer delineation without the expense and waste associated with drilling may be met with the cone penetrometer (CPT). Combined use of these technologies allows for the identification of permeable water bearing zones in stratified sediments and the collection of groundwater from selected zones. This approach proved to be very effective and provided data that are critical to characterization of a site which involves monitoring of five discrete subaquifers. The focus here is on one of many sites (Site 9) under investigation at Moffett Field Naval Air Station located at the southern end of San Francisco Bay, which acts as a submarine tracking facility in Mountain View, California.

One of the fundamental problems associated with the Moffett Field RI has been determining the distribution and correlatibility of the shallow aquifers. Aquitards are believed to be extensive and fairly persistent, while aquifer zones are seemingly sporadic. Determining the distribution of stratigraphically correlatable shallow aquifer zones was therefore one of five objectives identified in the characterization of Site 9 through the use of CPT/Hydropunch technologies.

Determining contaminant distribution and concentrations was also a primary objective that was closely linked to understanding the aquifer system. Since the areal distribution of groundwater contamination was being investigated, the permeable zones suspected of lateral continuity were targeted. The vertical distribution of contaminants was also a concern since two aquifer zones are known to be contaminated and two others are suspected of containing pollutants. It was theorized that the relationship of contaminant distribution within the two uppermost shallow aquifers would aid in determining the degree of aquifer interconnection.

A third data use objective associated with these issues was monitoring well siting. Several approaches for using CPT and Hp data to locate monitoring wells were used. One approach was simply determining if suitable aquifer material and contamination were present at a proposed location. Another approach used CPT/Hp data to estimate downgradient plume extent for location of monitoring wells on the downgradient edge. Since multiple contaminant sources exist at Site 9, the CPT/Hp data were also used to determine plume connection laterally. The use of the CPT/Hp data for determining contaminant plume distribution meant simply using the Hp-collected groundwater sample analysis as a data point. Using the data point approach satisfies one of the basic objectives for monitoring well installation (i.e., determining contaminant concentration at a point) and can result in the reduction of the number of wells.

The fourth overall objective associated with the CPT/Hp study at Site 9 was identifying the contaminant sources. The multiple source situation at Site 9 is complicated by an upgradient plume invading the site. Finally, the siting of aquifer test wells was a fifth objective. It was believed that the thickest aquifer sections within a depth correlatable interval were likely to be the most continuous laterally and therefore capable of the greatest capture zone during potential groundwater remediation.

SITE DESCRIPTION

Subsurface sediments at the site are unconsolidated and range from clay to gravel in size. The stratigraphy at Site 9, and at Moffett Field at large, is derived from an estuarine depositional environment subjected to several fluvial influxes of material from the alluvial fans that developed along the Santa Cruz highlands. Extensions of these fans subjected to wave action resulted in the redistribution of coarse alluvial material through shoaling. Periods of high sea level resulted in the deposition of low permeability bay mud. The permeable water bearing zones at Moffett result from bar sands, channel sands, and alluvial fan outwash. The shallow more permeable beds thin and become more discontinuous as one moves away from the alluvial fans northward toward the bay. These shallow aquifer zones occur over an interval of 10 to 55 feet (3 to 17 m) in depth and range in thickness from less than one to 15 feet (0.3 to 4.6 m). A fairly persistent aquitard occurs near the middle of this zone. Permeable beds that occur in the upper half of this sequence are known as the A_1 Aquifer, and those that occur in the lower half are known as the A_2 Aquifer. The occurrence of these beds is not consistent, and one or both may be absent at a location. Figure 1 presents a generalization of the shallow aquifer system.

Groundwater potentiometric head at Moffett increases with depth, resulting in an upward vertical gradient over most of the Base. The aquifers range from confined at depth (>300 feet) to semiconfined and unconfined at higher elevations. Much of the Base is below sea level and exhibits an upward gradient into the shallow A aquifers. Site 9, however, is above sea level and may have a downward vertical gradient between the A_1 and A_2 Aquifers.

Contamination at Site 9 includes fuel components solvents, and inorganics. Several suspected sources are within and upgradient of Site 9. These include underground storage tanks, paint waste sumps, a dry cleaning sump, an oil/water separator, and a vapor condensation recovery tank.

Previous investigations at this site included coring, borehole geophysical logging and collection of surface soil samples and shallow borehole samples, installation of a limited number of monitoring wells, groundwater and soil sample analysis, and a soil gas survey. The results of the soil gas survey and progressive downgradient positioning from the sources were used in determining the CPT and Hp locations.

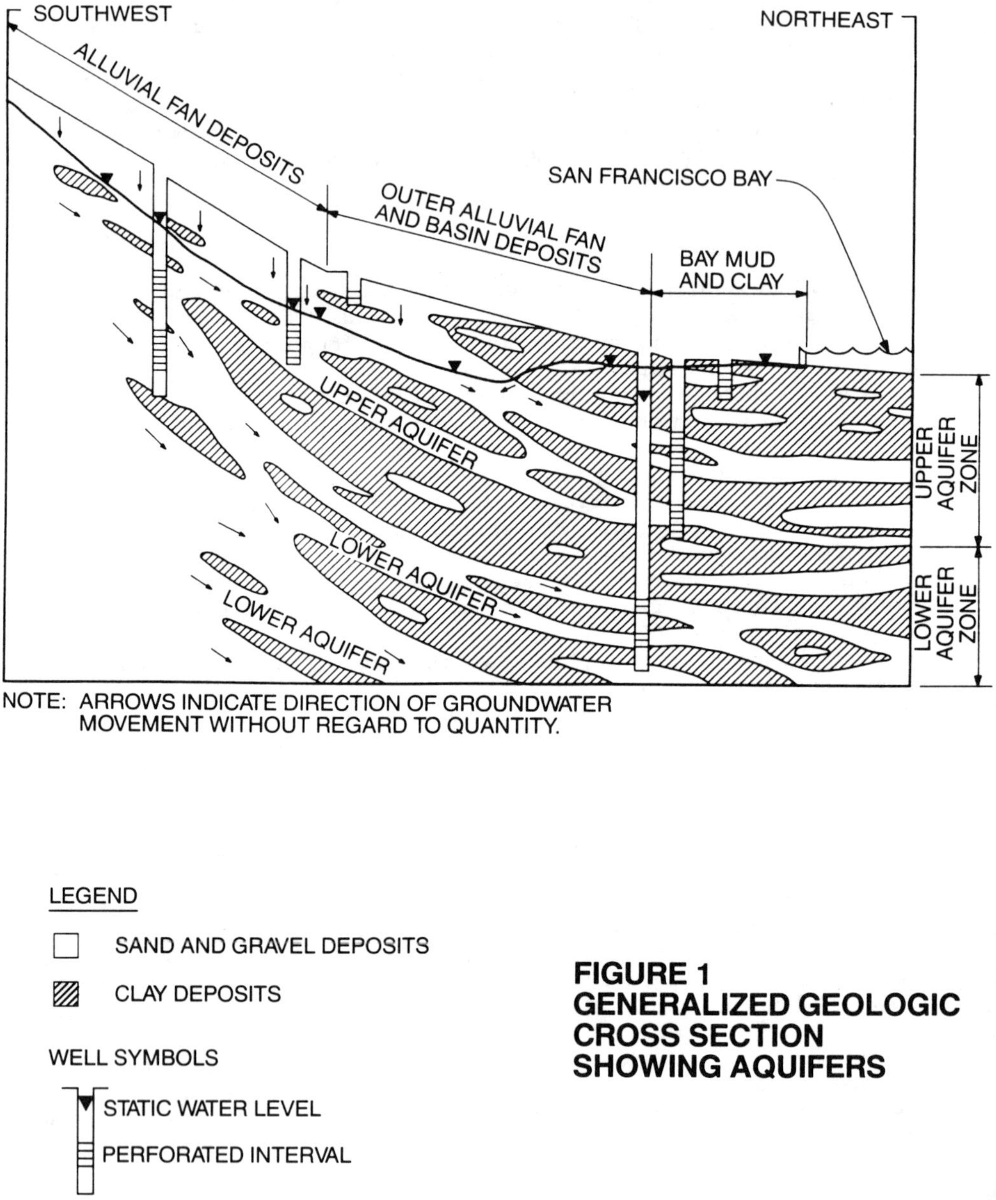

SOUTHWEST

NORTHEAST

ALLUVIAL FAN DEPOSITS

SAN FRANCISCO BAY

OUTER ALLUVIAL FAN AND BASIN DEPOSITS

BAY MUD AND CLAY

UPPER AQUIFER

UPPER AQUIFER ZONE

LOWER AQUIFER

LOWER AQUIFER

LOWER AQUIFER ZONE

NOTE: ARROWS INDICATE DIRECTION OF GROUNDWATER MOVEMENT WITHOUT REGARD TO QUANTITY.

LEGEND

SAND AND GRAVEL DEPOSITS

CLAY DEPOSITS

WELL SYMBOLS

STATIC WATER LEVEL

PERFORATED INTERVAL

FIGURE 1
GENERALIZED GEOLOGIC CROSS SECTION SHOWING AQUIFERS

THE CONE PENETROMETER

The Cone Penetrometer consists of a hydraulically driven steel probe that is equipped to measure the geotechnical properties of soil and data recording equipment. An electronic cone was used in this study. This cone is equipped with strain gauges, a pressure transducer, and an inclinometer. A 60° apex cone tip with a 10 cm^2 projected cross section area is the leading end of the probe, which is connected to a strain gage and measures the penetration resistance. The tip is also equipped with a porous stone piezoelement coupled to a pressure transducer. This provides measurement of pore pressure. The side of the probe is equipped with a sleeve that responds to the friction resistance incurred during penetration. The sleeve is connected to another strain gage within the probe. The inclinometer within the probe measures plumbness. Electronic signals from the pressure transducer, the strain gages, and the inclinometer are transmitted through a cable in the hollow push rods to analog and digital data recorders, which compose the data acquisition system. The system is housed in the cabin of a 20-ton (18 metric) truck. Data on cone tip resistance, sleeve friction resistance, friction ratio, depth, inclination, pore pressure, and differential pore pressure ratio are usually recorded. Two popular cone designs are presented in Figure 2. The cone used in this study included a porous filter in the cone face as shown.

The truck is rigged with a hydraulic jacking system used to drive the penetrometer. The load of the hydraulic ram is transferred to the punch rods by a hydraulic clamping system. The maximum thrust the push rods can withstand is 45,000 pound-force (lb_f) [200 kiloNewton (kN)]. A 45,000 lb_f (200 kN) thrust will penetrate depths up to 100 feet (30 m) in dense or still materials and up to 260 feet (80 m) in looser soils. Power is supplied by the power-takeoffs of the truck.

In this study, an interpretation of the CPT parameters was performed with computer software. An interpreted printout was provided that described the types of soils penetrated. The interpretation was based upon the relationship between the tip (cone) resistance and the friction ratio. The friction ratio is the sleeve friction (local friction) divided by the tip resistance, multiplied by 100, and expressed as a percent. The types of soils associated with various numerical relationships are available on soil classification charts, which plot cone resistance against friction ratio. Several charts are available with slightly different plotting of soil types based on that author's experience[1]. However, all of the charts exhibit a similar relationship between tip resistance, friction ratio, and soil types. In this relationship, sandy soils have high tip resistance and low friction ratios whereas clayey soils have low tip resistance and high friction ratios. Although both tip resistance and local friction increase with grain size, the rate of increase is not equal, which results in a numerical relationship beneficial to interpretation.

Interpretation based upon the relationship of these parameters has classically been accepted. Recently, however, the increased use of a porous filter piezoelement in the cone has brought a lot of attention to the role of pore pressure in CPT data and interpretation. In simple

FIGURE 2 PIEZOMETER CONE

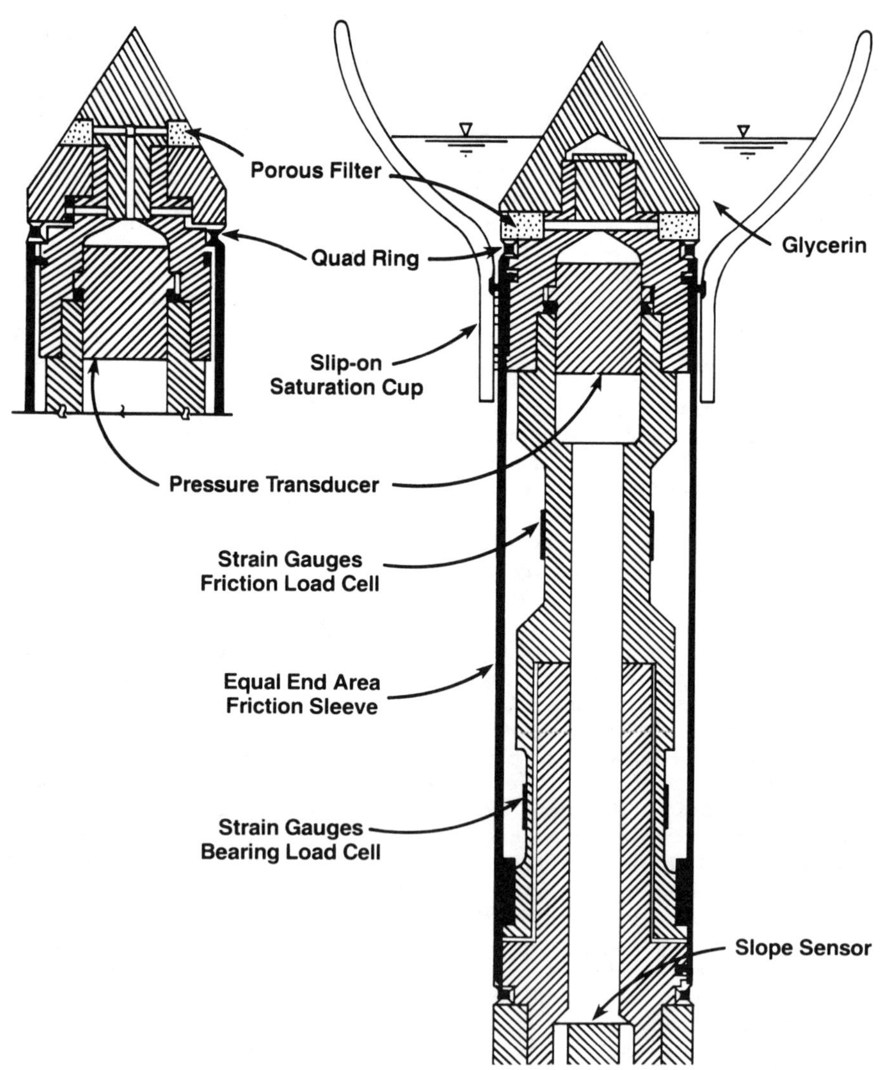

terms, excess pore pressure develops during penetration of low permeability undrained soils, such as clays. This situation results in data that are beneficial to stratigraphic interpretation because this excess pressure dissipates upon entering permeable zones, such as sand. The development of this potential is directly related to the grain size and saturation of the material. Drained material (e.g., above the potentiometric level, water table, or perched water zones) will not develop excess pore pressure. Slightly permeable material, such as silts to very fine sands, if saturated, may allow pore pressure dissipation as quickly as it builds with a resultant reading near zero. High permeability soils, such as sand, will allow rapid dissipation of excess pore pressure and often exhibit negative pore pressure. An alternate or complimentary means of CPT data interpretation involves comparison of the tip resistance and the pore pressure ratio. In its simplest form, the pore pressure ratio represents the excess pore pressure divided by the tip resistance, multiplied by 100, being represented as a percent. This ratio was not used in the software interpretation. Its relationship to the other parameters is depicted in Figure 3. The pore pressure and the ratio aid in conceptual understanding and refined interpretation of the CPT log.

THE HYDROPUNCH™

The Hydropunch™ groundwater sampling tool can be used with either cone penetrometer equipment or conventional drilling equipment to push or drive the sampler to the desired sampling depth. The sampler is approximately five feet (152 cm) long and 1.5 inches (3.8 cm) in diameter and is constructed entirely of stainless steel and Teflon. The Hp has a stainless steel drive point, a perforated section of stainless steel pipe for sample intake, and a stainless steel and Teflon sample chamber, which is capable of collecting 500 ml of groundwater. To drive the sampler, an adapter is used for coupling to either penetrometer push rods or standard soil sampling drill rods.

As the unit is pushed or driven through the soil, the sample intake port is shielded in water tight housing that prevents nonselect groundwater from entering the unit. The shape of the sampler and its smooth exterior surface prevents the downward transport of the surrounding soil and liquid as the tool is advanced. When the desired sampling interval is reached, the sampler is retracted upwards 12 to 18 inches (30.5 to 45.7 cm), exposing the sampling port to the water-bearing zone and permitting groundwater to flow through the screen into the sample chamber. A disposable polypropylene screen covers the sampling port to filter out sand particles.

As a groundwater sample is collected, the drive cone and the sample chamber are flush against the borehole wall, serving as packers that isolate the intake screen from groundwater above and below the zone being sampled. The sample is collected under in situ hydrostatic pressure with no aeration. After the sample chamber is filled, the Hp is pulled upwards. This results in a hydrostatic pressure in the unit greater than the aquifer, which closes the two Teflon check valves and retains the sample within the sample chamber. Upon retrieval, the upper check valve is replaced with a Teflon stop cock valve and a disposable tube. The groundwater sample is then decanted out of the top of the sampler, in much the same way as decanting a bailer. See Figure 4 for a sectional view of the Hp. The samples collected with

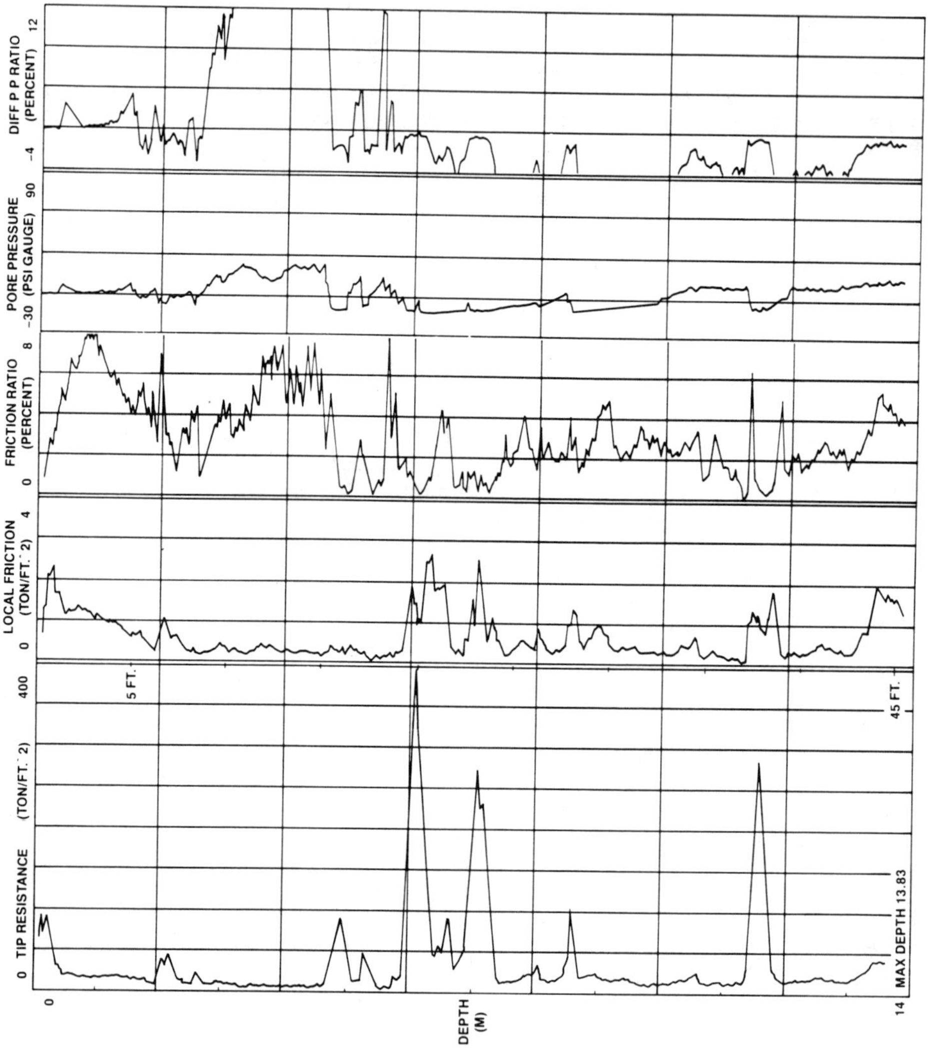

FIGURE 3 CPT GRAPHIC LOG

FIGURE 4 HYDROPUNCH

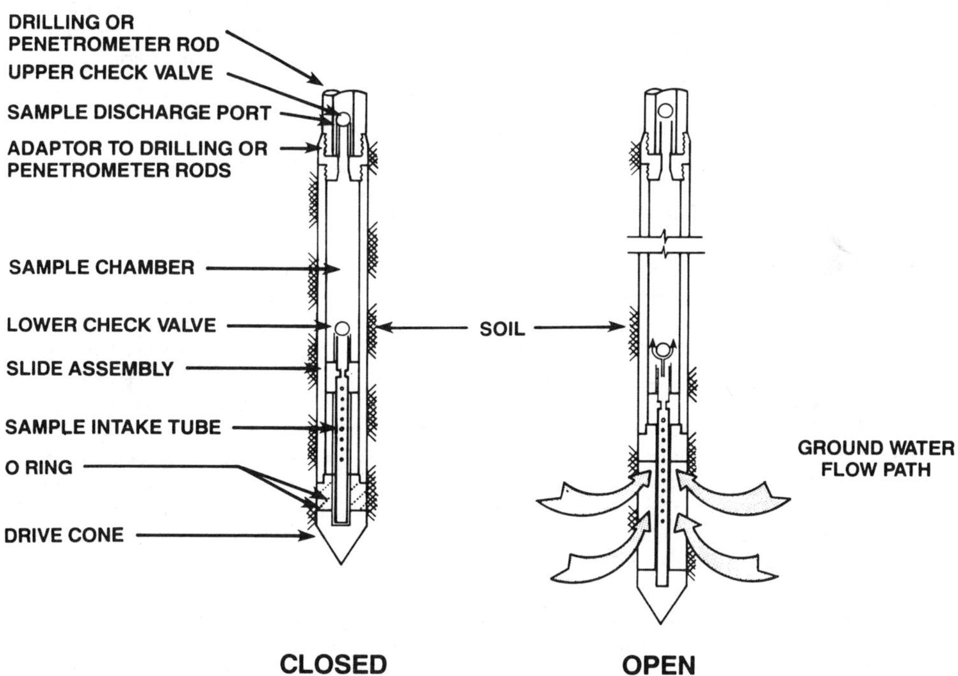

Once exposed, ground water flows through the intake tube and into the sample chamber.

the Hp were analyzed for Volatile Organic Compounds (VOCs) with selected samples being analyzed for Total Petroleum Hydrocarbons.

SITE CHARACTERIZATION

Results of 40 CPT test locations within the main body of Site 9 are presented here. An oblique view of the results is shown in Figure 5. The lithologies presented here have been simplified to be compatible with the scale. The CPT generates continuous data and the interpretation was provided at 0.2 feet (6.1 cm) intervals. Although such discrete interpretation can be valuable in understanding the geology at a single location, it brings too much complexity to an overall site view. Therefore, the lithologic columns have been simplified with respect to predominant grain size over large vertical intervals. Four grain size divisions have been chosen. Silt to silty clay, clayey silt to silt, sand, and sand and gravel. As will be discussed later, the software interpretation and discrimination between these soil types is not absolute and some overlap and cross-interpretation is unavoidable. This simplified reduction and segregation of the data seems to be the easiest way to view the data and is consistent with the data quality. Such data reduction is also consistent with one of the main data uses (i.e., understanding the potential for hydraulic communication between the A_1 and A_2 Aquifers). When viewed with respect to this, each category also represents potential ranges of hydraulic conductivity.

This interpretive presentation indicates a grain size distribution at Site 9 that is consistent with the grain size distribution known to exist on a larger scale at Moffett. As stated earlier, alluvial fans associated with the Santa Cruz highlands were a source of material as well as esturaine deposits. The aquifers show an overall increase in lateral extent and thickening towards the highlands, and inversely, the aquitards show a thickening toward the bay. A similar trend in grain size distribution is seen at Site 9 in that a greater amount of silt is seen towards the south end of the site. This trend which may indicate greater communication between A_1 and A_2, has potentially affected the contaminant distribution and may play a role if groundwater remediation is pursued.

As expected the CPT data showed the existence of preferential pathways and areas of low permeability that act as hydrologic barriers. CPT/Hp locations that showed low contaminant levels but were sited within suitable aquifer material and surrounded by higher contaminant concentrations are believed to be discontinuous laterally or pinch out into low permeability material. The A_1 Aquifer appears to have more sporadic discontinuities within Site 9 than the A_2 Aquifer. This is consistent with the increase in lateral persistence of deeper aquifer material observed basewide at Moffett. The southwest portion of Site 9 shows a northward trending flow path that is consistent with the northward potentiometric gradient direction. Also a narrow aquifer zone appears to enter the site from the area upgradient (south) of the site. The eastern part of the site exhibits a northwest oriented flow path that appears to be the result of a conductive flow path bounded on the east by low permeability material. See Figure 6 for these flow paths as evidenced by the contaminant distribution.

Much of the contamination within the A_1 Aquifer appears to result

FIGURE 5 CPT LITHOLOGY

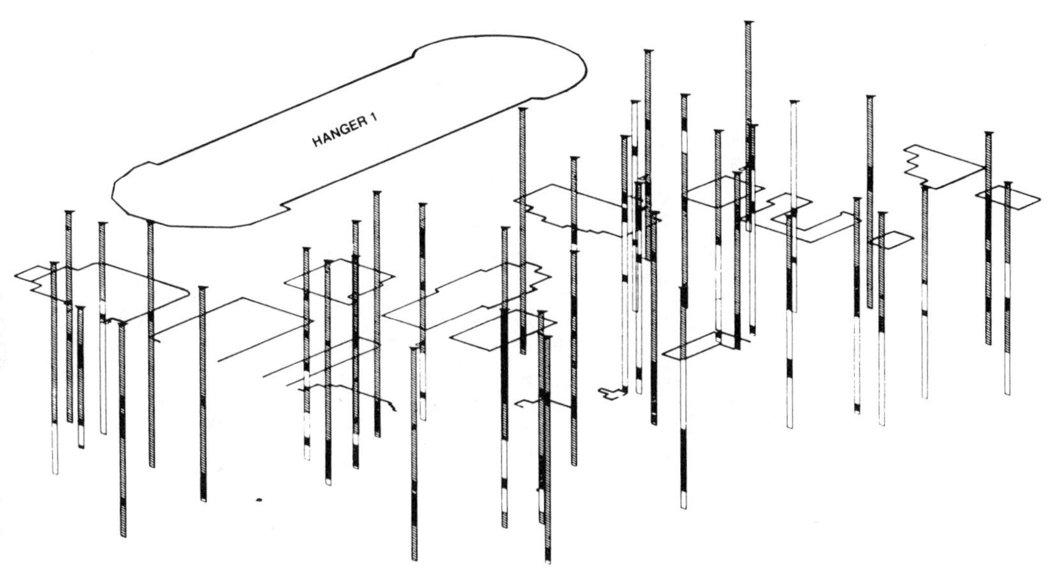

NOTE: CPT LITHOLOGY COLUMNS HAVE BEEN EXAGGERATED A FACTOR OF 10 FROM THE BASE MAP.

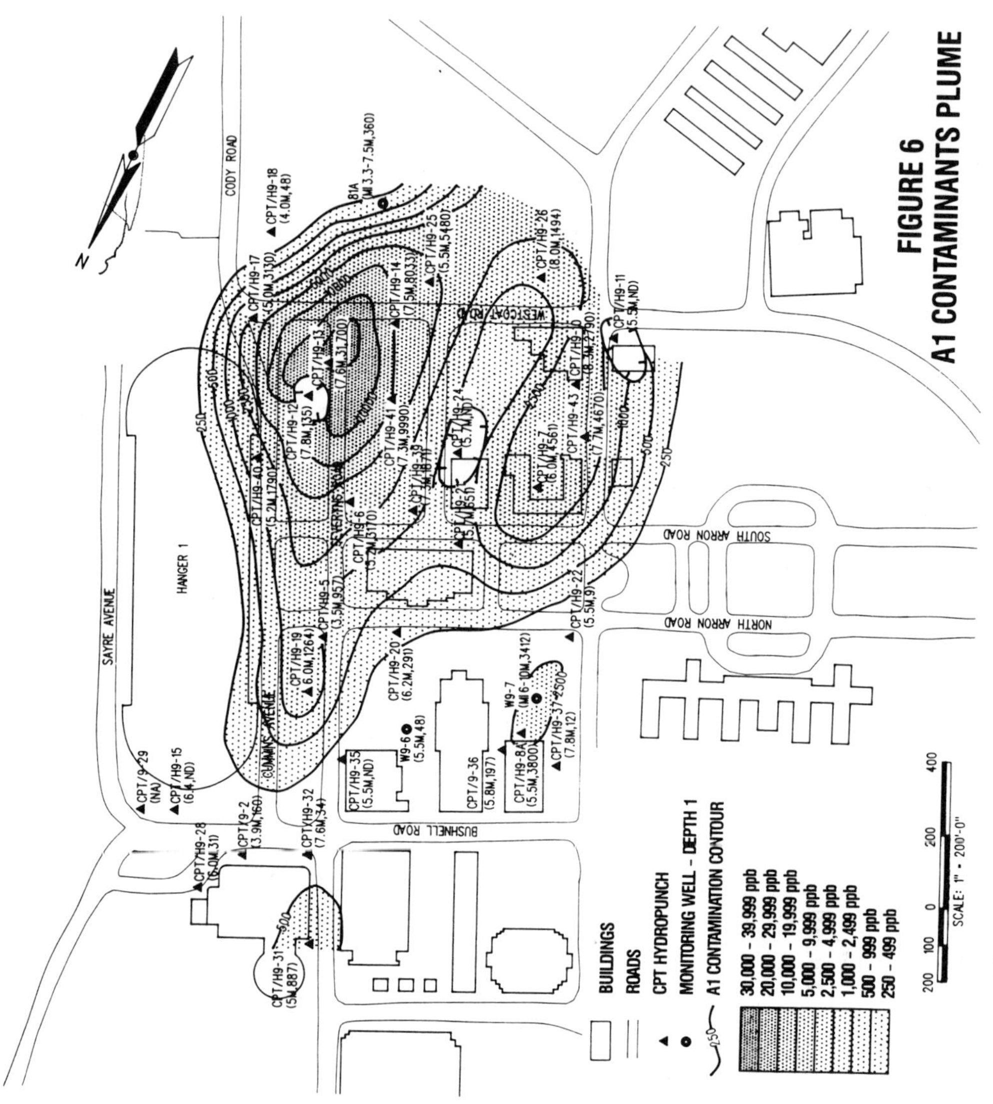

FIGURE 6
A1 CONTAMINANTS PLUME

from upgradient sources. However, a source (dry cleaners) upgradient of CPT/H9-13 is evidenced by the elevated VOC concentrations and anomalous presence of tetrachloroethene (PCE) locally. A source is also suspected near CPT/H9-8A. Up and downgradient concentrations indicate a very small localized plume.

The A_2 Aquifer CPT/Hp data indicate a fairly continuous northward trending plume with a westward extension that is probably a preferential flow path (See Figure 7). It is believed that the plume and aquifer material is continuous in the upgradient direction. However, it appears that the dry cleaner source may have contributed to the plume within Site 9. This conclusion s supported by the presence of PCE, lower reported upgradient VOC concentrations, and the coarsening of the A_1/A_2 aquitard toward the south end of the site. The presence of coarser material in the aquitard has probably resulted in increased permeability and allowed migration of contamination from the A_1 Aquifer to the A_2 Aquifer.

The CPT is a valuable tool but not without limitations. Within its working environment of nonlithified sediments, 200 feet (61 m) is close to its penetration limit, and it can be stopped by small gravel if there is little void space available for displacement. As would be expected, any cementation of sands or gravels can also prohibit penetration. Non retrieval of a soil boring is a limitation but is complemented by the fact that no soil waste is generated. Overall, the CPT is well refined and probably has less downtime than the average drilling.

A shortcoming associated with the CPT in this investigation was the computer software's interpretive capability. It may be that integration of more parameters into the program will resolve this, or refinement of the program may be needed. Some of the interpretation problems referenced in the literature appear to have been associated with this study. Noncorrelation between some specific intervals described by the CPT interpretation and equivalent depth intervals described on logs from nearby borings may partly be due to the drastic vertical and horizontal lithology changes that have been noted at Moffett.

The most significant difference in CPT versus boring log lithology was associated with sand and gravelly sand encountered in the borings. In instances where these materials were observed as being loose or very loose, they were sometimes interpreted by the CPT software as sandy silt to silt. This problem has been previously documented and is mentioned in some CPT research literature. This situation supports the need for a greater understanding of pore pressure relationships and incorporation into the interpretive software. Less significant interpretation variabilities were also observed. The material classified as clay by the CPT was logged in borings as silty clay, dry to slightly most, slightly plastic to clayey silt, slightly moist, firm, low plasticity. Material with a CPT software interpretation of clayey silt and silty clay was found in the borings to range from silty clay, slightly moist, soft to moderately stiff, moderate plasticity to clayey silt, slightly moist, nonplastic to sandy clay, moist, low to moderate plasticity. Material with a CPT interpretation of sandy silt and silt was found in the borings to range from sandy clay, wet, low to non-plastic, very loose to silt or a combination of sand and silt

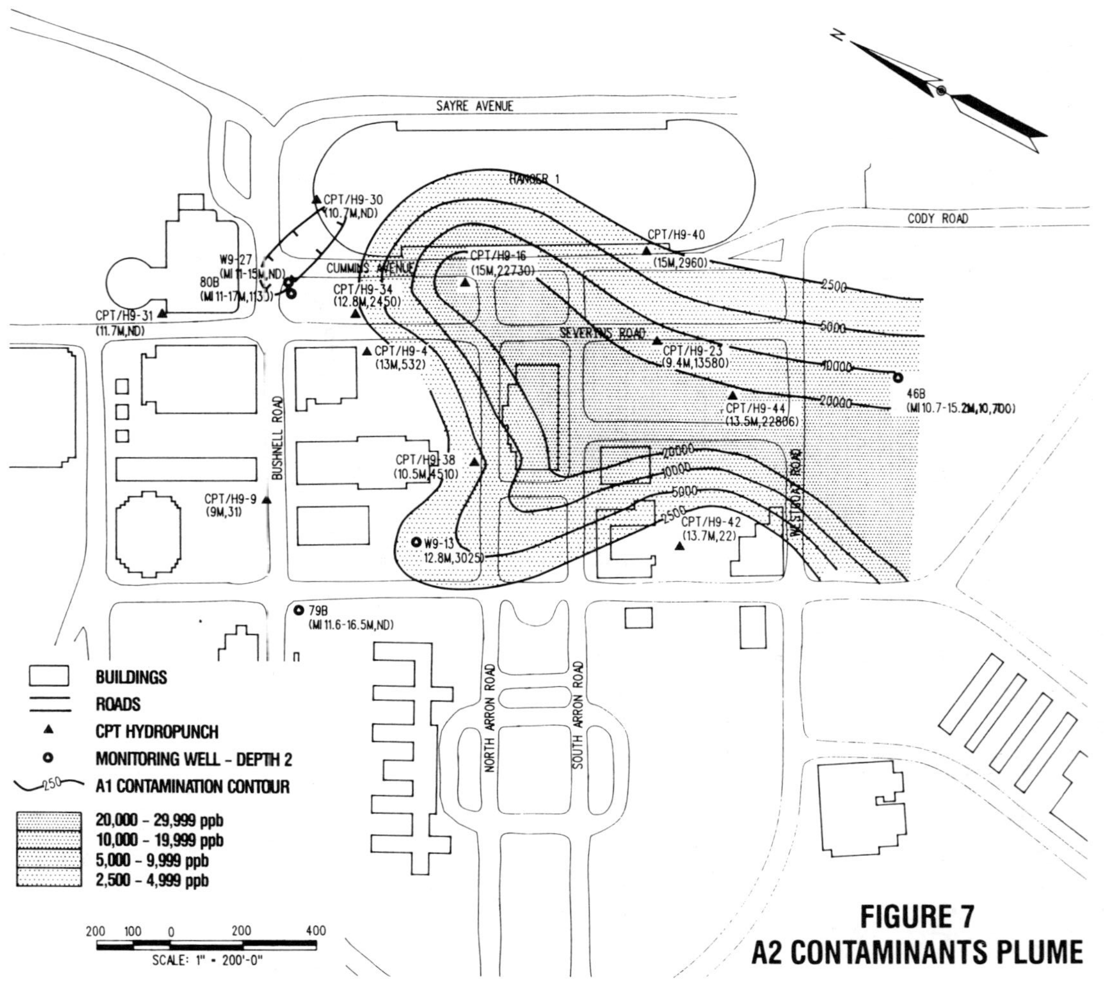

FIGURE 7
A2 CONTAMINANTS PLUME

to clayey silty sand, wet, fine to medium grain, and loose. As mentioned previously, the material interpreted as sandy silt and silt occasionally included much coarser material. Material with a CPT interpretation of silty sand was found to occur at transitions between sandy silt and silt zones and sand zones as interpreted by the CPT software. Specific correlative silty sand zones were not identified in the borings. This lithologic description was seldom used in the borehole logging. Material with a CPT interpretation of sand or sand and gravel was found to be in good agreement with the boring data. In general, the boring data confirmed most of the CPT interpretation.

Other data quality problems inherent to the CPT and interpretive software were not analyzed in this study, but are worthy of mention. One of the fundamental problems is associated with the trailing position of the friction sleeve relative to the cone tip. The friction sleeve's length and position relative to the tip results in some degree of depth inaccuracy. Although the distance it is set back from the tip is accounted for in the interpretations, the sleeve length results in data that represent an average resistance encountered over the entire sleeve length. This inaccuracy is compounded by the fact that tip resistance is influenced by the soil properties ahead of and behind the tip. The sensing distance increases with soil stiffness. Therefore, stiffer layers (e.g., sand) will require greater thickness for a full tip resistance response than softer layers such as clay. Schmertmann[1], and Treadwell[2], estimate a stiff bed thickness of 36 to 72 cm (14 to 28 inches) is required for a full tip response with a standard cone, whereas soft layer much thinner than 36 cm may fully respond. A potential result of this relationship is an interpretive log that is biased toward a clay predominance due to thin bedding and the tip's inability to fully respond in sand layers between clays. This is another situation where pore pressure data become valuable. Since pore pressure is measured at the porous element in the tip and response time is reported as 0.25 seconds or less, the response data are very depth discrete.

Another potential inaccuracy that has been reported in associated with excess pore pressure in low friction materials such as clays. Elevated pore pressures in such materials can increase the tip resistance and sleeve friction and thus influence data accuracy. A similar effect can occur with depth. Increasing overburden results in greater confining pressure that may not be accounted for in CPT software interpretation of tip resistance and sleeve friction.

CONCLUSIONS

The CPT/Hp combination proved to be a valuable site characterization tool. The CPT's most beneficial use was in selecting permeable zones for Hp sampling. Absolute lithology definition was complicated by thin bedding and loose sands. However, the overall lithologic trends and the thicker units were well identified. During the field effort, the CPT graphic logs (e.g., Figure 3) were relied upon, and interpretation was performed visually based on a conceptual understanding of the parameters. This approach was consistent with the data use and data quality associated with this stratigraphy screening method. Absolute stratigraphy definition will rely upon boring data.

The Hp worked well at almost all sampling locations. A few

permeable zones targeted with the CPT proved to be unsaturated or at too low hydraulic head to yield a sample. Under favorable conditions (i.e., in saturated conductive material and under at least five feet, 1.5 m of head) only a few minutes were required to fill the sampler. The data quality of analyzed Hp water samples versus samples from subsequent monitoring wells has not been investigated. The Hp does appear to be an effective method for groundwater sample retrieval in support of groundwater contaminant screening and plume mapping.

REFERENCES

[1] Robertson, P.K., and Campenella, R.G., 1986, "Guidelines for Use, Interpretation and Application of the CPT and CPTU", Hogentogler and Company, Inc., Gaithersburg, Maryland 20877.

[2] Schmertmann, J.H., July 1978, "Guidelines for Cone Penetration Test, Performance and Design", Federal Highway Administration, Report FHWA-TS-78-209, Washington, 145 pgs.

[3] Treadwell, D.D., 1975, "The Influence of Gravity, Prestress, Compressibility, and Layering on Soil Resistance to Static Penetration", Ph,D. Dissertation, Graduate Division of the University of California, Berkeley, 210 pgs.

Andrew I. Strutynsky and Timothy J. Sainey

USE OF PIEZOMETRIC CONE PENETRATION TESTING AND PENETROMETER GROUND WATER SAMPLING FOR VOLATILE ORGANIC CONTAMINANT PLUME DETECTION

REFERENCE: Strutynsky, A. I. and Sainey, T. J., "Use of Piezometric Cone Penetration Testing and Penetrometer Ground Water Sampling for Volatile Organic Contaminant Plume Detection," Current Practices in Ground Water and Vadose Zone Investigations, ASTM STP 1118, David M. Nielsen and Martin N. Sara, Eds., American Society for Testing and Materials, Philadelphia, 1992.

ABSTRACT: Piezometric Cone Penetration Testing (CPTU) and penetrometer ground water sampling were used to locate a volatile organic contaminant plume at an industrial site in southern Ohio. Nine CPTU tests (soundings) were performed to determine site hydrostratigraphy in real-time. On-site chemical analysis of penetrometer ground water samples provided near real-time detection of contaminants. These results were used to define subsequent monitor well locations. Using this investigation approach, drilling operations to set monitor wells began as penetrometer exploration ended. Program quality increased, while exploration costs decreased by using this combination penetrometer-drilling rig approach.

KEY WORDS: Piezometric Cone Penetration Test, Penetrometer Ground Water Sampling, Volatile Organic Contamination, Hydropunch, BAT, Hydrogeology

INTRODUCTION

Tricholoroethene (TCE), an industrial solvent, along with other volatile organic compounds, was identified in the ground water at a manufacturing site in rural southern Ohio. Contamination was initially found in four monitor wells, located

Andrew I. Strutynsky is the Vice President and Technical Director of STRATIGRAPHICS, 439 Taylor Ave., Glen Ellyn, IL 60137; Timothy J. Sainey is a Senior Project Geologist at ERM-Midwest, Inc., 450 W. Wilson Bridge Rd., Columbus, OH 43085.

---

around a suspected source. The immediate soils consist of mixed sand, silt, and clay, overlying granular strata, defined by CPTU measurements, as consisting of sand with gravelly layers, which overlie bedrock. The water table is located at El. 167.7 m (550 ft), or 2.4 to 6.1 m (8 to 20 ft) below the surface.

To define the limits of contamination, a soil gas survey was performed using a hand-held hammer drill to drive probes 1.5 m (5 ft) into the ground. Probing traced contamination to an off-site agricultural field and defined the general direction of contaminant movement (Figure 1). As the survey expanded, indications of volatile organic compounds abruptly stopped. It was suspected that a thicker clay unit was masking volatile organic compounds that might be present in the ground water. The limits of contamination still had to be determined, but additional exploration was subject to the following constraints:

- Contamination was determined to be moving off-site, where stratigraphy was uncertain.
- The off-site property owner did not want drilled boreholes and monitor wells in his field.
- Poor weather conditions prevailed during the late winter months of 1990.
- Agency deadlines and budgetary limitations influenced the scope and schedule.

The needs of the off-site property owner had to be met while still allowing exploration to proceed. His concerns were addressed by negotiating permission to set a maximum of nine wells to monitor the extent of contamination, with assurances of minimal surficial disturbance to future crop cultivation.

A possible solution to the constraints of the agreement was to use a drilling rig to split-spoon sample soils, and to sample ground water by using either a Hydropunch sampler or wellpoints. The exploration boreholes would be drilled and plugged until the extent of contamination had been determined. At this point, the monitor wells would be set. However, the cost of drilling and plugging numerous exploratory boreholes, and the overall ineffectiveness of drilling rigs to advance the Hydropunch in dense sands and gravels, made the use of a drill rig less than optimal.

Another solution was to use a cone penetrometer rig to explore subsurface conditions followed by conventional drilling to set monitor wells. This plan had the following advantages:

- The penetrometer rig has an enclosed work area so that bad weather does not significantly impact productivity.
- It is capable of collecting high quality, continuous stratigraphic information in real-time, both allowing for efficient penetrometer ground water sampling and allowing a drill rig to eventually set monitor wells without additional physical sampling.
- It can be used to hydraulically push ground water samplers into sands and fine gravels, under suitable conditions.
- Penetrometer operations do not generate possibly contaminated drill cuttings or fluids that require expensive disposal.

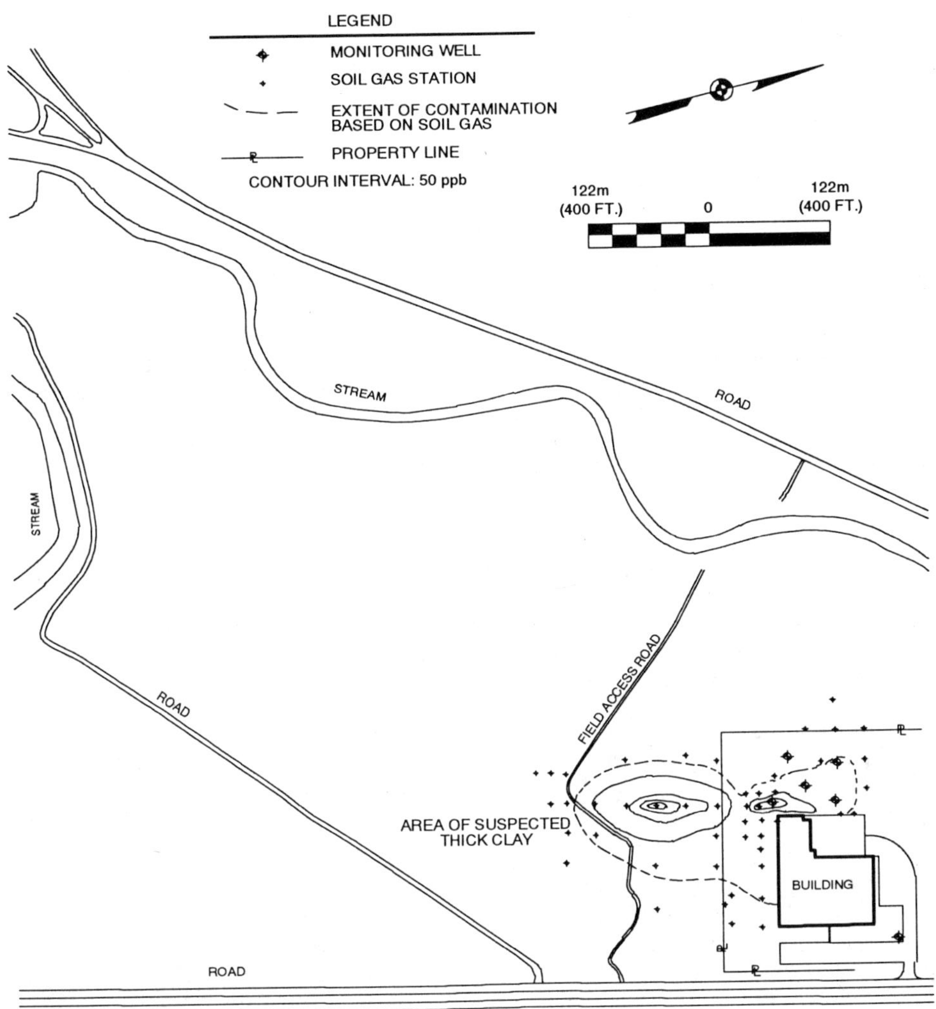

FIGURE 1 - EXTENT OF CONTAMINATION BASED ON SOIL GAS SURVEY, APRIL 1988

- The method does not result in large diameter holes that require extensive grouting to seal.
- Penetrometer operations are more rapid and thus less expensive than drilling operations.

Use of the penetrometer exploration/drill rig well installation approach was decided upon as the most cost- and performance-effective solution for this project. An analytical laboratory was set up on-site allowing for successive exploration points to be optimally chosen on the basis of contaminant plume detection, rather than on an arbitrary grid pattern.

PENETROMETER TECHNIQUE AND EQUIPMENT

Penetrometer methods are being used with increasing frequency, and they provide significant advantages for geo-environmental exploration. The technique uses a large hydraulic ram to push small (3.8 to 6.35 cm, 1.5 to 2.5 inch) diameter probes into the ground without drilling a borehole (Figure 2). Instrumented probes, called penetrometers, provide semi-direct and direct information on geotechnical, hydrogeological, and geochemical site conditions. Penetrometer samplers are also used to obtain physical samples of subsurface materials.

Penetrometer methods are used to their greatest advantage in sand, silt and clay deposits. Penetrometer profiling (Figure 3) is continuous and accurate, and measurements (sounding logs) are reliably interpreted for definition of aquifers and confining layers. Lateral continuity of layers is readily apparent from a series of adjacent penetrometer soundings.

Site disturbance and waste material disposal is minimized because no cuttings or drilling fluids are generated during penetrometer operations. Downhole equipment is steam cleaned during retrieval. Personnel exposure to contaminants is much less than exposures during drilling. Heaving sands pose little problem, as no open hole actually exists during penetrometer advance.

One hundred to 275 m (330 to 900 ft) of geotechnical soundings can be performed in a day, depending on access and project requirements. The special demands of geo-environmental investigations, such as ground water sample, pore pressure dissipation testing, decontamination and grouting, decrease daily footage, but productivity is still significantly higher than that using drilling rigs. Sounding depths in excess of 30.4 m (100 ft) can be achieved, depending on site stratigraphy.

The penetrometer is mounted at the end of a series of heavy-wall sounding rods. A hydraulic ram is used to push the penetrometer into the ground at a rate of 1.2 m (4 ft) per minute. Electronic signals from downhole sensors are transmitted by a cable, strung through the sounding rods, to a computer inside the penetrometer rig. Data are recorded at depth intervals of .95 to 1.9 cm (3/8 to 3/4 inch). A real-time data display is monitored for evaluation of test performance and for immediate definition of site conditions. At the end of a sounding, the penetrometer and sounding rods are retrieved and decontaminated. The open hole can be grouted where cross-contamination between water bearing formation may occur. The open hole was allowed to collapse at the end of each sounding for this project, due to the lack of confining layers at the site.

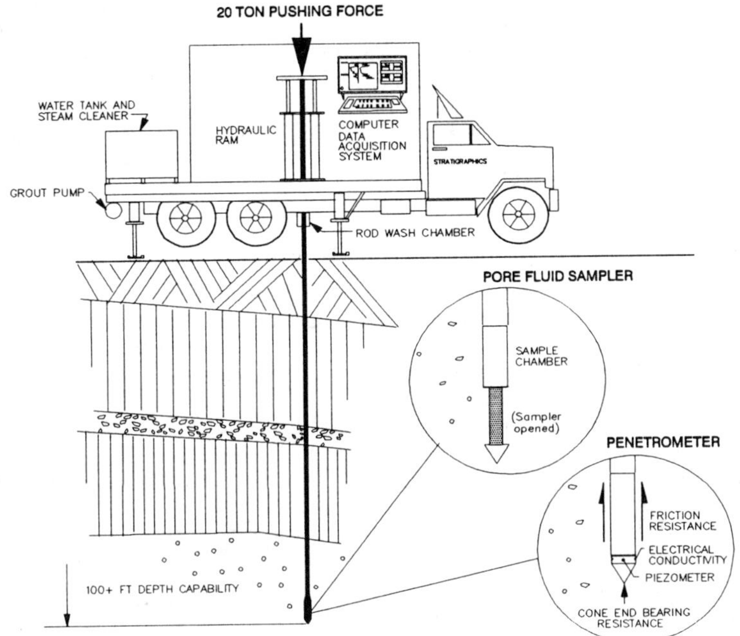

FIGURE 2 - PENETROMETER SUBSURFACE EXPLORATION SYSTEM

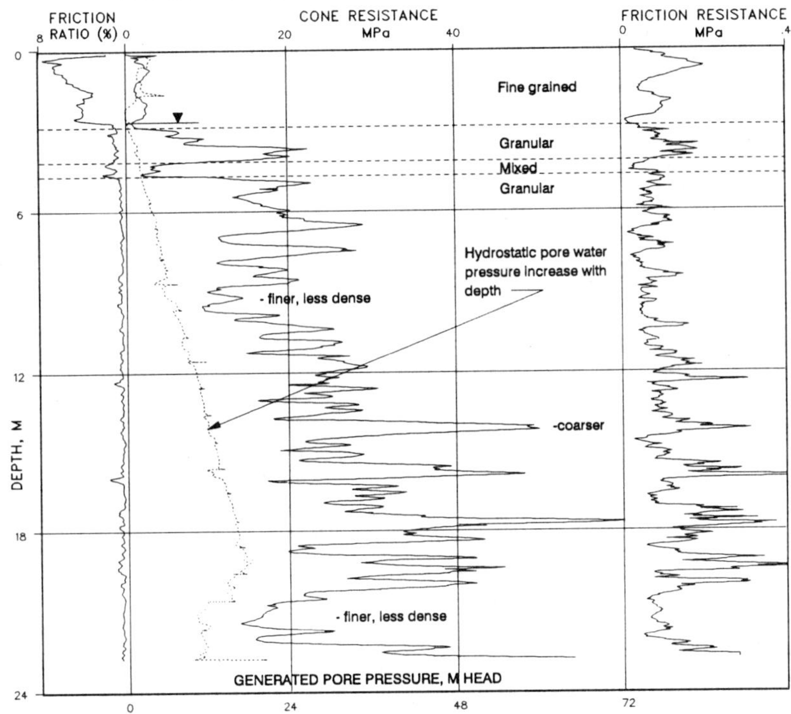

FIGURE 3 - CPTU SOUNDING LOG C-7

A special truck is used to house, transport, and deploy the penetrometer. Twenty tons of truck weight and ballast are used to counteract the thrust of the hydraulic ram. The work area is enclosed and includes heating and air conditioning. Computers, penetrometers, samplers, electrical power, lighting, compressed air, grout and water pumps, steam cleaner for equipment decontamination, 1045 L (275 gal) water tank, tools and spare parts are all included within the penetrometer truck, providing for self-contained operations.

Instrumentation

The basic electronic penetrometer consists of two separate soil shear resistance sensors, and is used to acquire information on soil strength and stratigraphy. Tests conducted with this penetrometer are called Cone Penetration Tests (CPT) and are specified by ASTM Standard D3441. Electronic CPT has been used for geotechnical engineering projects for more than 25 years, while less sophisticated, uninstrumented versions of the test (Dutch cone test) have been used since the 1930s.

Two laboratory grade, strain gage loadcells, mounted inside the penetrometer, are used to measure the soil shear resistance to penetration acting on the conical tip and along the cylindrical sides of the CPT penetrometer. CPT measurements are continuous, accurate and repeatable. The tip or cone end bearing resistance can respond to soil seams as thin as 5.1 to 10.2 cm (2 to 4 inches). The side or friction resistance measurement has a resolution of about 15.2 cm (6 inches).

A pressure transducer is added to the CPT penetrometer to additionally measure the soil pore water pressure response to penetration; this is called a Piezometric Cone Penetration Test (CPTU). CPTU piezometric data allow for the evaluation of soil saturation, water tables, potentiometric surfaces, and soil horizontal permeability in both aquifers and aquitards. The CPTU piezometric measurement has a resolution of about 2.5 cm (1 inch).

Another penetrometer has been used to measure the shear resistance, piezometric response, electrical conductivity and temperature of penetrated soils. This penetrometer is useful in detecting free hydrocarbon product on ground water. This test type is termed CPTU-EC. Additional details on penetrometer instrumentation can be found in Strutynsky, et al., 1985 and 1991.

Ground Water Samplers

The Hydropunch and BAT penetrometer samplers were used to sample ground water for this project. The Hydropunch sampler (Figure 4) consists of a stainless steel, shielded wellpoint and sample barrel assembly (Edge and Cordry, 1989); it is deployed using the heavy-wall penetrometer sounding rod. The shield prevents contamination of the sampler while pushing. When the shield is retracted, ground water flows under in situ pressure conditions into the 500 ml sample barrel.

A water level indicator can be lowered to the top of the Hydropunch in order to determine sampler filling. The sample is isolated in the barrel by two ball check valves. The entire sampler is retrieved to the surface to pour off the sample and for sampler decontamination. The tip of the Hydropunch sampler must be at least 1.2 m (4 ft) below the water table to provide the hydrostatic pressure needed to acquire a sample.

The BAT sampling system (Figure 5) consists of a wellpoint that is internally sealed with a septum (Torstensson, 1984). After pushing the wellpoint to depth, an evacuated 35 or 70 ml vial, also sealed with a septum, is wirelined down the casing. A double-ended hypodermic needle, mounted in an adapter below the vial, pierces both the wellpoint and the sample vial septa, and allows fluids to flow into the vial. The septa seal as the sample vial is retrieved to the surface, maintaining the sample at in situ pressure conditions. This procedure may be repeated to develop the wellpoint and to obtain increased sample volumes.

The BAT Enviroprobe and Mk. 2 wellpoints were used during this study. The Enviroprobe is shielded, while the Mk. 2 tip is unshielded. The Mk. 2 filter is exposed to possible cross-contamination during deployment. A thin-wall AQ drill rod is used to deploy the BAT samplers, as the heavy-wall penetrometer sounding rods are too small to pass the BAT sample vial.

PENETROMETER DATA INTERPRETATION

Correlations between penetrometer data and soil type have been developed from observational criteria on adjacent CPT soundings and drilled and sampled boreholes (Douglas and Olsen, 1981). CPT soil classifications reflect the shear response of a soil to penetration. Soil shear response is not entirely controlled by grain size distribution. However, soil types evaluated from CPT data generally agree with those classifications based on soil grain size distribution methods, such as the Unified Soil Classification System (USCS).

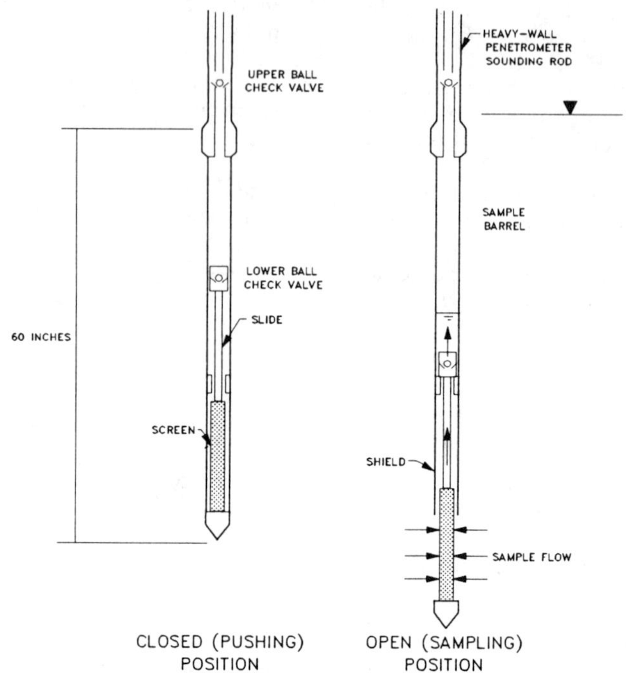

FIGURE 4 - HYDROPUNCH GROUNDWATER SAMPLER

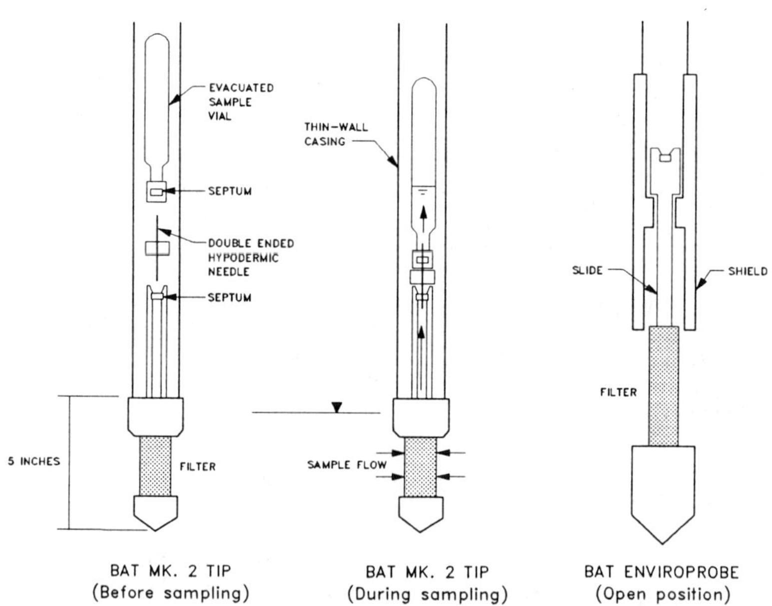

FIGURE 5 - BAT GROUNDWATER SAMPLING SYSTEM
Note: All dimensions approximate

The CPT cone end bearing resistance increases exponentially with increases in grain size. The cone end bearing resistance in dense sands ranges from about 15 to 30 MPa (150 to 300 tons per square foot (TSF)), while the cone end bearing resistance in a stiff clay ranges from about .7 to 1.5 MPa (7 to 15 TSF). The proportion of CPT friction to cone end bearing resistance, termed friction ratio, is related to the fines content of a soil. The friction ratio is low in sands and high in clays. CPT measurements are computer analyzed using CPT soil classification charts (Figure 6) to quickly define site stratigraphy.

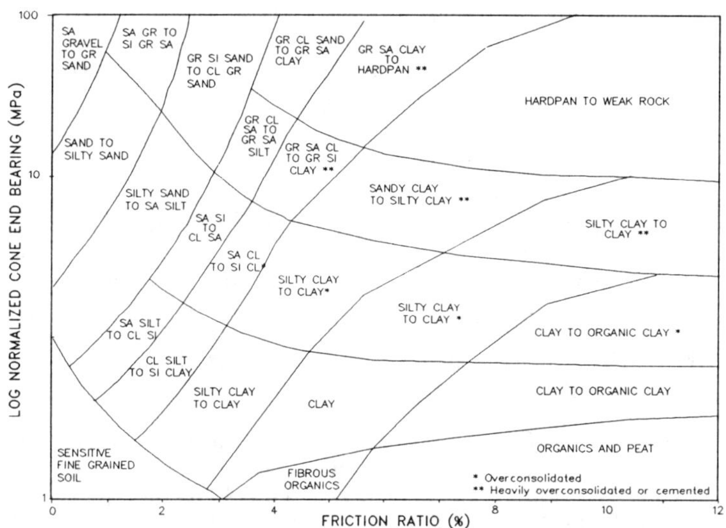

FIGURE 6 - CORRELATION CHART FOR CPT SOIL TYPES

The penetration of a saturated soil generates a localized pore water pressure field, in excess of equilibrium, around the penetrating probe. This generated pressure field dissipates quickly in soils of high permeability, so only the equilibrium pore water pressure field is measured in clean sands and gravels during Piezometric Cone Penetration Testing (CPTU). In low permeability soils, excess pore pressures require a significant amount of time to dissipate (Saines, et al., 1989). The dissipation of excess pore water pressure can be recorded as a function of time by pausing in the penetration process; this is termed a CPTU dissipation test. If the pauses are sufficiently long for all excess pressures to dissipate, measurements can be obtained of equilibrium potentiometric surfaces at multiple depths.

A CPTU dissipation test is somewhat similar to a falling head slug test, and can be used to calculate a value of soil horizontal permeability. However, tens to hundreds of meters of excess pore water pressure are induced in low permeability soils by penetrometer advance, as compared to several meters of head induced in a well during slug testing. The greater pressure changes during CPTU dissipation testing require soil compressibility effects to be included in analyses. The CPT cone end bearing resistance provides an index of compressibility for permeability computations.

ON-SITE LABORATORY

A Hewlett-Packard HP 589OA gas chromatograph (GC) with two electron capture detectors (ECD) was selected as the on-site analytical instrument for this project. It has the advantage of:

- Achieving laboratory results in the parts per billion (ppb) range
- Having a programmable oven which provides better separation of the eluting compounds for more accurate identification
- Qualifying and quantifying compounds as compared to internal reference standards

For the analysis of volatile organics, U.S. EPA SW-846 Method 3810 Headspace was used. Calibration procedures were performed using standard mixtures to establish the internal standard curve. Sample analysis consisted of pouring collected water samples into a 1.0 ml syringe, and injecting this sample into a nitrogen purged 40 ml septum vial. The sample was then heated in a 70 deg C water bath for 15 minutes; 100 ml of headspace was drawn from the center of the septum vial and injected into the GC for analysis.

For quality control (QC), every 10th unknown was run, per a modified U.S. EPA Method 3810. Modifications to the method, which were made by the on-site team, allow samples to equilibrate to 70 deg C, to use 40 ml vials with Teflon face septa, and to inject 100 ml of headspace gas. For each 10th unknown, four samples were injected to determine the quality of data. These consisted of a:

- Sample containing reagent water that had been carried through all stages of sample collection of unknown constituents
- Sample containing reagent water spiked with known amount of target analytes
- Sample containing unknown constituents
- Sample containing unknown constituents spiked with known amount of target analytes

In addition to the field analyses, select duplicate samples were sent to an off-site analytical laboratory for verification. Sample preparation, standards and QC were assisted by an analytical subcontractor.

PROGRAM RESULTS

CPTU soundings were performed at nine locations (Figure 7) during the period of February 6 to 13, 1990, for a total of 699 ft of test. Only partial days were worked on February 6 and 9, and no field work was done on February 10 when bulldozer support for access to locations in the wet, recently plowed field was not available. Twenty seven ground water sampling attempts, with 22 successfully recovered samples, were also performed during this period. The sequence of penetrometer operations was as follows:

1) The penetrometer rig was set up, and a CPTU sounding was performed to determine hydrostratigraphy. The field sounding log was analyzed immediately to determine ground water sampling depths.

2) The penetrometer rig was moved to provide about 1.5 m (5 ft) of offset between the sounding and sampling hole to avoid vertical cross-contamination of samples.

3) Where sands were especially dense, or contained a high gravel content, a prepunch tool was pushed to nearly the sampling depth in order to facilitate sampler deployment.

4) The sampler was pushed to depth. Three ground water samples were typically taken at each location at successively deeper penetrations (Figure 8), from shallow (4.6 to 9.5 m, 15 to 31 ft), to intermediate (11.3 to 14.3 m, 37 to 47 ft), to deep (20.7 to 27.4 m, 68-90 ft). Three Hydropunch samplers were available on site, so as a sampler was being deployed, the others were being decontaminated.

The CPTU soundings revealed site soil conditions to have general lateral continuity, as can be seen on a stratigraphic cross-section (Figure 9). Site stratigraphy is summarized in Table 1.

TABLE 1 -- CPTU soil stratigraphy

STRATUM NUMBER	DEPTH	DESCRIPTION	INCLUSIONS
1	0 m to (.6-.9 m) (0 m to 2.7-6.1 m in C-3, C-7 & C-8) (0 ft to 2-3 ft (0 ft to 9-20 ft in C-3, C-7 & C-8))	Very stiff to firm sandy clay to clayey silt, with increasing silt content with depth	
2	.6-.9 m to 22.9-28.6 m (2-3 ft to 75-94 ft) (completion depth)	Medium dense to dense sand to silty sand, with layers of gravelly sand and sandy gravel.	a) a seam of cl. silt at 25.6 m (84) ft at C-5 b) sa. silt from 13.8 to 4.7 m (15.5 ft) at C-7 c) cl. sand from 24.4 to 8.1 m (26.6 ft) at C-9

A highly detailed characterization of the site geology was obtained from the CPTU sounding logs. Split-spoon sampling during drilled investigations often results in poor recovery in sands and gravels, which make detailed description impossible. In contrast, the CPTU data provided a continuous record of the material being penetrated, thus allowing even minor changes to be recorded accurately (Figures 3, 8 and 9). The high resolution of the CPTU data was important in finding zones of higher permeability for future ground water remediation and also verified the lack of confining layers at the site. The assumption of a thick, surficial clay zone interfering with the soil gas survey was verified by soundings C-3, C-7 and C-8 (Figures 3, 8 and 9).

Groundwater conditions were measured using CPTU piezometric data (Figures 8 and 9). Water table measurements were also taken using a water level indicator lowered into the open hole left after a sounding. The potentiometric surface at one ground water sampling depth was measured by allowing the sounding rod string to fill through the Hydropunch sampler.

The Hydropunch sampler was successfully used in obtaining 22 ground water samples out of 27 attempts. In general, the Hydropunch sampler worked well, especially in regards to depth capacity. Some disadvantages of the Hydropunch sampler were:

- Lack of definitive feedback as to whether the shield opened
- Minor galling and seizing of sampler parts as is common with unlubricated stainless steel assemblies
- Slight bending of the sample barrel during pushing with forces in excess of 90 kilo Newtons (10 tons)

The BAT ground water samplers were used during two sampling attempts at the site. The thin-wall deployment casing buckled while pushing the Mk. 2 wellpoint through the shallow gravelly sands at location C-2, which precluded taking a sample. The Enviroprobe was successfully used at location C-5 and illustrated some of the features of the BAT system, including the retrieval of multiple samples and rapid feedback as to wellpoint shield opening.

Ground water samples were analyzed at the on-site laboratory for four organic volatiles that had been identified in the ground water. TCE was found to be the most prevalent contaminant. Several samples were sent to an off-site laboratory for verification of the field results. On-site analytical results were consistently lower than results generated in the off-site laboratory. However, the on-site laboratory results were sufficiently accurate to guide the location of successive exploration locations.

CONCLUSIONS

The approach of using a truck-mounted penetrometer rig, CPTU soundings, penetrometer ground water samplers and an on-site GC laboratory to conduct geo-environmental site characterization studies, in combination with drill rigs to set monitor wells, proved to be highly advantageous in terms of:

- Minimal site disturbance and generation of wastes

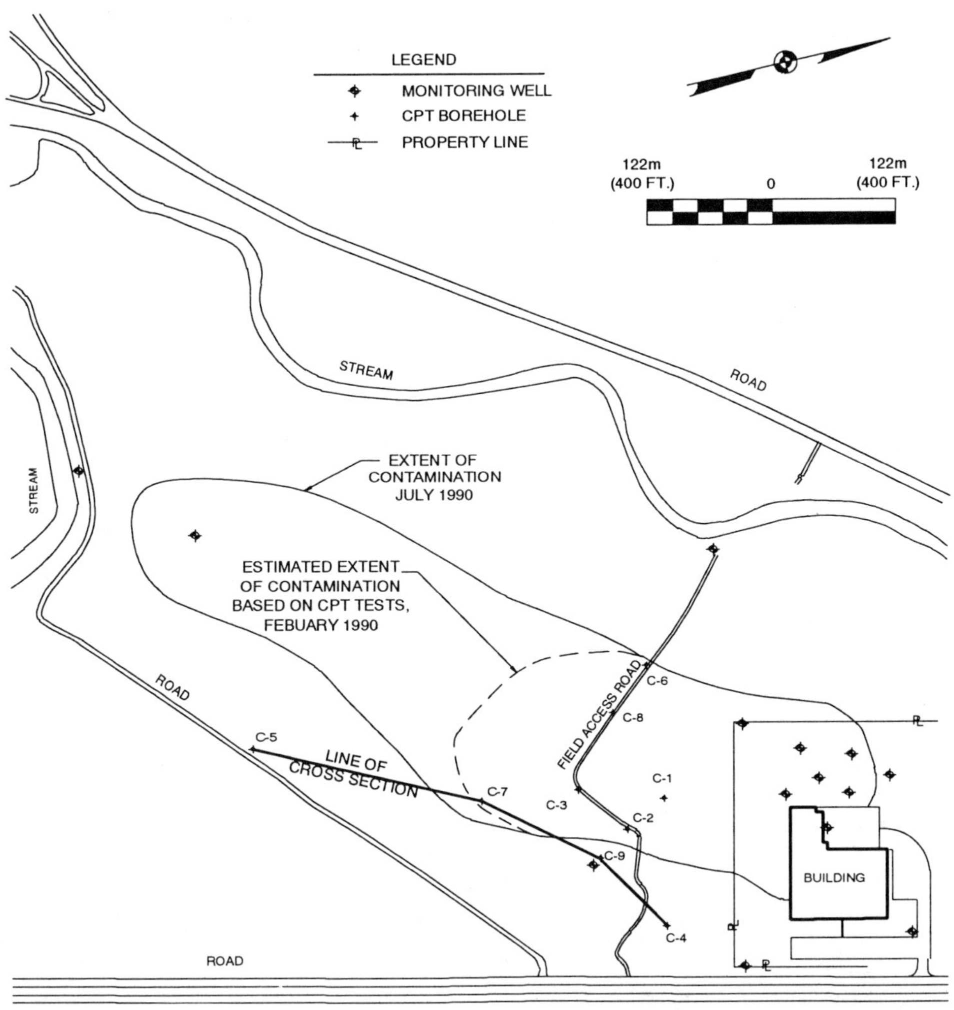

FIGURE 7 - LOCATION OF CPT TESTS, MONITOR WELLS, LINE OF CROSS SECTION AND EXTENT OF TCE CONTAMINATION AT THE TOP OF GROUNDWATER

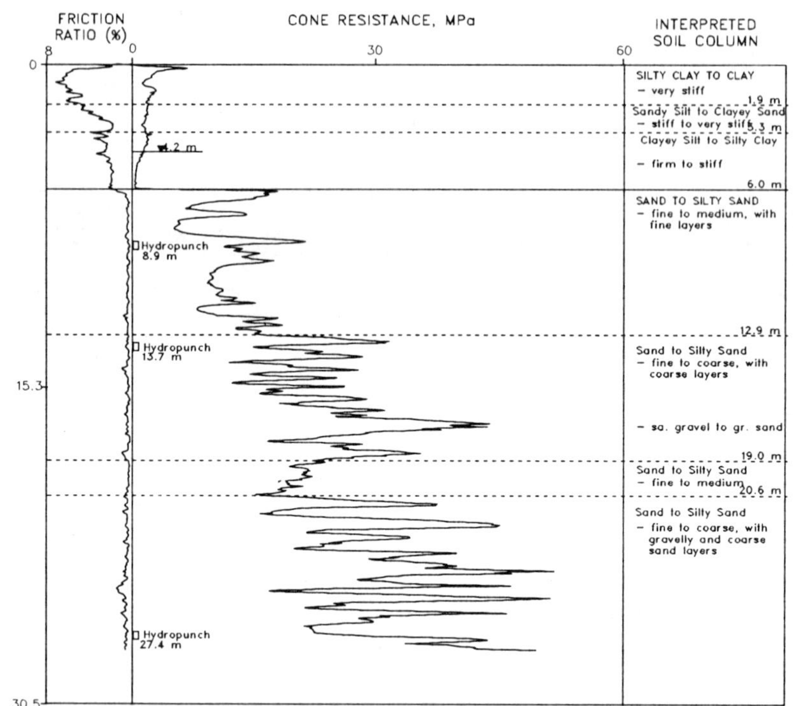

FIGURE 8 - INTERPRETED CPT SOUNDING LOG C-3 WITH SAMPLING DEPTHS

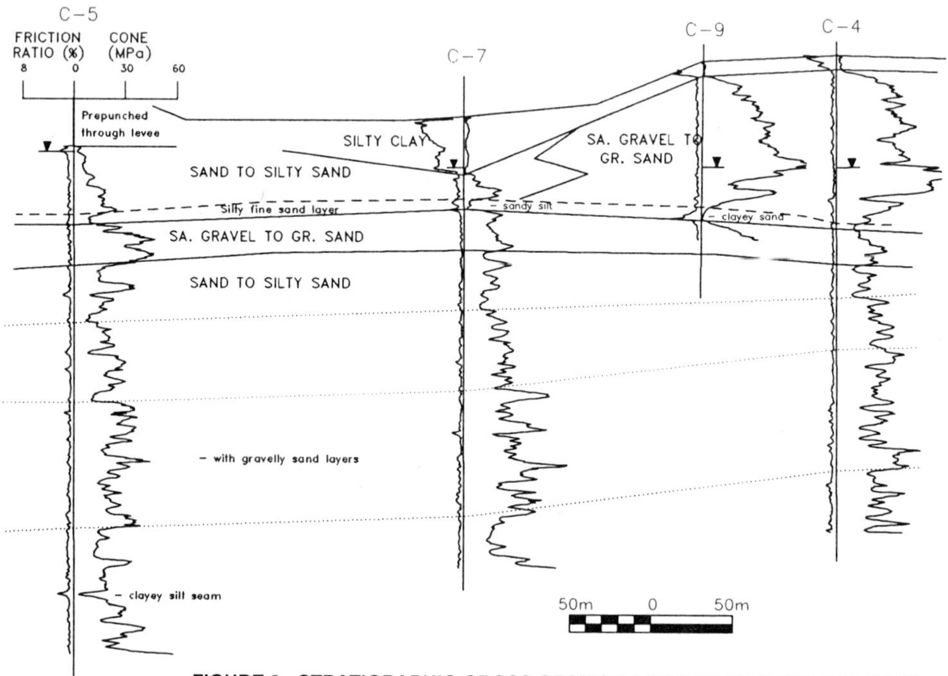

FIGURE 9 - STRATIGRAPHIC CROSS-SECTION DEVELOPED FROM CPT DATA

- Collection and analysis of high quality, high resolution hydrostratigraphic data, with little or no downtime

- Optimal positioning of drill rig installed monitor wells, with little additional split-spoon sampling or geotechnical laboratory testing

- Meeting deadlines and budgetary constraints

Based on the results of the penetrometer investigation, monitor wells were installed using drill rigs along the downstream perimeter of the contaminant plume. Little split-spoon sampling was performed during well installation as stratigraphy had already been defined to a high degree by the CPTU sounding data. Several rounds of monitor well sampling have verified the accuracy of penetrometer ground water sampling and the use of field analytical testing.

REFERENCES

[1] Edge, R.W., K. Cordry, 1989. The Hydropunch: An In Situ Sampling Tool for Collecting Ground Water from Unconsolidated Sediments. Ground Water Monitoring Review, Vol. IX (3), pp 177-183.

[2] Douglas, B.J., R.S. Olsen, 1981. Soil Classification Using Electric Cone Penetrometer. Cone Penetration Testing and Experience, Proceedings, St. Louis, MO, October 1981, ASCE, pp 209-227.

[3] Saines, M., A.I. Strutynsky and G. Lytwynyshyn, 1989. Use of Piezometric Cone Penetration Testing In Hydrogeologic Investigations. Presented at the First USA/USSR Hydrogeology Conference, Moscow, USSR.

[4] Strutynsky, A.I., B.J. Douglas, L.J. Mahar, G.F. Edmonds, and E. Hencey, 1985. Arctic Penetration Test Systems. Civil Engineering in the Arctic Offshore, ASCE, pp 162-168.

[5] Strutynsky, A.I., R. Sandoford, and D. Cavaliere, 1991. Use of Piezometric Cone Penetration Testing with Electrical Conductivity Measurements (CPTU-EC) for Detection of Hydrocarbon Contamination in Saturated Granular Soils. Accepted for publication, Ground Water and Vadose Zone Investigations, ASTM.

[6] Torstensson, B-A, 1984. A New System for Ground water Monitoring. Ground Water Monitoring Review, Vol. IV (4), pp 131-138.

AUTHORS

Andrew I. Strutynsky, P.E., the Technical Director of STRATIGRAPHICS, is responsible for all penetrometer operations at the company. He designs custom penetrometer equipment and performs research and development on new techniques. He was a speaker at the First International Symposium on Penetration Testing (1988), and conducted a Piezometric Cone Penetration Test (CPTU) workshop at the Third Outdoor Action Conference (1989), sponsored by the NWWA. He has over 10 years of subsurface exploration experience, eight of which have been exclusively devoted to penetrometer exploration projects. He has worked throughout the United States, in the Far East, the Middle East, Central America, Europe, and the High Arctic. He has performed penetrometer research for a major oil company and both the U.S. Geological Survey and the National Science Foundation. He has authored various papers on penetrometer equipment and interpretation of penetrometer data. He has the B.S. and M.S. degrees from the University of Illinois.

Timothy J. Sainey, P.G., is Senior Project Geologist at ERM-Midwest, Inc. With more than 17 years of geologic experience, his specialties include OVA/GC instrumentation, soil gas surveys and subsurface geophysical testing methods. While at ERM, Mr. Sainey has managed numerous environmental assessment, site remediation and closure projects, involving soil and ground water sampling, well construction, installation and monitoring, and the evaluation of remedial actions, Previously, he managed a geotechnical laboratory that specialized in ASTM testing. Also, he worked as an exploratory geologist for an oil company. Mr. Sainey received his B.S. and M.S. in geology from Ohio University in 1969 and 1973.

Ground-Water Monitoring Wall Design and Installation

Louise V. Parker[1]

SUGGESTED GUIDELINES FOR THE USE OF PTFE, PVC AND STAINLESS STEEL IN SAMPLERS AND WELL CASINGS

REFERENCE: Parker, L. V., "Suggested Guidelines for the Use of PTFE, PVC and Stainless Steel in Samplers and Well Casings," Current Practices in Ground Water and Vadose Zone Investigations, ASTM STP 1118, David M. Nielsen and Martin N. Sara, Eds., American Society for Testing and Materials, Philadelphia, 1992.

ABSTRACT: This paper reviews the literature on PTFE, PVC and stainless steel well casings and samplers and develops a series of suggested guidelines for use of these materials in sampling programs. Three factors are considered: the resistance of these materials to chemical reaction and thus degradation by the environment, the effect these materials have on the integrity of ground water samples, and the physical strength of these materials. Specifically, factors involved in the corrosion of stainless steel and the ability of aqueous solutions of organics to degrade PVC are discussed. Also considered are sorption and leaching of organic and metal species by these materials. The differences in the behavior of flexible vs. rigid PVC are emphasized.

KEYWORDS: PVC, PTFE, Teflon®[2], stainless steel, samplers, well casings.

Much attention has been focused on the issue of polyvinylchloride (PVC) vs. polytetrafluoroethylene (PTFE, Teflon®) vs. stainless steel (SS) well casings. This concern arose in part from considerations raised by the EPA's "Resource Conservation and Recovery Act (RCRA) Ground-Water Monitoring Technical Enforcement Guidance Document" (TEGD) [1]. The TEGD raised two issues: 1) the resistance of these materials to chemical reaction (such as oxidation of metals and swelling of plastics by chemical solvents) and thus degradation, and 2) the ability of these materials to affect the chemistry of ground-water samples either by sorption of analytes of interest or by leaching contaminants. With respect to leaching, either the analytes of interest or compounds that interfere with analysis may be leached. Also, desorption of sorbed contaminants may occur if water quality improves.

This paper examines what has been learned from studies on well casings and samplers and uses these findings to establish suggested guidelines for future ground-water sampling programs. In order to keep this paper concise, not all of the available literature was included in this

[1]Research Physical Scientist, U.S. Army Cold Regions Research and Engineering Laboratory, 72 Lyme Road, Hanover, N.H. 03755-1290.

[2]E. I. Du Pont De Nemours, Inc., Wilmington, Delaware.

review. For example, laboratory studies that did not appear to include proper controls and in-situ studies that compared the chemistry of samples taken from different but adjacent wells were not included. The in-situ studies often looked at only one or two sites and had conflicting results. Given the lack of control that is inherent in this type of study, it is not surprising that in-situ studies often yield conflicting results. In such cases, differences in the sample chemistry between wells may arise from tapping different parts of the aquifer or may be due to (inadvertent) differences in well construction. When assessing laboratory studies, the reader also should be alert to whether losses due to biodegradation and biotransformation can be ruled out because measures were taken to prevent these changes.

This paper first examines the resistance of casing and sampler materials to chemical degradation, describes the effects that these materials can have on sample integrity via sorption and leaching, and finally discusses the physical strength of these materials, with respect to use as casings and screens in monitoring wells.

RESISTANCE OF SS, PVC AND PTFE TO CHEMICAL ATTACK

PTFE is known for its excellent resistance to corrosive reagents and dissolution by solvents. While PVC also has relatively good resistance to chemical attack, it is subject to degradation by several solvents, including low molecular weight ketones, aldehydes, amines, and chlorinated alkanes and alkenes [2]. While it is known that these neat (or pure) solvents can degrade PVC, the effect of aqueous solutions of these and other solvents on the structural integrity of rigid PVC is not generally known. Berens [3] predicted the permeation of organic chemicals through PVC pipe based on gravimetric sorption data for samples of PVC exposed to a range of activities of organic solvents.[3] Berens [3] noted that the polymer-solvent interaction parameter (χ) in the Flory-Huggins equation can be used as a measure of the swelling or solvent power of a particular solvent. In general, if χ is less than 0.5, the solvent will completely dissolve the polymer. For values > 0.5, the larger the number, the less the swelling power. In these studies, Berens observed that the rate of penetration in PVC is governed by the activity of the penetrant in the environment of the PVC and not by the penetrant's absolute concentration or physical state, and that the rate of penetration changed many orders of magnitude as the activity of the solvent in the environment changed. He noted that at ambient temperatures, rigid PVC will be softened only by strong solvents or swelling agents (for $\chi = 0$) at activities > 0.5. For less aggressive solvents, even higher activities are needed. For aqueous solutions, the solvent activity can be approximated by dividing the concentration of the compound in solution by its solubility. Thus for aqueous solutions of solvents to cause swelling or softening of rigid PVC, the solvent must be able to completely dissolve PVC and must exist in the environment at a concentration that exceeds one-half its solubility in water. Thus, Berens [3] concluded that softening of PVC is only possible in the presence of nearly undiluted solvents or swelling agents for PVC.

---

[3]Both vapor sorption and liquid immersion experiments were conducted. Initial tests involved samples of powders, sheets and films of simple PVC resin or resin plus a tin stabilizer. Confirmation tests were conducted using a powder of a commercial pipe formulation and samples cut from PVC pipe. Immersion tests used polyethylene glycol as the activity depressant.

For mixtures of organics such as gasoline, Berens [3] predicted that gasoline would not appreciably swell PVC. Schmidt [4] confirmed these findings when he subjected sections of Type I PVC well screen to several different grades of gasoline for six months and found no detectable change in the screen when examined by using a scanning electron microscope.

Olson et al. [5] also studied the permeation of two organics (1,1,1-trichloroethane and toluene) through PVC pipe from the neat solvent and from saturated aqueous solutions. Their findings agreed well with those of Berens [3]. Olson et al. [5] concluded that PVC pipe was an effective barrier to low activity (< 0.25) levels of swelling solvents, but becomes permeable to swelling solvents at high or unit activities.

While stainless steel is subject to corrosion under some conditions, it is considered to have good to excellent corrosion resistance [6]. It is more resistant to corrosion than many other metals or alloys, including carbon steel and low-carbon steel and galvanized steel [7], and silicon red brass, Anaconda red brass, and common yellow brass [6,8]. Typical forms of corrosion to which stainless steel well casings may be exposed include oxidation (rusting), selective corrosion (loss of one element of an alloy or dezincification), bimetallic corrosion by creation of a galvanic cell, surface pitting, and stress corrosion [6,7,9]. Factors that can affect corrosion include low pH, high dissolved oxygen content, presence of hydrogen sulfide (H_2S), dissolved solids, high carbon dioxide (CO_2), and high chloride [7,10]. Several different alloys of stainless steel are available; the most common alloys for casings are types 304 and 316 [7]. Stainless steel 304 (SS 304) has good resistance to corrosion under aerobic conditions, while stainless steel 316 (SS 316) has improved resistance to sulfur species and sulfuric acid solutions, thus performing better under reducing conditions [2,7]. However, both types of stainless steel are subject to corrosion under long-term exposure to corrosive environments [2,7].

Several laboratory studies conducted by our laboratory [11,12,13, 14] have found problems with the rusting of both SS 304 and 316. Parker et al. [11] observed that approximately half the stainless casings showed signs of surface rust, and in some cases a hydrous iron oxide precipitate was visible in the glass containers. Hewitt [13,14] noted that when rusting did occur on stainless steel casings, it usually occurred on the weld. Because of this enhanced potential for corrosion in the vicinity of a weld, Aller et al. [7] recommended using threaded stainless steel casings rather than stainless casings with welded joints.

Often corrosion of metals is mediated or enhanced by the presence of microorganisms. Many aerobic bacteria utilize hydrogen and thus can take part in the cathodic reaction in a galvanic cell [6]. Under anaerobic conditions, sulfate-reducing bacteria can actively cause corrosion of stainless steel [6]. Iron bacteria, which oxidize ferrous iron to ferric iron, can cause problems with encrustation, which can lead to total blockage of well screens [6].

Clearly PTFE is the most inert to chemical attack. However, in many circumstances the geochemistry of a site will not be corrosive and thus stainless steel can be used. In most instances rigid PVC can be used unless it will be exposed to a nearly saturated solution of a PVC solvent or the pure solvent itself.

MATERIAL EFFECTS ON SAMPLE INTEGRITY

Sorption of Metals from Aqueous Solution

Masse et al. [15] outlined the factors involved in sorptive losses of metals on containers: 1) chemical form and concentration of the metal;

2) characteristics of the solution, such as pH, total dissolved solids, salinity, hardness, complexing agents, dissolved gases (especially oxygen, which can influence the oxidative state), suspended matter (competitor in the sorption process) and microorganisms (e.g., uptake of trace elements); 3) properties of the container, such as chemical composition, surface roughness, surface cleanliness, surface area/volume ratios, and history of the container (e.g., previous exposure to heat and chemicals); 4) external factors such as temperature, contact time, exposure to light, and agitation. Most of these factors are also applicable to materials in samplers and well casings.

Comparisons of SS 304, SS 316, PVC, and PTFE well casings indicate that with respect to sorption of metals (arsenic, chromium and lead) in solution, PTFE is clearly the least sorptive, while the two stainless steel casings were the most sorptive [11]. No loss of arsenic or chromium was noted in solutions exposed to either polymer casing after 72 hours. Presumably these elements existed as anions in this experiment, and negatively charged species do not associate strongly with plastic surfaces such as PTFE and polyethylene, which are known for their cation exchange capacity [15]. Parker et al. [11] also noted a tremendous variability in the concentrations of cadmium and chromium in samples exposed to both stainless steel casings and in the concentration of lead in samples exposed to SS 316. This variation was attributed to surface oxidation of the stainless steel casings. All surfaces sorbed lead; after four hours losses were approximately 10% for PVC and 20% for the two stainless casings. However, even with PTFE, significant losses of some metals can occur rapidly. Masse et al. [15] noted a 10% loss of silver in distilled water (pH 8.5) within 30 minutes.

Recently Hewitt [14] conducted a similar study under low dissolved oxygen conditions. As in the two previous studies [11,12], PTFE was the least active surface, and the two stainless surfaces were the most active. As might be expected, there was no oxidation of the stainless steel casings under these low oxygen conditions.

Sorption of Organics From Aqueous Solution

Many of the factors involved in sorption of metals on containers that were outlined previously [15] are also applicable to sorption of organics by samplers and well casings. Early work by Barcelona and coworkers [16] found that flexible PVC tubing was highly sorptive of chlorinated organics from solution. For example, they found that loss of chloroform was 80% after only one hour in the presence of PVC tubing. They also found that Teflon® (PTFE) tubing was less sorptive than any of the other tubings they tested (silicone rubber, linear polyethylene, and polypropylene). However, even PTFE sorbed substantial amounts of several of these organics quickly, within 10-60 minutes. For example, after 60 minutes, total sorption of the halocarbon mixture was 237 $\mu g/m^2$ for PTFE and 622 $\mu g/m^2$ for (flexible) PVC. This study has frequently been cited as a justification for using Teflon® vs. any type of PVC. However, rigid PVC is much less sorptive of organics than flexible PVC (or Tygon®[4]) tubing [16,17,18]. Two studies [11,18] conducted under sterile conditions have shown that the rate and extent of sorption of organics from aqueous solution is greater for PTFE than for rigid PVC. For example, Parker et al. [11] found that loss of trichloroethylene (TCE) was 10% after eight hours for PTFE, while losses for PVC were only 6% after one week and 12% after six weeks. In comparison, after six weeks, loss for PTFE was 60%. Another

[4]Norton Performance Plastics, Wayne, New Jersey.

study [19], conducted under nonsterile conditions, looked for loss of five chlorinated hydrocarbons and found more rapid and extensive losses of two compounds with PTFE vs. rigid PVC, and equivalent losses for the two materials for two other compounds. There appears to be some problem with the results for the fifth compound since concentrations of samples exposed to PTFE exceeded the initial concentration for much of the experiment. In all these studies, sorption of the organics was attributed to absorption within the polymer matrix. Uptake of organic compounds by polymers is considered to proceed first by sorption/dissolution in the polymer surface, followed by diffusion into the polymer matrix [18,19,20,21]. For PTFE, Parker et al. [11] were able to correlate the loss of hydrophobic organic constituents at equilibrium, or the equilibrium constants (K_{eq}), with the substances' octanol water partition coefficient (K_{ow}) values, although this correlation overestimated losses for more polar organic substances. Serota et al. [20] and Polack et al. [22] developed a similar relationship between the permeability or sorption of organics in polyethylene and their hexane/water partition coefficients.

The reduced losses associated with rigid vs. flexible PVC may be attributed to the higher density and greater crystallinity of rigid PVC [16]. Also, diffusion would be slower in rigid PVC because of decreased mobility within the polymer [23].

For stainless steel, losses of organics have been observed under nonsterile conditions [24] but not under sterile conditions [11,18]. However, the losses observed in the study conducted under nonsterile conditions [24] appear to be due to reductive hydrogenolysis, as the authors suggested, since only the most highly halogenated compounds were affected. This reaction is mediated by transition metals and transition metal complexes. Several transition metals (chromium, manganese, iron and nickel) are components of stainless steel.

While one would not expect partitioning or absorption of organics into stainless steel, one might expect to see some losses due to adsorption. The fact that Parker et al. [11] did not note any losses of organics with K_{ow} values of 10^3 indicates that adsorption would have to involve highly hydrophobic contaminants (K_{ow} of ~10^4 or higher). However, extremely hydrophobic compounds are also less likely to be found in ground water because of their high affinity for adsorption and absorption.

Leaching of Metals into Aqueous Solution

With respect to the leaching of metals, PTFE is clearly superior. Hewitt [12] looked for leaching of nine metals (arsenic, barium, cadmium, chromium, copper, lead, mercury, silver and selenium) from SS 304, SS 316, PTFE and PVC well casings after 1, 5, 20 and 40 days of exposure. Generally, PTFE leached the least amount of these metals, and either SS 304 or SS 316 (or both) leached the most. ANOVA revealed that after 24 hours, levels of barium were significantly elevated for samples exposed to all four well casings, including PTFE. In the same amount of time, statistically significant amounts of copper were leached from the SS 316 casings but none of the others. While ANOVA revealed that levels of lead were only significantly higher for PVC after 24 hours, substantial amounts did leach from all four casings. Mean concentrations for these four samples were all at least two times that for the control sample. The reason ANOVA did not reveal significant differences for the other casings was due to the large variability in this data. A similar problem with large variability in the chromium data also occurred. Thus, ANOVA did not reveal any significant differences in the concentration of chromium between samples exposed to casings and the controls after 24 hours. However, at that time, the mean concentration was six times the control value for PVC, eight times the control value for SS 304, and thirty times the control value for SS 316.

After 24 hours, cadmium concentrations were below the detection limit for samples exposed to PTFE and SS 304, while the mean concentration was 0.136 µg/L for samples exposed to PVC and 1.125 µg/L for samples exposed to SS 316. As in the previous study [11], the variance was greatest among replicate samples exposed to stainless steel. This variance was attributed to corrosion of the stainless steel casings. In some instances the initial increases seen after 24 hours were followed by loss, indicating subsequent sorption of the leached analytes. However, leaching of these metals (barium, cadmium, chromium, copper and lead) from all four casing materials was always well below levels set for these metals by the National Sanitation Foundation [25,26] for leachates from PVC well casings.

A similar study, conducted under low dissolved oxygen conditions, produced very similar results [14]. There was no statistically significant difference in the values of cadmium, chromium, copper, lead, iron or nickel concentrations in samples exposed to PTFE vs. the control values, although copper levels were 1.4-2.4 times the control values. In most cases, the greatest amount of leaching of metals was from the two stainless steel casings.

In the United States, organotin compounds are used in potable water and monitoring pipe, while in other parts of the world lead stabilizers are commonly used. A number of older studies have shown that PVC pipe will leach these stabilizers [27,28,29,30,31,32]. In the case of organotin stabilizers, these compounds react with the polymer during processing, and thus it is the reaction products, alkyltin chlorides, which leach [32]. The amount of these stabilizers that is leached is highest initially, drops off with additional flushing with water, and then levels off [27,28, 30,32]. In the case of lead stabilizers, this phenomenon has been attributed to the dissolution of a lead-rich layer on the inner surface of the new pipe followed by minimal additional leaching once this layer has been removed [29]. This layer could be removed by pretreatment with either alcoholic sodium hydroxide solution or a mixture of dilute nitric and perchloric acids [29]. However, pretreatment with distilled water or 20% detergent solution was not as effective [29]. One study indicates that the amount of stabilizer leached can vary between pipes of similar composition but made by different manufacturers, and even along the same length of pipe [27]. The specific stabilizer used and its concentration, the amount of filler, the presence of impact modifiers, and the extrusion temperature have all been found to affect the degree of leaching of the stabilizer [27,30,31,33]. Thus manufacturers can apparently reduce the leaching of stabilizers from PVC by carefully selecting the amount and type of stabilizer and other components, and by controlling processing temperatures. Treating the finished product with either alcoholic sodium hydroxide or a dilute solution of mineral acid would also appear to significantly reduce leaching of stabilizers. However, this requires further study before specific recommendations can be developed.

In addition to leaching associated with sampler and casing materials, leaching of metals from paints and greases used on sampling pumps can also occur [34].

Leaching of Organics into Aqueous Solution

With respect to leaching of organics, perhaps the compounds most often cited as troublesome for PVC are plasticizers. Again, it is important to differentiate between flexible and rigid products. Flexible PVC has been found to leach various phthalate esters (plasticizers) into a variety of fluids (blood, plasma, saline and distilled water) [16,35,36, 37,38,39]. Estimates of plasticizer contents in flexible PVC products range from 30-50% by weight [7,37]. Barcelona et al. [16] looked at various flexible tubings including PVC and Teflon® (also silicone rubber, linear polyethylene and polypropylene) and found that all of the tubings except for Teflon® leached significant amounts of organic carbon in 30 minutes, and that flexible PVC and silicone rubber leached the most. The migration of plasticizers from the tubing into a liquid has been found to depend upon the amount of "free" plasticizer on the inner walls of the tubing, the rate at which the "free" plasticizer is renewed by diffusion from the bulk of the PVC, the solubility of the plasticizer in the liquid, and the partition coefficient of the plasticizer between the liquid and PVC tubing wall [39].

In comparison, rigid PVC contains virtually no plasticizers (<0.01%) [2] and thus would not be expected to leach large quantities of these materials. Plasticizers are not normally used with PTFE resins, although various inorganic fillers may be added to provide variations in physical properties [40]. Curran and Tomson [17], who compared the leachates from several materials (polypropylene and polyethylene) including Teflon®, Tygon® (flexible PVC) and rigid PVC, found that the least amount of contamination leached was from the Teflon® but that rigid PVC was a very close second. Miller [41] was unable to find any organic constituents leached from rigid PVC using gas chromatography (flame ionization detector) following extraction and a thousandfold concentration.

The leaching of ppb levels of alkyltin compounds (stabilizers) from PVC pipe has already been discussed. While leaching of vinyl chloride monomer (VCM) from PVC pipe was a problem in the past [42,43], standards set for potable pipe and well casings by the National Sanitation Foundation (NSF) in 1977 have substantially reduced this problem [7]. At that time, the NSF set the maximum limit of RVCM in PVC pipe at 10 ppm , which limited leached concentrations of VCM to 1 or 2 ppb [2,7]. A more recent standard [25] set the limit at 3.2 mg/kg.

In addition to the components that may leach from well casing or sampler materials themselves, lubricants and joining compounds may also leach and thus degrade water quality. The problems associated with the leaching of solvents from solvent-bonded joints are well documented in the literature [32,41,44,45,46,47]. For this reason, PVC that has solvent-bonded joints should not be used when sampling for organics. There are no similar problems associated with PTFE because it is not subject to solvation and thus cannot be solvent-bonded.

Several laboratory studies have indicated that PTFE and flexible and rigid PVC will desorb sorbed organics [11,16,41] For different compounds, the amount desorbed did not parallel the amount sorbed [11,41]. For example, after three days, Parker et al. [11] found higher concentrations of tetrachloroethylene and t-dichloroethylene in leachates from PTFE casings, even though more p- and m-dichlorobenzene had been sorbed by the PTFE casings. They concluded that smaller molecules may be more easily or rapidly desorbed. These results agree well with findings of Berens and Hopfenberg [48]. Berens and Hopfenberg found that diffusivity of organic molecules in PVC and other polymers decreases exponentially with increasing diameter of "spherical" penetrant molecules. Serota et al. [20] noted that while adsorption/desorption are affected by the relative affinity of the organic molecule for the aqueous solution vs. the polymer, the diffusion process is regulated by the effective size of the penetrant molecule, extent of the plastic-penetrant interaction, and structural characteristics of the polymer. Further rate studies of this process are needed.

STRENGTH OF CASING MATERIALS

Finally, in selecting a material, other considerations, such as the physical strength of the material, clearly may be overriding. Of the three materials considered in this paper, stainless steel is the strongest.

While PVC is not as strong as stainless steel, it can be used down to depths of 1200-2000 ft (366-610 m) [49] and thus can be used in most ground-water monitoring applications. Recommended depths for PTFE range from 225-375 ft (68.6-114 m) [49]. Problems associated with the bending of PTFE casings and the partial closing of PTFE well screens at depth, and measures to compensate for these problems, are discussed in detail in several references [7,49,50]. One additional concern that recently has been raised is that because of the antistick property of PTFE, it is difficult to achieve an annular seal with neat grout cement [49]. This failure of grout to bond to PTFE results in formation of a "micro annulus" that can channel surface water down the outside of the casing [49]. These issues are discussed in more detail by Nielsen and Schalla [49].

RECOMMENDATIONS

None of the materials discussed in this review (PVC, PTFE and stainless steel) are the ideal "inert" material. All are either subject to chemical degradation or affect sample integrity, or both. Thus, selection of a material for a casing or sampler will depend on the conditions of exposure, its intended use, and the duration of expected exposure. The selected material must have the durability and strength necessary for the specific application.

Casings, samplers, or pumps that remain in the well are subject to long-term effects of the environment. In these instances, stainless steel cannot be used if the environment is corrosive. Conditions such as low pH, high dissolved oxygen content, presence of H_2S, dissolved solids, high CO_2, and high chloride indicate that corrosion of stainless steel may be a problem. A corrosion probe can be used to determine whether corrosion of stainless steel will be a problem [4].

With respect to possible degradation of rigid PVC by solvents, only high activities of solvents or swelling agents of PVC will cause softening and thus degradation of PVC. If neat solvents are expected, the solvent interaction parameter, or χ value, can be used to determine if the solvent is able to cause softening [3]. For aqueous solutions of solvents, PVC casings would be ruled out only for solvents that dissolve PVC and are present at concentrations greater than half their solubility in water.

With respect to material-sample interactions, contact time is critical. The contact time for bailers would be only a few minutes. In this case either rigid PVC, PTFE or stainless steel can be used. However, for well casings and many samplers, the contact time is more likely to be considerably longer (minutes to hours). For these samplers and well casings, material interactions can be important. In these instances, PTFE is the best material if one is interested only in analyzing for metals. Rigid PVC is a reasonable second choice, but stainless steel should be avoided. For sampling organics, rigid PVC and stainless steel are the best choices. While stainless steel is best in that it is least likely to affect the integrity of samples that will be analyzed for organics, in most instances the contact time is sufficiently short so that PVC can be used. PVC should be used if analyzing for both organics and metals.

With respect to leaching and sorption of organics by flexible tubings, PTFE is better than flexible PVC. However, since PTFE can be highly sorptive of some organics, contact time between sample and sampler line should be kept as brief as possible to reduce losses of organics. There is not much information available on the leaching or sorption of metals by flexible PVC. Presumably, PTFE also would be better than flexible PVC for sampling metals.

Since desorption of sorbed organics has been shown to occur from both PVC and PTFE, dedication of PVC and PTFE sampling equipment is strongly recommended when sampling for organics. Cleaning should be adequate for decontaminating sampling equipment made of stainless steel. Thus all-stainless samplers need not be dedicated if sampling for organics.

Only threaded PVC, PTFE, and stainless steel casings should be used. Solvent bonding should be avoided in PVC samplers used for organics.

ACKNOWLEDGMENTS

Funding of this work was provided by the U.S. Army Toxic and Hazardous Materials Agency, Martin H. Stutz, Project Monitor. The author thanks Alan Hewitt and Dr. Thomas Jenkins (research chemists with the author's laboratory) for their useful review comments.

REFERENCES

[1] U.S. Environmental Protection Agency, Resource Conservation and Recovery Act (RCRA) Ground-Water Monitoring Technical Enforcement Guidance Document, U.S. Government Printing Office, Washington, D.C., 1986.

[2] Barcelona, M.J., Gibb, J.P., and Miller, R.A., A guide to the selection of materials for monitoring well construction and ground water sampling, U.S. Environmental Protection Agency Report No. EPA-600/2-84-024, U.S. Government Printing Office, Washington, D.C., 1984.

[3] Berens, A.R., "Prediction of organic chemical permeation through PVC pipe," Journal of the American Water Works Association, 1985, pp. 57-65.

[4] Schmidt, G.W., "The use of PVC casing and screen in the presence of gasolines on the ground water table," Ground Water Monitoring Review, Vol. 7, No. 2, 1987, pp. 94-95.

[5] Olson, A.J., Goodman, D., and Pfau, J.P., "Evaluation of organic solvents through PVC, Asbestos/Cement and Ductile Iron Pipes," Journal of Vinyl Technology, Vol. 9, No. 3, 1987, pp. 114-118.

[6] Lloyd, J.W. and Heathcote, J.A., Natural inorganic hydrochemistry in relation to ground water: An introduction, Clarendon Press, Oxford, England, 1985.

[7] Aller, L., Bennett, T.W., Hackett, G., Petty, R.J., Lehr, J.H., Sedoris, H., Nielsen, D.M., and Denne, J.E., Handbook of suggested practices for the design and installation of ground-water monitoring wells. National Water Well Association, Dublin, Ohio, 1989.

[8] Campbell, M.D. and Lehr, J.H., Water Well Technology, McGraw-Hill, New York, 1973.

[9] Driscoll, F.G., Ground Water and Wells, Johnson Division, St. Paul, Minnesota, 1986.

[10] Moss, R., Design of casings and screens for water production and injection wells. American Petroleum Institute, 1964.

[11] Parker, L.V., Hewitt, A.D., and Jenkins, T.E., "Influence of casing materials on trace-level chemicals in well water." Ground Water Monitoring Review, Vol. 10, No. 2, 1990, pp. 146-156.

[12] Hewitt, A.D., Leaching of metal pollutants from four well casings used for ground-water monitoring. Special Report 89-32, U.S. Army Cold Regions Research and Engineering Laboratory, Hanover, NH, 1989.

[13] Hewitt, A.D., Influence of well casing composition on trace metals in ground water. Special Report 89-9, U.S. Army Cold Regions Research and Engineering Laboratory, Hanover, NH, 1989.

[14] Hewitt, A.D., Potential influences of common well casings on metal concentrations in well water with low dissolved oxygen. CRREL Report 91-13, U.S. Army Cold Regions Research and Engineering Laboratory, Hanover, NH, 1991.

[15] Masse, R., Maessen, F.J.M.J., and DeGoeij, J.J.M., "Loss of silver, arsenic, cadmium, selenium and zinc traces from distilled water and artifical sea-water by sorption on various container surfaces." Analytica Chimica Acta, Vol. 127, 1981, pp. 181-193.

[16] Barcelona, M.J., Helfrich, J.A., and Garske, E.E., "Sampling tubing effects on ground water samples," Analytical Chemistry, Vol. 57, 1985, pp. 460-464.

[17] Curran, C.M. and Tomson, M.B., "Leaching of trace organics into water from five common plastics," Groundwater Monitoring Review, Vol. 3, 1983, pp. 68-71.

[18] Gillham, R.W. and O'Hannesin, S.F., "Sorption of aromatic hydrocarbons by materials used in construction of ground-water in sampling wells," in Ground Water and Vadose Zone Monitoring, ASTM STP 1053, American Society for Testing and Materials, Philadelphia, 1990, pp. 108-122.

[19] Reynolds, G.W. and Gillham, R.W., "Adsorption of halogenated organic compounds by polymer materials commonly used in groundwater monitors," in Proceedings of Second Canadian/American Conference on Hydrology. Hazardous Wastes in Ground Water: A Soluble Dilemma, National Water Well Association, Dublin, Ohio, 1986, pp. 125-132.

[20] Serota, D.G., Meyer, M.C., and Autian, J., "Effects of structure on permeability of substituted anilines from aqueous solution, through polyethylene," Journal of Pharmaceutical Sciences, Vol. 61, No. 3, 1972, pp. 416-419.

[21] Yasuda, H. and Stannett, V., "Permeability Coefficients." Polymer Handbook, Wiley Interscience Publications, 1975, pp. III-229 to III-240.

[22] Polack, A.E., Nunez, L.J., and Autian, J., "Transport of solutes into polyethylene bottles from aqueous solutions; empirical relationships of the data," International Journal of Pharmaceutics, Vol. 3, 1979, pp. 157-175.

[23] Berens, A.R., "The diffusion of gases and vapors in rigid PVC," Journal of Vinyl Technology, Vol. 1, No. 1, 1979, pp. 8-13.

[24] Reynolds, G.W., Hoff, J.T., and Gillham, R.W., "Sampling bias caused by materials used to monitor halocarbons in groundwater," Environmental Science and Technology, Vol. 24, No. 1, 1990, pp. 135-142.

[25] National Sanitation Foundation, National Sanitation Foundation Standard No. 14 for plastic piping system components and related materials, National Sanitation Foundation, Ann Arbor, Michigan, 1989.

[26] National Sanitation Foundation, National Sanitation Foundation Standard No. 14 for plastic piping system components and related materials, National Sanitation Foundation, Ann Arbor, Michigan, 1980.

[27] Packham, R.F., "The leaching of toxic substances from unplasticized PVC water pipe. Part I. A critical study of laboratory test procedures," Water Treatment and Examination, Vol. 20, No. 2, 1971, pp. 108-124.

[28] Packham, R.F., "The leaching of toxic substances from unplasticized PVC water pipe. Part II. A survey of lead levels in PVC distribution systems," Water Treatment and Examination, Vol. 20, No. 2, 1971, pp. 144-151.

[29] Packham, R.F., "The leaching of toxic substances from unplasticized PVC water pipe. Part III. The measurement of extractable lead in PVC pipes," Water Treatment and Examination, Vol. 20, No. 2, 1971, pp. 152-164.

[30] Gross, R.C., Engelbart, B., and Walter, S., "Aqueous extraction of lead stabilizers from PVC compounds," Society of Plastic Engineering, Technical Paper 20, 1974, pp. 527-531.

[31] Dietz, G.R., Banzer, J.D., and Miller, E.M., "Water extraction of additives from PVC pipe," in Proceedings of the Safety and Health with Plastics National Technical Conference, Society of Plastic Engineering, 1977, pp. 25-29.

[32] Boettner, E.A., Ball, G.L., Hollingsworth, Z., and Aquino, R., Project summary: Organic and organotin compounds leached from PVC and CPVC pipe, EPA-600/S1-81-062, U.S. Environmental Protection Agency, U.S. Government Printing Office, Washington, D.C., 1982.

[33] Wong, M.K., Gan, L.M., Koh, L.L., and Lum, O.L., "Some further studies on factors affecting the leaching of lead from unplasticized poly (vinyl chloride) pipes," Water Research, Vol. 24, No. 4, 1990, pp. 451-455.

[34] Houghton, R.L. and Berger, M.E., "Effects of well-casing composition and sampling method on apparent quality of ground water," in Fourth National Symposium on Aquifer Restoration and Groundwater Monitoring, National Water Well Association, U.S. Environmental Protection Agency, and National Center for Ground Water Research, 1984.

[35] Jaeger, R.J. and Rubin, R.J., "Plasticizers from plastic devices: Extraction, metabolism, and accumulation by biological systems," Science, Vol. 170, 1970, pp. 460-462.

[36] Jaeger, R.J. and Rubin, R.J., "Migration of a phthalate ester plasticizer from polyvinyl chloride blood bags into stored human blood and its localization in human tissues," The New England Journal of Medicine, Vol. 287, No. 22, 1972, pp. 1114-1118.

[37] Junk, G.A., Svec, H.J., Vick, R.D., and Avery, M.J., "Contamination of water by synthetic polymer tubes," Environmental Science and Technology, Vol. 8, No. 13, 1974, pp. 1100-1106.

[38] Christensen, D., Neergaard, J., Nielsen, B., Faurby, V., and Nielsen, O.F., "The release of plasticizers from PVC tubing," Proceedings of the 6th International Congress on Pharmacology, Vol. 6., Pergamon, Oxford, 1976.

[39] Fayz, S., Herbert, R., and Martin, A.M., "The release of plasticizer from polyvinyl chloride haemodialysis tubing," Journal of Pharmacy and Pharmacology, Vol. 29, 1977, pp. 407-410.

[40] Voter, R.C. "VI. Ethylene and fluoroethylene polymers in high polymers, Vol. XII," in Analytical chemistry of polymers, Part I. Analysis of monomers and polymeric materials. Plastics-resins-rubbers-fibers, Interscience Publishers, Inc., New York, 1959, pp. 165-196.

[41] Miller, G.D., "Uptake and release of lead, chromium, and trace level volatile organics exposed to synthetic well casings," in Proceedings of the Second National Symposium on Aquifer Restoration and Ground Water Monitoring, National Water Well Association, U.S. Environmental Protection Agency, and National Center for Ground Water Research, 1982, pp. 236-245.

[42] Banzer, J.D., "The migration of vinyl chloride monomer from PVC pipe into water," in Proceedings of the Safety and Health with Plastics National Technical Conference, Society of Plastic Engineering, Vol. 77, 1977, pp. 30-34.

[43] Dressman, R.C. and McFarren, E.F., "Determination of vinyl chloride migration from polyvinyl chloride pipe into water," Journal of the American Water Works Association, January, 1978, pp. 29-30.

[44] Boettner, E.A., Ball, G.L., Hollingsworth, Z., and Aquino, R., Organic and organotin compounds leached from PVC and CPVC pipe, U.S. Environmental Protection Agency, Report Number EPA-600/1-81-062, 1981.

[45] Sosebee, J.B., Geiszler, P.C. Jr., Winegardner, D.L., and Fisher, C.R., "Contamination of ground water samples with poly (vinyl chloride) adhesives and poly (vinyl chloride) primer for monitor wells," in Hazardous and Industrial Solid Waste Testing: Second Symposium, American Society for Testing and Materials, Philadelphia, 1982, pp. 38-50.

[46] Desrosiers, D.G., and Dunningan, P.C., "The diffusion of chloroform and carbon tetrachloride from rigid PVC pipe and rigid CPVC pipe into water," Journal of Vinyl Technology, Vol. 5, No. 4, 1983, pp. 187-191.

[47] Cowgill, U.M., "The chemical composition of leachate from a two-week dwell-time study of PVC casing and a three-week dwell-time study of fiberglass reinforced epoxy well casing," in Ground-water

Contamination: Field Methods, ASTM STP 963, American Society for Testing and Materials, Philadelphia, 1988, pp. 172-184.

[48] Berens, A.R. and Hopfenberg, H.B., "Diffusion of organic vapors at low concentrations in glassy PVC, polystyrene, and PMMA," Journal of Membrane Sciences, Vol. 10, 1982, pp. 283-303.

[49] Nielsen, D.M. and Schalla, R., "Design and installation of ground-water monitoring wells," in Practical Handbook of Ground-Water Monitoring, D.M. Nielsen, editor. Lewis Publishers,Chelsea, MI, 1991, pp 239-331.

[50] Dablow, J.F. III, Persico, D., and Walker, G.R., "Design considerations and installation techniques for monitoring wells cased with Teflon® PTFE," in Ground-water Contamination: Field Methods, ASTM STP 963, American Society for Testing and Materials, Philadelphia, 1988, pp., 199-205.

Stephen D. Reynolds[1] and Dawn A. Zemo[2]

METHODOLOGY FOR THE IN-FIELD DESIGN OF MONITORING WELLS IN HETEROGENEOUS FINE-GRAINED FORMATIONS[3]

REFERENCE: Reynolds, S. D. and Zemo, D. A., "Methodology for the In-Feild Design of Monitoring Wells in Heterogeneous Fine-Grained Formations," Current Practices in Ground Water and Vadose Zone Investigations, ASTM STP 1118, David M. Nielsen and Martin N. Sara, Eds., American Society for Testing and Materials, Philadelphia, 1992.

ABSTRACT: As part of the remedial investigation for a southern California Superfund site, clusters of depth-staggered monitoring wells were installed in heterogeneous unconsolidated alluvium. It was imperative that filter packs and well screens be properly designed such that ground-water samples and hydraulic head data would be suitable for use in litigation. The highly variable geometry and texture of the sediments precluded the use of a generic well design and suggested the need for a method of in-field design of monitoring wells.

In order to provide timely well design parameters, representative core samples of intervals to be screened were analyzed on-site. Simultaneous with the geophysical logging of the core boring, samples were sieved and the grain-size distribution data were used to design the filter packs and well screens. This approach provided a means by which each individual well could be tailored to the specific zone of interest, without delaying the pace of drilling and well construction or incurring costs above those normally associated with the more conventional approach employing a generic well design.

KEYWORDS: well design, turbidity, sieve analysis, continuous wireline sampling

BACKGROUND

Environmental investigations often require the installation of monitoring wells in formations that are significantly finer grained and more heterogeneous than those traditionally described as aquifers. In such formations the use a single generic design for all wells at a site invariably results in ground-water samples whose analytical results are

[1]Geologist with the California Dept. of Toxic Substances Control, Alternative Technology Div., P.O. Box 806 Sacramento, CA 95812-0806

[2]Geologist with Geomatrix Consultants, 100 Pine St., Suite 1000, San Francisco, CA 94111

[3]The official policies of the California Department of Toxic Substances Control or the United States Environmental Protection Agency are not necessarily reflected by the views or concepts presented in this manuscript.

readily impeachable due to the effects of inefficient wellbore hydraulics and suspended sediment (turbidity).

As part of the on-going remedial investigation for a southern California Superfund site, clusters of depth-staggered monitoring wells were installed to obtain litigation quality data needed to assess if a potential upgradient source was contributing to ground water contamination.

Review of previous efforts [1-6] to install monitoring wells on- and off-site found that 15 of the 20 existing wells have problems, including: deviated casing, clogged screens, and foaming and turbid samples. In addition, water-level data and analytical results from these wells have been called into question, resulting in costly resampling and legal disputes. Previous investigators reported difficult drilling conditions, principally hole stability problems, that were deemed responsible for the problematic wells. The authors' review noted that a single generic well intake design was used in each of the previous investigations, even though the geology is extremely heterogenous.

The combined factors of the highly variable nature of the formation; depth of installation, 65 to 98 meters (m) (215 to 320 ft.), hole stability problems, stringent requirements for sample quality of 5 nephelometric turbidity units (NTU)or less; and limited budget presented challenging problems for monitoring well design and installation. In order to meet the performance requirements, the authors concluded that wells needed to be individually designed for each interval to be monitored. In addition, the design and construction procedure could not result in a significant increase in cost over more conventional approaches.

HYDROGEOLOGY

Geologic Setting

The site is on the southern limb of the northwest-southeast trending Coyote Hills anticline which is located in the central block of the Los Angeles Basin, southern California (Figure 1). The project involved drilling and sampling two formations, the La Habra and Coyote Hills, as described by Yerkes [7]. Both formations exhibit a southerly dip of 10 to 15 degrees. In the project area the La Habra Formation is limited to the upper 45 m (147 ft.) while the Coyote Hills Formation is exposed at the surface in the northern portion of the project area and extends to depths in excess of 300 m (1000 ft.) Sedimentary structure revealed in drill cores consisted of alternating coarsening- and fining-upward sequences [8].

Based on the results of this investigation and on previous work [1-6] at the site, the Coyote Hills Formation is comprised of laterally discontinuous deposits of unconsolidated gravel, sand, silt, and clay. The La Habra Formation is very similar to the Coyote Hills Formation, but contains a higher percentage of silt and clay strata. Evaluation of all the continuous core collected during the authors' study indicates that the sands are predominately subarkosic, ranging in texture from sand with gravel(SW/GP) to fine-grained sand with silt and clay (SM/SC). Silts are nonplastic (ML), containing up to 25% fine-grained sand. Clays are low plasticity (CL), containing variable amounts of silt up to 20%. Sands comprised 72% of core samples with the remainder of the samples being 11% silt, 6% clay, and 11% minus 76 millimeter (mm) (3 in.) gravel [8].

The internal structure and texture of the sediments in conjunction with their depositional pattern strongly suggest deposition in a transitory delta to alluvial plain environment exhibiting an overall regressive pattern. This depositional model is consistent with observations made by the previous investigators [1-6] who noted an inability to correlate lithologic units even over short distances, less

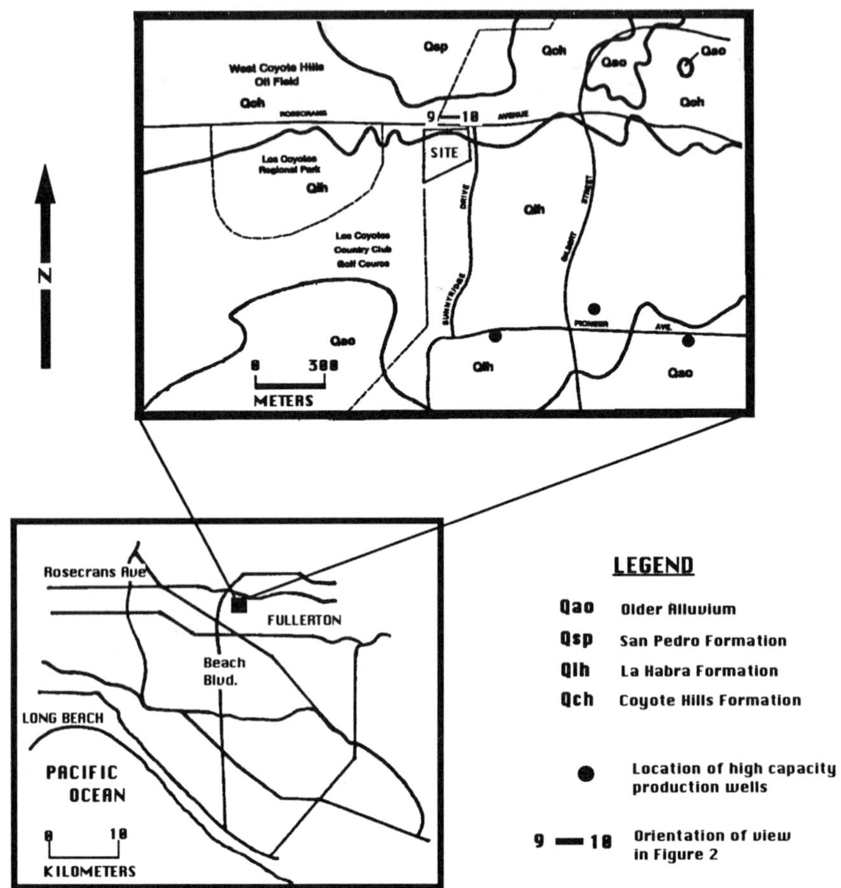

FIG. 1 -- Location map

than 30 m (100 ft.). Figure 2 illustrates the lithologic variability.

Occurrence and Movement of Ground Water

Existing reports indicated that the regional aquifer was unconfined and that the water table was between 61 m and 76 m (200 to 250 ft.) below ground surface (bgs). In addition, the reports noted perched ground water at depths between 12 m and 30 m (40 to 98 ft.) bgs. During the authors' investigation the water table was encountered at depths ranging from 61 m to 63 m (200 to 210 ft.) bgs and perched ground water was not encountered. Information provided by the county water district[4] indicated that high capacity, 3.8×10^6 liter per day

[4]Nira Yamichika, Orange County Water District, personal communication, 1988.

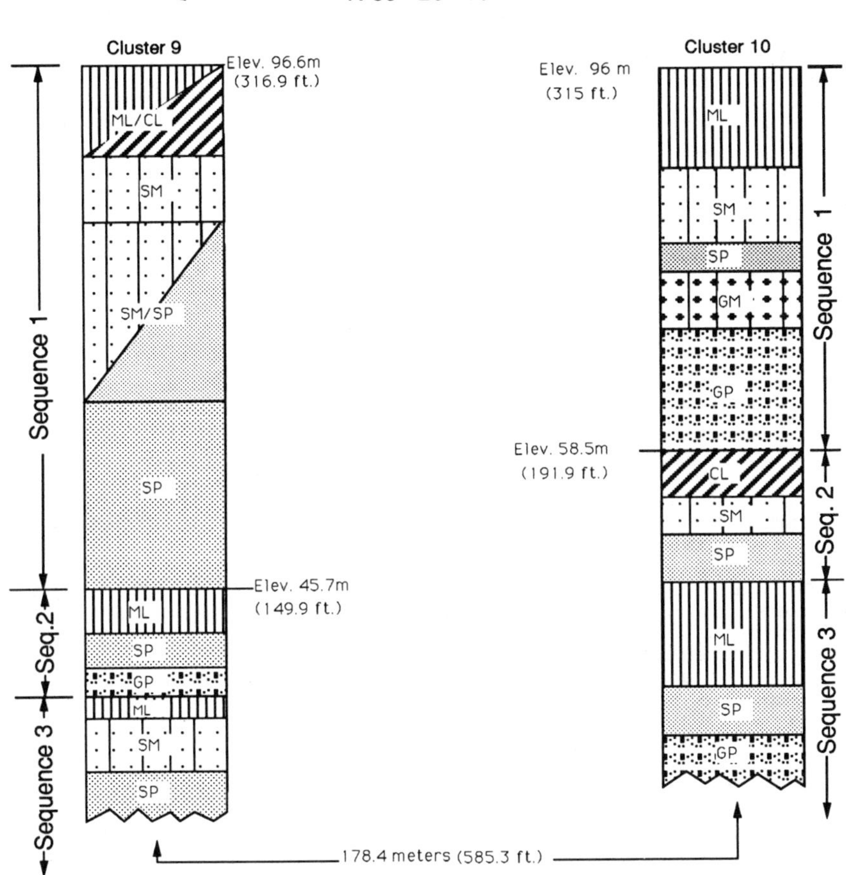

FIG. 2 -- Generalized geologic logs illustrating the lateral variability of lithology.

(one million gallons per day), production wells were in continuous operation within approximately 0.8 kilometers (0.5 miles) south of the site (Fig. 1).

Efforts by previous workers to contour existing water level data resulted in anomalous contours exhibiting an incongruent assemblage of mounds and troughs. These results indicated a range of flow directions from southeast to west-southwest at horizontal gradients of 0.02 to 0.15 [6]. The incongruent water level contours suggested the presence of vertical gradients. Evaluation of the existing water level data using an assumed downward vertical gradient between 0.04 and 0.06 resulted in a reduction of the anomalies.

Well construction data for the wells used to obtain the anomalous water level data had screen lengths of 1.5 m to 18 m (5 to 60 ft.) and were screened over various intervals from the water table to approximately 31 m (100 ft.) below the water table. These findings were consistent with the presence of vertical gradients. Wells screened over widely varying intervals would, in the presence of vertical gradients, be monitoring very different potentiometric surfaces. Therefore, water level data from these wells would not reflect the actual water table and thus would generate anomalous water-table contour maps.

Considering the proximity of the production wells, the location of the site on a topographic high, the well construction data, and the contours exhibited by the corrected water-level data. The presence of downward vertical gradients was considered likely. Accordingly, wells installed during this investigation were designed with short, three meter (10 Ft.), screened intervals and placed in depth staggered clusters in order to resolve and quantify the horizontal and vertical components of ground-water flow.

FIELD METHODS

Sampling Method Selection and Results

Rigorous design requirements necessitated obtaining soil samples at depths in excess of 98 m (320 ft.) in order to identify the most appropriate monitoring intervals and to design well intakes. The most common sampling methodology employs drive sampling at 1.5 m (5 ft.) intervals.

This approach was considered impractical because the frequent handling of sampling tools would exacerbate the anticipated borehole stability problem and be very time consuming and thus costly.

Rather, a sampling method was needed that would not require the repetitive exchange of drilling and sampling tools, be faster than more conventional approaches, and be capable of detecting the presence of perched ground water.

The authors concluded that a wireline coring system would best meet the project requirements. The chosen sampling system consisted of a Christensen 94 mm (3.7 in.) wireline soil core barrel, also referred to as the punch core system (Fig. 3), circulating 6.9 bar (100 psi) filtered air. Air was chosen as the drilling fluid to permit detection of perched ground water. This system was used to continuously sample the deepest hole at each well cluster location. Core samples from the deep hole were used to design all wells in the cluster.

It was readily apparent that one of the significant

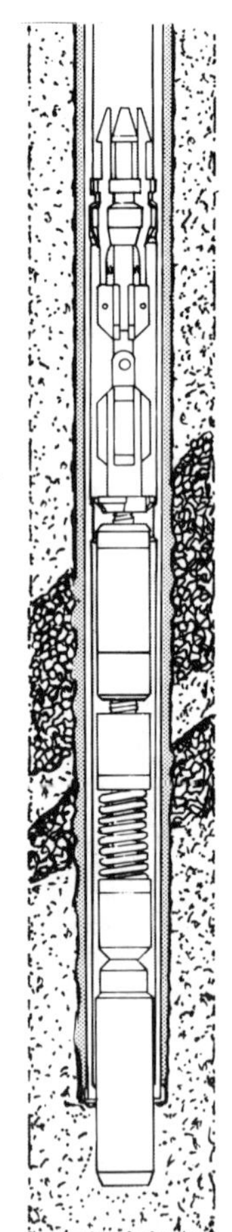

FIG. 3 -- 'Punch Core' system. (not to scale)

advantages of continuous wireline sampling was the completeness and accuracy of the geologic log. Core recovery ranged from 30% in sand with gravel (SW/GP) to 90% in fine-grained sand, silt and clay (SP, SM, ML, CL) as compared to a theoretical maximum of 30%, 0.46 m/1.5 m (1.5 ft./5.0 ft.), for collecting drive samples at 1.5 m (5 ft.) intervals.

The completeness of information on subsurface conditions greatly improved the authors' ability to select appropriate screen intervals and samples that best represented those intervals for on-site sieve analysis and filter pack design.

Geophysical Logging

To complement the coring, a suite of borehole geophysical logs was run. The suite consisted of three-arm caliper, natural gamma ray, spontaneous potential; 0.41-m and 1.6-m (16 and 64 in.) normal resistivity, and 1.8-m (6 ft.) lateral and single point resistivity.

The spontaneous potential and resistivity logs showed excellent distinction between the lithologic units and correlated well with the textural variations observed in the core. However, the water table was not readily discernable on the resistivity logs. The gammma ray logs had limited application due to radioisotopes contained within the ubiquitous mica. This resulted in both sands and clays generating a similar response on the gamma ray log. The caliper log proved useful in correcting borehole volumes and correlating areas of lost core with 'washed-out' zones.

Formation Analysis and Well Intake Design

Sieve analyses were done at the drill site, Figure 4, concurrent with geophysical logging. In all instances sieve analyses and well intake design were completed prior to or at the same time as the geophysical logging.

FIG. 4 -- Photograph of field sieving apparatus

Samples were selected from the finest grained stratum within the screen interval. Samples were thoroughly dried using a two-burner camp stove and weighed to the nearest 0.1 gram (0.004 oz.) on a beam balance. An assortment of appropriately sized sieves was assembled and the samples were shaken using a manually driven mechanical shaker for approximately 10 minutes. The contents of each sieve and the pan were weighed on the beam balance and cumulative percent retained was determined. These data were then plotted in the field logbook. Figure 5 illustrates a representative analysis from well cluster nine.

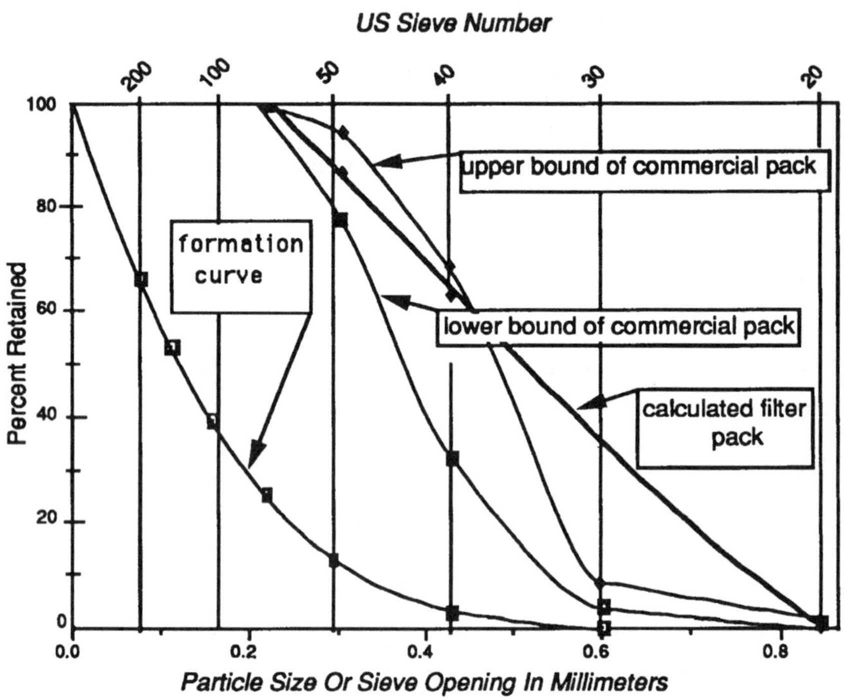

FIG. 5 -- Plot of formation and filter pack sieve analyses.

Results of sieve analyses were employed in essentially the classic well intake design method described by Driscoll and others [9, 10-12]. The authors elected to modify the conventional method in that a less conservative D_{70} (70% retained sieve size) formation multiplier of 7 was used rather than the more conservative multiplier of 4 or 5, as would have been indicated by the conventional method. The resulting increase in the coarseness of the filter pack enhances the effects of well development and permits the use of a larger well screen slot size. This approach relies on the development process as the means to stabilize the formation and prevent an unacceptable amount of fines migrating into the well. As previously mentioned, control of fines was critical as these wells were required to provide litigation-quality samples, which meant sample turbidity values of 5 NTU or less [13].

The ideal filter pack was compared with a range of commercially available filter packs and the commercial pack that best matched the ideal pack was selected (Fig. 5). Well screen slot size was then selected on the basis of filter pack gradation. Well screen slot sizes of 0.254 mm (0.01 in.), 0.508 mm (0.02 in.), 0.762 mm (0.03 in.), and 1.02 mm (0.04 in.) were needed for this project, reflecting the heterogeneity of the formations.

Well Construction Materials

In order to meet project goals and keep costs down, a hybrid approach for both well casing and annular seals was employed. The well casing consisted of 102 mm (4 in.), Type 304 stainless steel below the water table and Schedule 80 polyvinyl chloride (PVC) above the water table. All casing was internal-external flush threaded and fitted with centralizers. Well screens were of the continuous slot (wire wrapped) design, 3 m (10 ft.) in length, and constructed of Type 304 stainless steel.

Annular seals consisted of Volclay grout below the water table and cement-bentonite grout blended to Haliburton [14] specifications above the water table. The hybrid seal design provided a means whereby a suitable seal could be rapidly placed in that portion of the borehole experiencing greatest stability problems.

Construction

Following geophysical logging, the pilot borehole was reamed to its total depth with a 251 mm (9 7/8 in.) tricone bit using the mud rotary method. The circulation system included a shale shaker and desanders set in series. During reaming, drilling mud viscosity and weight were monitored and maintained at a maximum 35 seconds per liter (s/l) and 1.13 kilograms per liter (Kg/l) (9.4 lbs.per gal.), respectively.

Concurrent with geophysical logging and reaming, the driller was provided with the specifications for filter pack gradation and well screen slot size. These materials were obtained and the well screen, casing, and centralizers were decontaminated prior to completion of reaming. Upon completion of reaming, the drilling mud was thinned to 29 s/l and 1.08 Kg/l (9 lbs.per gal.). Immediately following the thinning of the drilling mud, the casing string was run into the hole. The filter pack and annular seals were placed using the tremie method to preclude segregation and bridging. The position of both the filter pack and annular seal was monitored by frequently sounding to locate the top of the filter pack or annular seal. The annular seal was placed to within a meter of the ground surface. The remaining meter of annulus was filled with a sand-cement mix that was contiguous with the surface completion (well pad).

The use of the mud rotary drilling method for well construction was extensively deliberated. The key factors in the decision were depth of installation, maintenance of hole stability, and minimization of formation damage (as compared to augers or drive casing). There was concern about the potential for cross contamination via the drilling mud. However, the authors were of the opinion that a proper drilling mud control program would minimize if not alleviate the potential for cross contamination by developing a uniform and effective mud cake, minimizing filtrate loss, and minimizing recirculation of cuttings. Drilling mud consumption was monitored during reaming. As reaming progressed drilling mud was added to the system in an amount proportional to the increasing borehole volume. This provided a qualitative check on filtrate loss.

Well Development

Well development was done using a boom truck rather than the drill rig to allow the drilling rig to proceed with drilling. Development was completed in two steps. Initially the well was surged with a vented surge block and all materials brought into the well were removed by bailing. The surge block was initially positioned above the well screen. Surging was conducted using a 0.3- meter (1 ft.) stroke at approximately 30 strokes per minute and was continued until sediment could no longer be bailed from the well. The surge block was then lowered approximately one meter (3 ft.) and the process was repeated. This process continued until the entire screened interval had been surged and sediment could no longer be bailed from the well. Following surging, the well was pumped with a submersible pump at four times the anticipated purge rate until pH, temperature, and turbidity readings stabilized. All wells except one met the goal of 5 NTU or less. One well stabilized at 8 NTU, which was attributed to the presence of an immiscible compound detected during chemical analysis.

Water produced during well development was held in a storage tank for chemical analysis. Depending upon the analytical results, the produced water was either discharged to sewer, or mixed with drill cuttings and drilling mud to be incorporated into landfill cover.

Time for complete development varied from 10 to 15 hours per well. The length of time is attributed to several factors; non-conservative filter pack design, the high percentage of fines, and the low transmissivity of this part of the formation. The low transmissivity greatly reduced the amount of formation water available for development. Several wells were bailed dry, thus increasing the time required for development.

FINDINGS AND CONCLUSIONS

Head data from the well clusters confirmed the presence of a downward vertical gradient of between 0.08 and 0.09. The magnitude of the vertical gradient is significant in comparison to an average horizontal gradient of 0.015 over the 20.2 hectare (50 acre) project area. Existing water table contour maps of the site were redrawn with water level data corrected for the vertical gradient. Previous water-level anomalies, attributable to the presence of downward vertical gradients, either disappeared or became so subtle that they could reasonably be attributed to variations in lithology. The new contours indicate a more southerly direction of ground water flow, which is consistent with the location of the municipal supply wells.

The monitoring wells have been sampled nine times since their installation. Turbidity values have remained below 5 NTU except for the well with the immiscible compound. Analytical results for all ground-water samples have indicated the presence of toluene, xylene, and ethylbenzene at single digit part per billion levels. The lack of turbidity in conjunction with extremely low detectable contaminant levels over an extended period of time, demonstrates that these wells have consistently met the stringent requirements for providing data suitable for use in litigation.

Comparison with cost data for the previous well installation efforts at this site, in which more conventional soil sampling and well construction techniques were employed, showed that the time savings during sampling offset the higher per-foot costs of the wireline technology. Well installation costs ranged from approximately 15% lower to 5% higher (corrected to 1989 dollars) than the previous efforts. No additional costs were incurred for specified filter pack and well screens as the authors utilized commercial 'off-the-shelf' materials that the drilling contractor stocked or could obtain from suppliers on a next day basis.

Long term data are being gathered to evaluate whether further cost savings will be realized by a reduced need for well redevelopment, and by higher quality samples that will reduce the need for resampling and eventual time-consuming legal disputes.

The authors conclude that this in-field method of well design is appropriate for a wide range of well installation conditions, particularly those involving heterogeneous sediments, initial installation of wells at a site, and/or litigation-sensitive ground water sampling and analysis. The described well design and installation method has been demonstrated successful, cost-effective, and provides a preferred alternative to the prevalent generic methods of monitoring design and installation.

ACKNOWLEDGEMENT

The authors wish to acknowledge Mr. Bud Broadus of CJ Drilling Supply, Inc. The project could not have succeeded without his invaluable advice and insight.

REFERENCES

[1] Fox, R. W., "Log of Test Borings," Memorandum Report California Department of Transportation, Division of Construction and Research to California Department Health Services, 1981.

[2] Stephens, E. E., "Geotechnical Investigation of the McColl Site," Memorandum Report, Alternative Technology Section to Site Cleanup and Emergency Response Section, 1982.

[3] Radian, "Preliminary Hydrogeologic Evaluation; McColl Site Fullerton, California," Technical Memorandum prepared for the California Department Health Services, 1982.

[4] Radian, "McColl Phase II, Hydrogeologic Aspects of the McColl Site Fullerton, California," Technical Memorandum prepared for the California Department Health Services, 1983.

[5] Reynolds, S. D., "Summary Report of Exploratory Drilling at the McColl Site," Alternative Technology Section to Program Management Section, 1984.

[6] CH2MHILL, "Final Draft Groundwater Investigation Report - McColl Site, Fullerton California," Final Draft Report prepared for USEPA Region 9, 1987.

[7] Yerkes, R. F., "Geology and Oil Resources of the Western Puente Hills Area, Southern California," USGS professional paper 420-c, U.S. Govt. Printing Office, Washington D.C., 1972.

[8] Zemo, D. A., "Monitoring Well Installation Report, McColl Site, Fullerton, California," Ecology & Environment FIT Report for USEPA Region 9, 1989.

[9] United States Bureau of Reclamation, Ground Water Manual, John Wiley & Sons, New York, 1981.

[10] Driscoll, F. G., Groundwater and Wells, Johnson Division, Signal Environmental Systems Inc., St.Paul, Minnesota, 1986.

[11] National Water Well Association, Selection and Installation of Well Screens and Gravel Packs (Anthology), National Water Well Association, Dublin, Ohio, 1988.

[12] California Department Health Services, The California Site Mitigation Decision Tree Manual, Toxic Substances Control Division, 1986.

[13] United States Environmental Protection Agency, RCRA Ground-Water Monitoring Technical Enforcement Guidance Document, Office of Solid Waste, OSWER-9950.1, September 1986.

[14] Haliburton Services, "Lightweight Additives, Prehydrated Bentonite," in Cementing Technical Data, Haliburton Services Company, Duncan Oklahoma, 1985, pp. 8-12.

George A. Gillespie

AN EFFECTIVE MONITORING SCREEN FOR FINE SAND AQUIFERS

REFERENCE: Gillespie, G. A., "An Effective Monitoring Screen for Fine Sand Aquifiers," Current Practices in Ground Water and Vadose Zone Investigations, ASTM STP 1118, David M. Nielsen and Martin N. Sara, Eds., American Society for Testing and Materials, Philadelphia, 1992.

ABSTRACT: Collecting sediment free samples of low turbidity from monitoring wells installed in fine sand aquifers has been a difficult problem. Wells are often screened in aquifers containing thin lenses of silts and clays having a formation effective size of 0.1 mm (0.004") or less. Fine sand silty formations are common to the California Bay area where wells, at depths of 6 to 15 meters (20 to 50 feet), are used to monitor landfills. Most of these wells use a conventional design consisting of a 5 cm (2 in.) or 10 cm (4 in.) diameter slotted or wire wrap screen with a filter pack externally placed around the screen to retain the fine sediment. Many of the wells are not useful because they do not provide clear, sediment free samples at reasonable sampling rates. Often, the wells are purged dry before the required purge volumes are removed.

In July, 1988, EMCON Associates installed a series of preassembled, dual walled screens filled with a fine filter pack around the Palo Alto, California landfill to monitor these fine sand aquifers at two different depths. Numerous tests were conducted to determine turbidity, pumping, and recovery rates for three wells that used 0.43 x 0.28 mm (40 x 60 # US Sieve) filter packs. The filter pack passed a #40 Sieve and was retained on a #60 Sieve.

Well Products Technical Director
Johnson Filtration Systems Inc.
PO Box 64118
St. Paul, MN 55164

This paper describes well development procedures, methods and measurements, and test results. Based on test results over a simulated four year sampling period, the following results have been achieved.

1. After the wells were stabilized, turbidity readings were consistently less than 20 Nephelometric Turbidity Units (NTU). Sediment was almost nil.

2. Using effective well development techniques, sampling and well recovery rates in fine sand formations can be improved so wells can be sampled in less than one hour.

3. There is no evidence of plugging of the dual walled screens during a simulated four year sampling program.

KEYWORDS: Turbidity, sediment, filter pack, dual walled screens, slotted screens, wire wrap screens, Channel Pack®, purge volume.

BACKGROUND: Proper monitoring of groundwater quality in landfills or other hazardous waste sites often requires completing monitoring wells in silty fine sand aquifers where 15% or more of the formation is finer than a 0.15 mm (#100) US Sieve size. Adequate retention of these fine particles requires a fine filter pack uniformly placed around the well screen. During well development this pack must not be displaced because fine materials will continue to migrate into the well.

There are numerous reported reasons why the conventional well design may not be effective.

1. Filter packs of 0.84 x 0.42 mm (20 x 40 # US Sieve) or finer are difficult to place properly around the screen. Bridging can occur, and the pack may segregate during placement [1].

2. Well development may displace the filter pack so the proper envelope is not maintained around the screen [2].

3. These wells can only be purged at low pumping rates. Often the wells purge dry which require extra sampling time.

4. Samples taken from many wells are turbid which requires extra settling or filtering time, and may affect the chemistry of the water sample.

Consequently, many wells do not meet the EPA Technical Guideline of 5 NTU or less turbidity when the well is being sampled [3].

PALO ALTO LANDFILL SITE: These problems have been reported at many sites throughout the San Francisco Bay area. One of these sites, the Palo Alto Landfill, has used PVC slotted pipe monitoring wells since 1976 to monitor water quality in two vertically separated and distinct aquifers. Some of the wells yielded turbid samples after the required purge volumes were removed. Since several wells were apparently screened across the two aquifers, EMCON Associates was selected in 1988 to develop a new series of monitoring wells to sample each aquifer separately [4].

The Palo Alto Landfill's two thin aquifers, located at depths of 7.5 m (2 ft.) and 14.3 m (47 ft.), consist of fine sands with thin layers of silt and clays referred to as young and old Bay muds. Table 1 shows the screen settings and aquifer characteristics of three wells. EMCON Associates conducted slug tests on several wells to determine aquifer permeability. The aquifer characteristics listed were developed from EMCON Associates's October 1988 report [5].

PRODUCT DESIGN: EMCON Associates chose a dual walled screen design for the ten new wells to be installed. Their decision was based on reduced well development costs and clarity of water samples at another test site in the Bay area [4]. The basic design (Figure 1) consisted of a fine filter pack completely encased between two cylindrical stainless steel wire wrapped screens.

The screens were made of Vee-shaped surface wires welded to longitudinal ribs or channels. The channels were used to insure there was no displacement or excessive settling of the filter pack during transport and installation. The dual walled screens, called Channel Pack®, were developed and furnished by Johnson Filtration Systems Inc.

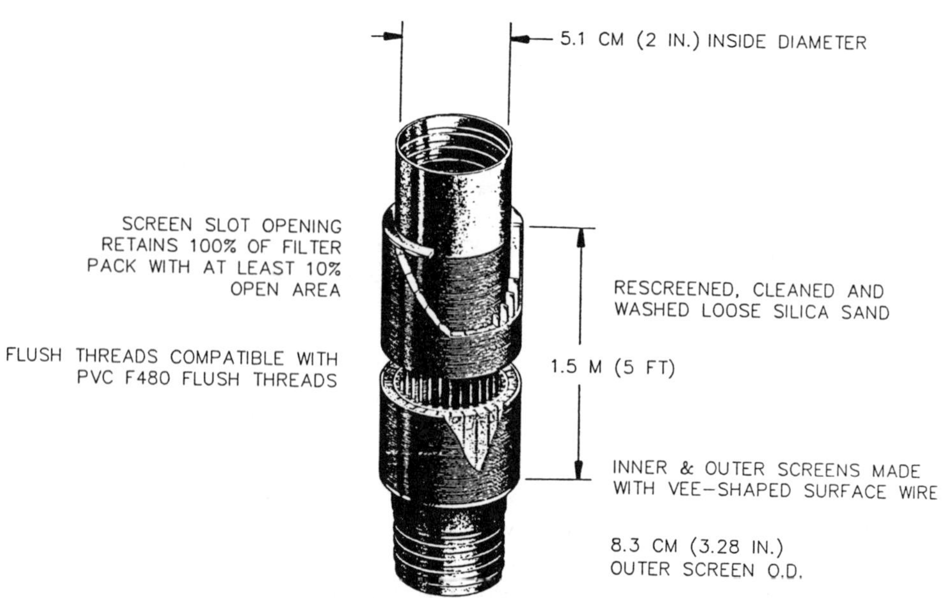

FIGURE 1-- DUAL WALLED SCREEN DESIGN
By Johnson Filtration Systems Inc.

Based on the formation sieve analysis, each dual walled screen was factory assembled with either a 0.84 x 0.42 mm (20 x 40 # US Sieve) or a 0.42 x 0.26 mm (40 x 60 # US Sieve) filter pack encased between the two screens. The screen slot openings were designed to completely retain the filter pack and to provide at least 10% open area, which is considered the minimum for adequate hydraulic performance in monitoring wells [2]. The filter media consisted of round, resieved, and washed silica sand. It was placed in a loose condition with no chemical additives used.

The dual walled screens were installed in July 1988 through a 20.3 cm (8 in.) hollow stem auger and placed at the bottom of the different wells. PVC casing was used as riser pipe to the surface. The 1.52 m (5 ft.) long screen was placed opposite aquifers of 0.3 m (1 ft.) to 2.7 m (9 ft.) thickness.

A coarse filter pack (#12 x 20 US Sieve) was used to fill the annulus between the screen and borehole. The annulus was sealed by conventional means. Installation was quick and easy as the coarse filter pack was placed in about one third of the time needed to install a fine filter pack. An estimated two to three hours installation time was saved on each well.

EMCON Associates developed the wells by bailing and using a surge block similar to one shown in Figure 2. The wells were bailed several times to remove the sediment. Several wells would bail dry shortly after one purge volume was removed.

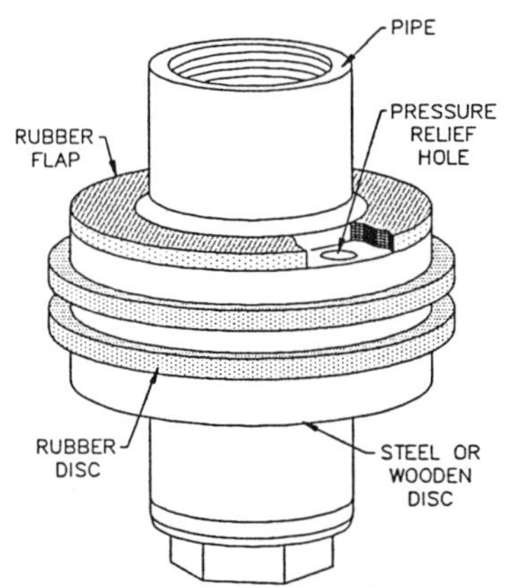

FIGURE 2-- TYPICAL SURGE BLOCK TOOL
By Johnson Filtration Systems Inc.

TEST PROGRAM - The dual walled screen filled with fine filter pack was introduced by Johnson Filtration Systems Inc at the May 1988 NWWA Outdoor Action Conference. Due to limited field data, there were several concerns raised about its performance. The key questions were:

1. What turbidity results could be expected in fine sands?
2. Could wells be developed to provide adequate flow rates for sampling?
3. Would migration of fine sand and silt plug off a dual walled screen resulting in low pumping rates which could require abandoning the well?

In August 1988 Johnson Filtration Systems Inc received permission from EMCON Associates and Palo Alto Public Works to conduct a series of tests to answer these questions. Wells G10, G12, and G15 were selected because of similar depth, purge volume, and aquifer characteristics as shown in Table 1.

Table 1 -- Summary of Well and Aquifer Characteristics

	Well Designation		
Well Characteristics	G10	G12	G15
Well Depth (meters)	4.5	14.3	14.8
Static Water Level (meters)	3.7	3.6	3.3
Purge Volume (liters)	21.9	21.0	23.3
Top of Screen (meters)	12.9	12.5	12.9
Bottom of Screen (meters)	14.4	14.1	4.4
Aquifer Characteristics			
Thickness (meters)	0.90	1.10	0.90
Effective Size (millimeters)	.09	.06	.06
Average Size (millimeters)	.58	.34	.46
Percent Passing of #200 Sieve	10 %	16%	10%
Percent Passing of #100 Sieve	15 %	26%	22%
Permeability (cm/sec x 10^{-4})	---	1.30	---

The project was divided into the following phases:

1: WELL DEVELOPMENT - Redevelop two of the three test wells and evaluate the results versus the one well which had no further work done.

2: TURBIDITY - Determine the well turbidity in NTU units after well redevelopment and purging of the wells.

3: PUMPING RATES - Identify the pumping rates of the wells for each purging event.

4: RECOVERY RATES - Determine recovery rates for each well after a purging event.

Phase 1: WELL REDEVELOPMENT - The wells had been developed and bailed by EMCON Associates using a surge block similar to Figure 2. The two test wells were redeveloped by mechanical means using the swab tool shown in Figure 3. Other methods, such as airlift or water jetting were not authorized.

A. Well G10: Conduct the various tests without any well redevelopment work.

B. Well G12: Swab this well once using the 3/4 in. pipe size (PS) tool shown in Figure 3.

C. Well G15: Swab this well twice using the same tool.

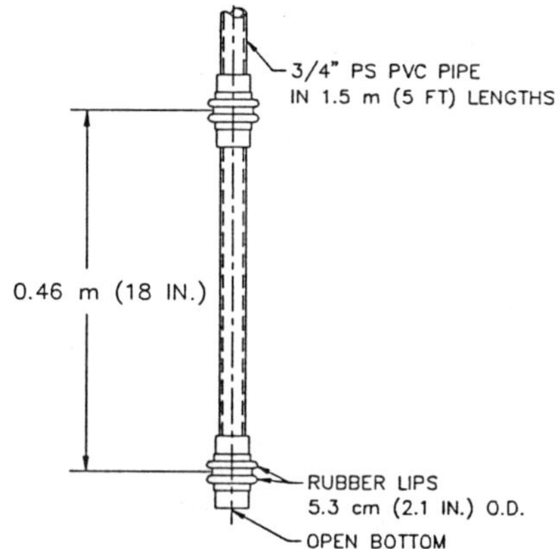

FIGURE 3-- SWAB TOOL USED ON PALO ALTO TEST WELLS

The wells were surged by swabbing the well starting at the top of the water level. A series of upward and downward strokes were used to push the standing water in the casing back and forth across the screen into the formation. The water level gradually decreased until the well was essentially dry. The entire length of each screen was swabbed at about 45.7 cm (18 inch) intervals. About 30 minutes was spent surging a well.

Well G12 and G15 were surged in August 1988; and Well G15 was surged again in November 1988.

Phase 2: TURBIDITY - All ten wells at the site were first checked for sediment using an Imhoff cone in August 1988. When the wells were bailed there would be a small amount of sediment which became almost nil when one or more purge volumes were removed. After a well was swabbed, a significant amount of sediment entered the well indicates the well was not fully developed. Table 2 lists the volume of sediment taken from several wells in August and the results for swabbing Well G15 in November.

Table 2 -- Sediment Recovered After Swabbing the Wells

Well Number	Filter Pack Size	Sediment (ML)
G12	40 x 60	35
G14	20 x 40	169
G15	40 x 60	22
		10

Wells G14 and G17 are in the upper aquifer, which has a slightly higher permeability [5]. More sediment was pulled through their screens because a 20 x 40 filter pack was used in these wells.

Sediment from several wells was analyzed to determine the size range of materials passing through the dual walled screens. Table 3 summarizes results for samples taken from Wells G14 and G15 in August 1988. Virtually all the passed sediment is less than .05 mm (.0016 in.) size.

Table 3 -- Sediment Size and Distribution for Wells

Particle Size	*Number of particles/100 ML of Water* G14 Value	G14 Percent	G15 Value	G15 Percent
.025 to .050 mm	3,433,694	97.1%	8,949,604	97.8%
.050 to .100 mm	83,080	2.4%	159,791	1.8%
.100 to .200 mm	15,108	0.4%	36,320	0.4%
> 200 mm	2,906	0.1%	4,358	0.1%

The particle counts were conducted on samples taken after swabbing and bailing the well. Well G14 had 7 ML/L of sediment after 75.7 liters (20 gallons) were bailed. Well G15 had 22 ML/L of sediment in the initial purge volume bailed.

Turbidity readings were taken when the wells were sampled each quarter. In addition, test pumping was done several times to collect additional data for turbidity levels and the pumping and recovery rates. Almost no sediment was evident. Some cloudiness was observed if the turbidity exceeded 100 NTU.

TURBIDITY RESULTS - Figures 4, 5, and 6 show the turbidity results for 24 consecutive readings over a 2 year period.

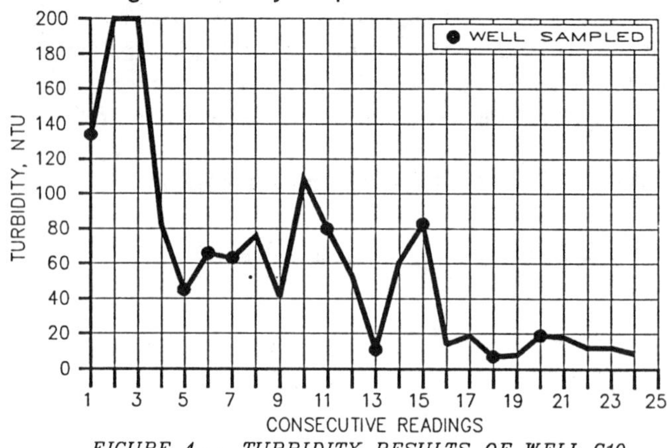

FIGURE 4-- TURBIDITY RESULTS OF WELL G10

Well G10's readings are higher and more erratic for the first 15 readings. During this period increased cloudiness was observed when Well G10 purged dry. Beginning with reading 16, which was taken in May 1990, the turbidity in Well G10 has been consistently less than 20 NTU.

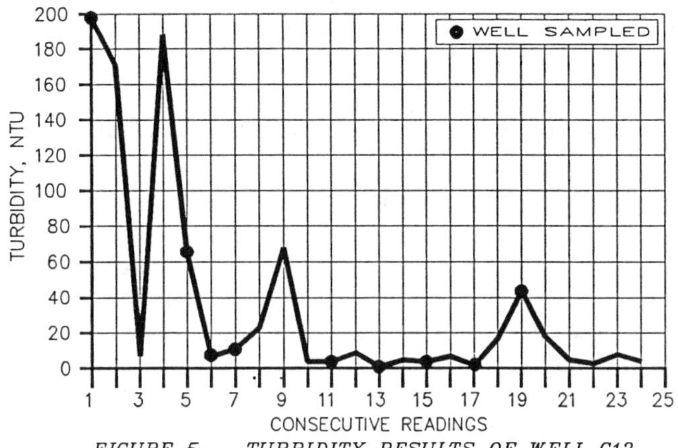

FIGURE 5-- TURBIDITY RESULTS OF WELL G12

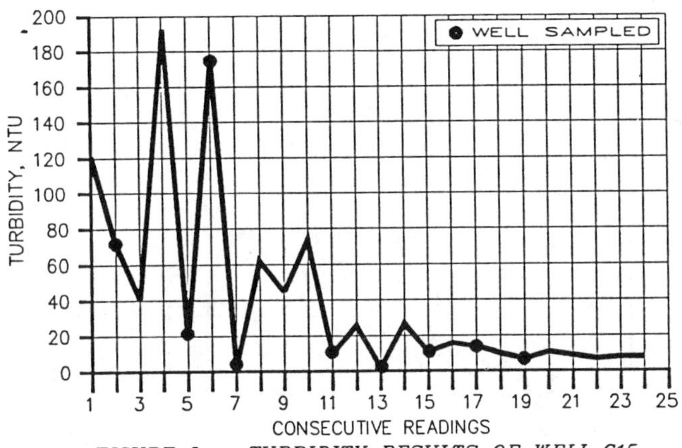

FIGURE 6-- TURBIDITY RESULTS OF WELL G15

The initial turbidity readings in Wells G12 and G15 were much higher because the wells were still being stabilized by well redevelopment work. The readings in these wells stabilized to less than 50 NTU after 11 readings, and virtually all later readings are less than 20 NTU, with many less than 10 NTU.

Phase 3: PUMPING RATES - Periodic pumping tests were conducted in the selected wells. The basic approach was to remove three or more purge volumes or purge the well dry. A purge volume was defined as the volume of water inside the (2 in.) casing and screen. Initially, hand pumps or bailers were used. Dedicated pumps were installed by EMCON Associates in late 1988 [4]. After an initial purge volume was removed, the time was recorded for removal of the additional required purge volumes or pumping the well dry. The flow rate was then calculated for each well.

The calculated purge volume (PV) was determined by:

$$PV\ (\text{gal}) = C\ (WD - SWL) \qquad (1)$$

Where:

PV	=	Purge Volume (Gal)
C	=	0.16312, a constant to convert cubic feet to gallons for liquid in a 2 inch pipe
SWL	=	Static Water Level (Ft.)
WD	=	Well Depth (Ft.)

PUMPING RATE RESULTS - Figure 7 shows that there are significant differences in the pumping rates even though the wells are of similar size and depth, and the aquifer permeabilities are comparable.

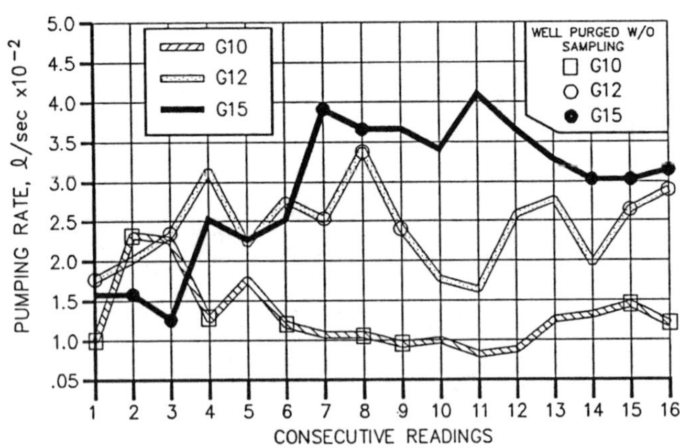

FIGURE 7-- PUMPING RATES AFTER ONE PURGE VOLUME IS REMOVED

Well G10 pumped dry after removing less than two purge volumes. The technician had to return to the well later to collect water samples.

Table 4 shows the average pumping rate of Well G12 was almost twice Well G10's rate. The water level in G12 was drawn down to about 9.1 m (30 feet) when the pumping stopped. Swabbing this well significantly improved its sampling rate.

Table 4 -- Summary of Pumping Rates for the Test Wells

Value (ℓ/sec x 10^{-2})	Pumping Rate		
	G10	G12	G15
Average Rate	1.22	2.43	2.91
Lowest Rate	0.82	1.64	1.26
Highest Rate	2.33	3.41	4.10
Sampling Time (hr)	1-1/2 to 2 hrs	3/4 to 1 hr	3/4 to 1 hr

The pumping rates were determined after removal of an initial purge volume, which represents the standing water in the casing and screen.

Well G15, which was swabbed twice, had an average pumping rate about 20% greater than Well G12. Maximum drawdown water levels were 7.6 m (25 ft.). Prior to the well redevelopment work, this well would go dry shortly after one purge volume was removed. There was concern whether it could be used because its recovery rate was very slow - about 13 hours to recover 90% of its water level.

Swabbing the wells with a surge tool as shown in Figure 3 improved the well performance. The vertical movement of the tool created changes in pressure and flow direction in the well casing. This broke down bridges or skin effects on the borehole wall left from the drilling process. The seal between the surge tool and the screen, and surging the entire screen length are important factors that affect well performance [7].

Figure 3 is an improved version of a surge tool that isolates sections of the screen by flexible rubber seals. This localizes the flow of water across the well bore to move the formation fines through the filter pack and into the well. The fines do not plug the screen because the vee shaped wires have excellent release characteristics, and the significantly higher amount of open area results in improved well development when compared to a slotted screen [6].

The reduced sampling time for Wells G12 and G15 confirmed that the well development of dual walled screens was effective. Well G10 takes 1-1/2 to 2 hours to sample, which EMCON Associates and Wahler Associates feel is the typical time for conventional slotted wells of similar size and depth in the Bay area. The other two wells were sampled in less than an hour due to their improved pumping rates. Well G15 was sampled on several occasions in about thirty minutes. Based on a $60/hr. sampling rate and one hour reduced sampling time. The savings are $1200 per well for quarterly sampling events over a five year period.

Phase 4: WELL RECOVERY RATES -Numerous measurements were taken over time after pumping was stopped. Typical measurement periods were 1, 3, 5, and 30 minutes; and in some cases several hours later. Even if the well is dry, the initial recovery rate would be higher due to the storage water in the dual walled screen and external filter packs. Consequently, the recovery level after thirty minutes was selected to determine the recovery performance of the formation water flowing into the dual walled screens. A steady decline or sudden drop off in the recovery water level after thirty minutes would indicate plugging of the dual walled screen. This time was selected because 90% recovery in a half hour normally allowed sampling to be done the same day.

WELL RECOVERY RESULTS - . Figures 8, 9, and 10 illustrate the standing water levels in Wells G10, G12, and G15 taken thirty minutes after pumping was stopped.

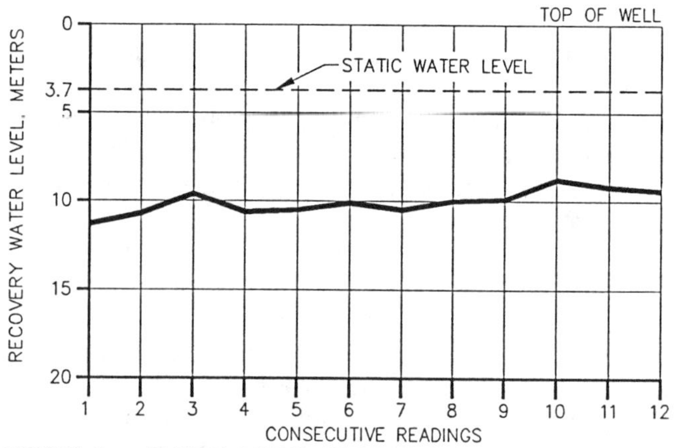

FIGURE 8-- WATER LEVEL IN WELL G10 AFTER 30 MINUTES

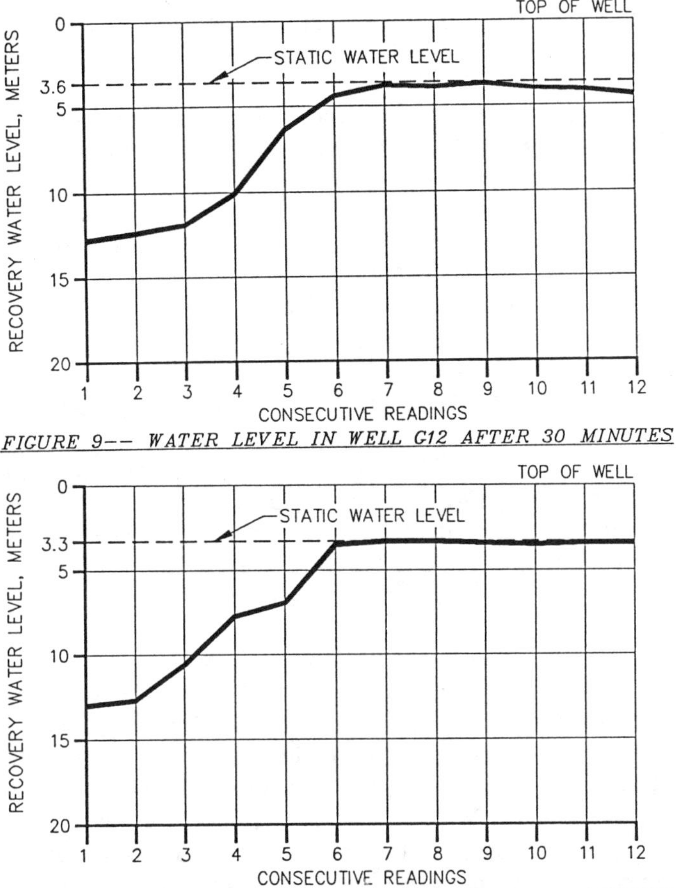

FIGURE 9-- WATER LEVEL IN WELL G12 AFTER 30 MINUTES

FIGURE 10-- WATER LEVEL IN WELL G15 AFTER 30 MINUTES

The recovery performance for Well G10 is less than the other wells. Since this well goes dry, a greater volume of water must flow into the well to return to the static water level. Table 5 shows the drawdown, average recovery level, and percent recovery of the wells at the thirty minute interval.

Table 5 -- Recovery Levels After 30 Minutes

Value	G10	G12	G15
Static Water Level	3.7m	3.6m	3.3m
Drawdown level	12.8m	9.1m	7.6m
Average Recovery Level	9.7m	4.0m	3.4m
Percent Recovery	34%	92%	97%

In Table 5, the static water levels are the 1988 values. These have declined somewhat due to the drought. The average recovery level and percent recovery are based on the last seven readings.

Wells G12 and G15 recover almost completely in thirty minutes while Well G10 takes several hours to reach the 90% recovery level.

Figures 9 and 10 show the improvement of the recovery rates for Wells G12 and G15 due to the swabbing of the wells. These increased rates indicate well redevelopment was effective.

All of the wells have a consistent recovery rate after thirty minutes for the past seven readings. This indicates there is no detectable plugging of the dual walled screen over the sampling period, which is equivalent to about four years based on quarterly sampling of the wells.

CONCLUSIONS: A 1987 test program in Wisconsin concluded that silt sized filter packs and screen with very fine openings (0.05 mm) are needed to retain fine grained sands [1] . The Palo Alto results demonstrate this is not the case, because the dual walled screen, using commonly available wire wrapped screens and filter packs, consistently achieved excellent retention of fine, silty sands. The comprehensive field tests conducted at the Palo Alto landfill confirm that dual walled screens are an effective well design for fine sand, silty aquifers as evidenced by:

1. Turbidity - 20 NTU or less was achieved. Several times the EPA guideline of 5 NTU was met. Filter pack size and uniformity, screens designed with high open area and Vee shaped openings, and 100% retention of the filter pack were critical factors to achieve the improved turbidity results.

2. Well Redevelopment - The 100% improvement in the pumping rates demonstrates that wells can be properly developed through a dual walled screen design . This results in lower sampling costs and fewer wells being purged dry. The well development results confirm that surging or swabbing is a more effective method than bailing or pumping wells to improve the hydraulic conductivity of the borehole [6].

3. Plugged Filter Packs - The consistent pumping and recovery rates for all three wells indicates that the dual walled screen does not plug off over time. Instead, the filter pack in both the screen and borehole have stabilized the fine silty sands as evidenced by little or no sediment being visible at any stage during purging of the wells.

The dual walled screen design has also been used effectively in other monitoring applications. The applications include heaving or running sands, reduced water or cuttings disposal, and leachate collection in active landfills.

ACKNOWLEDGEMENTS: The author gratefully acknowledges Edgar Ugarte and Chad Centola of Palo Alto Public Works for allowing these tests to be conducted at their landfill site; Julian C. Isham, Rex Heppe, and Steve Lofholm of EMCON Associates for their excellent support and guidance in conducting the tests; Alberto Gurevich, Pat Williamson, and Peter Michel of Wahler Associates for their fine work in providing turbidity and pumping rates for 1989 sampling tests, and finally Ruth Vandelac ,Ed Wolfe, and Daiv Sowder of Johnson Filtration Systems Inc for their excellent and timely help in preparing this paper.

REFERENCES

[1] Paul, Duane G., Palmer, Carl D., and Cherkauer, Douglas S. "The Effect of Construction, Installation and Development on the Turbidity of Water in Monitoring Wells in Fine - Grained Glacial Till" Groundwater Monitoring Review, Winter 1988 Issue, pp 73-83.

[2] Schalla, Ronald and Walters, Wallace H. "Rationale for the Design of Monitoring Well Screens and Filter Packs" Groundwater and Vadose Zone Monitoring, ASTM STP 1053. D. M. Nielsen and A. I. Johnson Eds, pp 64-75.

[3] "RCRA Groundwater Monitoring Technical Enforcement Guidance Document (TEGD)" September 1986 U.S. Government Printing Office A870-172-437, pp 86-89.

[4] Heppe, Rex S., Isham, Julian C., and Wheeler, Mark C., of EMCON Associates "State-of-The-Art Groundwater Monitoring Well Network at the Palo Alto Landfill" paper presented at 1989 Hazmacon Conference in Santa Clara, CA.

[5] EMCON Associates, "Implementation of Interim Groundwater Monitoring Plan" October 1988 report to Palo Alto Public Works Department.

[6] Driscoll, Fletcher G., Groundwater and Wells 2nd Edition, 1986, Johnson Division, St. Paul, MN pp 497-522.

[7] Kill, David L., "Monitoring Well Development - Why and How" Groundwater and Vadose Zone Monitoring, ASTM STP 1053 D. M. Nielsen and A. I. Johnson Eds, pp 82-90.

Lawrence D. Beard[1]

DESIGN AND CONSTRUCTION OF DEEP GROUNDWATER MONITORING WELLS

REFERENCE: Beard, L. D., "Design and Construction of Deep Groundwater Monitoring Walls," Current Practices in Ground-Water and Vadose Zone Investigations, ASTM STP 1118, David M. Nielsen and Martin N. Sara, Eds., American Society for Testing and Materials, Philadelphia, 1992.

ABSTRACT: Deep groundwater monitoring wells for the Colbert Landfill Superfund Project in Spokane County, Washington were constructed using 6.4-cm (2.5-in.) diameter polyvinylchloride (PVC) well casing, instead of 5.1-cm (2.0-in.) diameter casing, to provide additional strength and a larger inner well diameter for groundwater sampling equipment. Two well screen/filter pack combinations were used to provide the convenience of preconstruction purchase of well construction materials, and the flexibility to adjust the well intake design for site-specific hydrogeologic conditions. High percent-solids bentonite grout was used for the annular seal in lieu of a bentonite pellet seal to avoid material bridging during installation. Well casing alignment was maintained by applying a tensile force to the casing during the placement of annular fill materials rather than using casing centralizers. Also, threaded temporary steel casing was used for borehole support (versus welded temporary steel casing) to reduce boring and installation time and expense. This monitoring well design and construction approach maintained construction quality standards and resulted in significant cost savings by eliminating common deep well construction problems.

KEYWORDS: monitoring well design, monitoring well construction, bentonite grout, well casing, well alignment

INTRODUCTION

Suggested practices for groundwater monitoring well design are provided in various publications [1], [2], [3]. These publications provide useful guidance on the design of groundwater monitoring wells, but do not address the design considerations that must be evaluated to avoid construction difficulties during installation of deep groundwater monitoring wells [those wells exceeding about 70 m (230 ft) in depth]. Common construction difficulties encountered during installation of deep groundwater monitoring wells include material bridging and maintaining acceptable vertical alignment of the well casing. Material bridging usually occurs during installation of the bentonite pellet seal. Common bridging locations are between the temporary and permanent well casings,

[1]Lawrence D. Beard is a senior engineer at Landau Associates, Inc., P.O. Box 1029, Edmonds, WA 98020-9129.

at casing centralizers, and within the tremie pipe (if used). Monitoring well alignment difficulties often result from short-radius bends (kinks) that develop during installation of the well casing as it deflects under its own weight. This problem occurs with greater frequency as well depth, and resulting casing axial load, increases. These construction difficulties increase well installation time and cost, and sometimes compromise the integrity of the completed well.

The monitoring well design and construction procedures used at the Colbert Project site minimized the construction difficulties associated with deep monitoring wells, while maintaining high construction quality standards. These monitoring well design and construction features include:

- Use of 6.4-cm (2.5-in.) diameter Schedule 80 PVC casing
- Use of two well screen/filter pack designs to address variable hydrogeologic conditions
- Use of a high percent-solids bentonite grout (Pure Gold Grout, American Colloid Co.) for the annular seal
- No use of casing centralizers above the screened section of the well
- Application of an upward tensile force to the well casing during placement of annular materials
- Use of threaded temporary steel casing for boring advancement.

Thirty monitoring wells were constructed using these design features and construction techniques at the Colbert Project site, and no significant construction difficulties were experienced. Monitoring wells were installed to depths exceeding 125 m (410 ft), and a total of about 2,400 linear meters (8,000 ft) of well casing was installed. All wells meet the project well alignment requirements of allowing insertion of groundwater sampling equipment and the overall well performance criteria of producing groundwater samples with turbidity levels below 5 NTU (nephelometric turbidity units).

BACKGROUND

The Colbert Landfill is an inactive 40-acre municipal solid waste landfill located about 16 km (10 miles) north of Spokane, Washington. The landfill is situated in a broad, fluvially modified glacial valley, about 1 km east of the Little Spokane River, as shown on Figure 1.

The subsurface lithology in the Colbert Landfill vicinity consists of unconsolidated gravel, sand, and silt deposits overlying granitic bedrock. Two aquifers are present in the unconsolidated sediments. A perched sand and gravel aquifer of limited areal extent (upper aquifer) is present at a depth of about 25 m (80 ft) below ground surface. A silt aquitard separates this perched aquifer from regional groundwater flow; the aquitard thins from west to east, and pinches out near the east edge of the landfill. The underlying sand and gravel regional aquifer (lower aquifer) thickens from east to west and extends to a

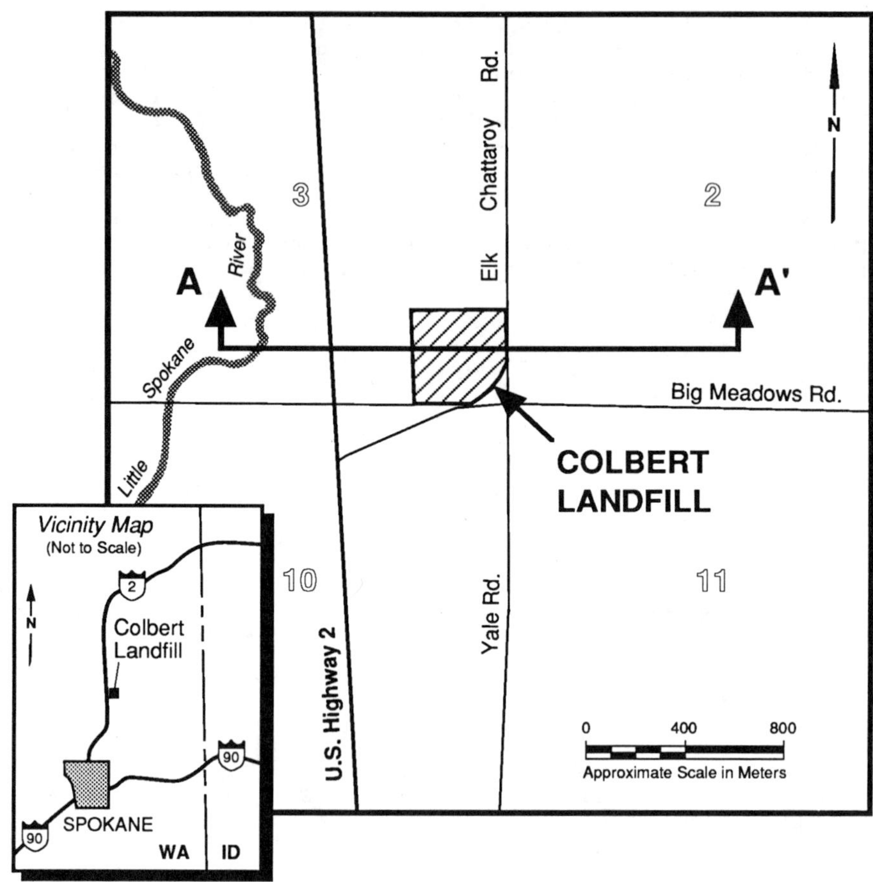

FIG. 1 -- Site plan and location of cross section A-A'.

maximum depth of about 125 m (410 ft) below ground surface. Granitic bedrock forms the Lower Aquitard for the regional aquifer. A generalized geologic cross section of the Colbert Project site is shown on Figure 2.

Spent chlorinated solvents, primarily 1,1,1-trichloroethane and methylene chloride, were disposed of at the Colbert Landfill between 1975 and 1980. Subsequent contamination occurred in the two aquifers, and the combined contaminant plumes involve about 6.4 km^2 (2.5 mi^2). The Colbert Project site was placed on the U.S. Environmental Protection Agency (EPA) National Priorities List (NPL) in August 1983. A remedial investigation/feasibility study (RI/FS) was completed in 1987.

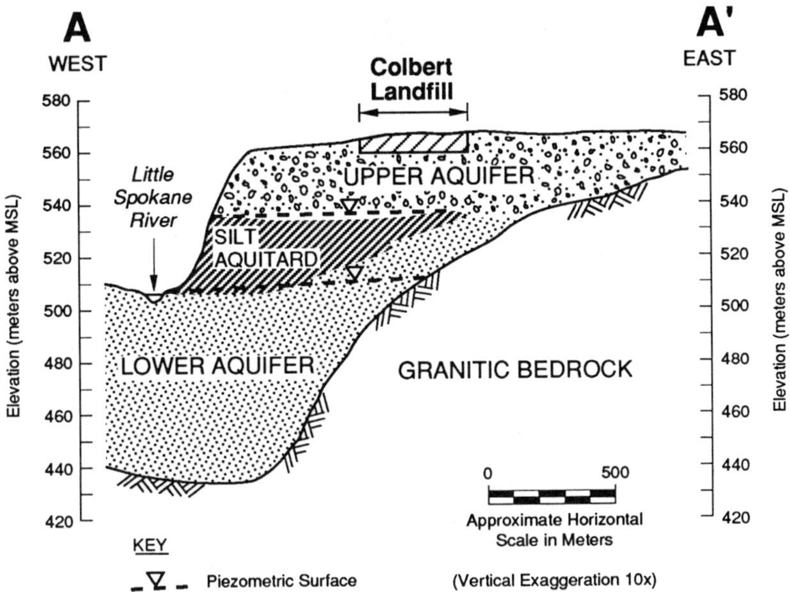

FIG. 2 -- Generalized geologic cross section.

The RI/FS recommended a pump-and-treat system for remediation of groundwater at the site. EPA's 1987 record of decision (ROD) supported this recommendation. Landau Associates, Inc. was retained by Spokane County, the potentially responsible party (PRP) implementing the cleanup, to design the remedial system.

A significant amount of additional site characterization was needed prior to design and construction of the remedial system. Additional site characterization included groundwater monitoring at the base of the lower aquifer because the dense nature of the chlorinated solvents disposed of at Colbert Landfill posed the potential for density-dependent contaminant migration. The depth to the base of the lower aquifer was not determined during the RI/FS, but available data indicated a potential aquifer depth of 150 m (500 ft) below ground surface, and a corresponding aquifer thickness of up to 100 m (330 ft). Experience at sites with similar hydrogeologic conditions indicate that previously described construction difficulties often occur when monitoring wells are constructed to these depths using 5.1-cm (2-in.) diameter PVC casing, frequently placed casing centralizers, and a bentonite pellet annular seal. The well design and construction methods discussed in the following sections of this paper were developed to avoid these difficulties.

MONITORING WELL DESIGN

The groundwater monitoring wells installed as part of the Colbert Project include design components applied by Landau Associates on other projects and the use of Pure Gold Grout (an American Colloid Co. product), a high percent-solids bentonite grout that became available shortly before the start of field activities in August 1989. A typical monitoring well design diagram for the Colbert Project is shown on Figure 3, and the components of this design are presented in the following subsections.

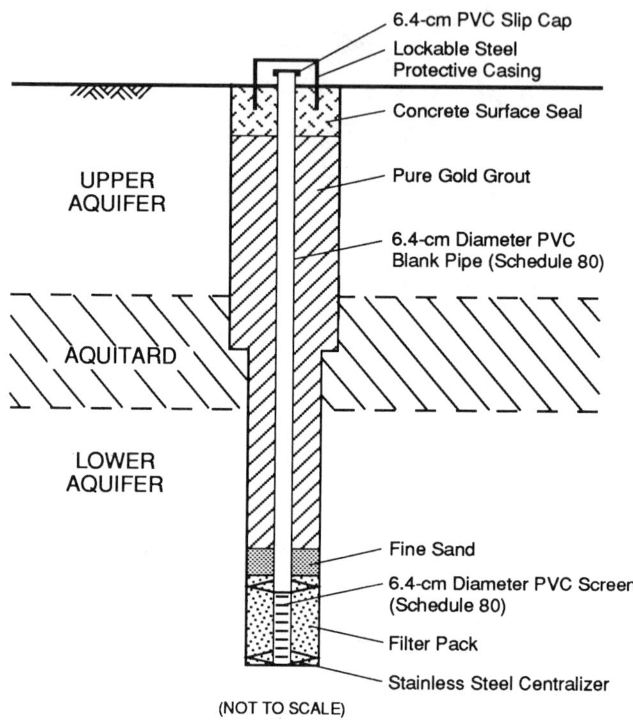

FIG. 3 -- Typical monitoring well design.

Well Casing

Monitoring wells for the Colbert Project were constructed using 6.4-cm (2.5-in.) diameter Schedule 80 PVC well casing. The primary advantage of 6.4-cm well casing over more commonly used 5.1-cm (2-in.) diameter casing is the reduced probability of difficulties during insertion of post-construction groundwater sampling equipment. Most groundwater sampling pumps are designed for 5.1-cm diameter wells, and minimal clearance between the sampling pump and well casing is common. Consequently, if small radius bends are present in the well casing (which are common in deep monitoring wells), the insertion of groundwater sampling equipment into a 5.1-cm diameter well to sampling depth may be difficult or impossible. The additional 1.3-cm (nominal) inner well diameter provided by 6.4-cm diameter casing reduces the probability that small-radius casing bends will adversely affect the use of groundwater sampling equipment.

The advantages of additional annular space are, in part, offset by the additional presampling purge volume associated with a larger diameter well casing. Increased purge volume usually results in increased sampling time and cost, and increased purge-water treatment and disposal costs (when applicable). A 6.4-cm diameter well provides additional annular space at a moderate 56 percent increase in unit volume over 5.1-cm diameter casing. By comparison, a 10.2-cm (4-in.) diameter casing, the most commonly used alternative to 5.1-cm diameter casing, provides a 300 percent increase in unit volume over 5.1-cm diameter casing.

The strength characteristics of 6.4-cm diameter casing also provide advantages to 5.1-cm and 10.2-cm diameter casings. As a result of a low-dimension ratio (outside casing diameter/wall thickness), pipe stiffness and hydraulic collapse pressure are greater for 6.4-cm diameter casing, as shown in Table 1. Consequently, casing failure is less likely to occur when 6.4-cm diameter casing is used.

TABLE 1--Physical Properties of Schedule 80 PVC Well Casing [4]

Outside Diameter					
Nominal (cm) [in.]	Actual (cm) [in.]	Minimum Wall Thickness (cm)	Dimension Ratio[a]	Hydraulic Collapse Pressure (KPa)	Pipe Stiffness[b] (KPa)
5.1 [2.0]	6.032 [2.375]	0.554	10.9	6461	13238
6.4 [2.5]	7.302 [2.875]	0.701	10.4	7653	15445
10.2 [4.0]	11.430 [4.500]	0.856	13.3	3406	6481

a. Outside Diameter/Wall Thickness

b. At 5 Percent Deflection

One potential disadvantage of 6.4-cm diameter casing is the boring diameter required for well installation. Although well construction regulations vary from state to state, a 5.1-cm (2-in.) annular space is commonly required between the well casing and borehole. Based on the annular space between the outside of the well casing and outside of the temporary steel casing, a 5.4-cm annular space is achieved when installing a 5.1-cm nominal diameter casing within a 15.2-cm (6-in.) nominal diameter boring. However, only a 4.8-cm (1.9-in.) annular space is present when a 6.4-cm nominal diameter casing is installed within a 15.2-cm boring. Thus, a 20.3-cm (8-in.) diameter borehole (the next larger standard size) would be required if the 5.1-cm annular space requirement were rigorously applied. The Washington State Department of Ecology (Ecology) granted a well construction variance for the Colbert Project, allowing installation of 6.4-cm diameter well casing within a 15.2-cm diameter borehole. The variance was granted largely because the flow characteristics of the annular sealant (Pure Gold

Grout) are superior to that of bentonite pellets, allowing placement of an adequate annular seal within a slightly smaller annular space.

Filter Pack

An artificial filter pack was placed within the screened section of all groundwater monitoring wells constructed for the Colbert Project. The installation of an artificial filter pack:

- Filters out fine aquifer materials, reducing groundwater sediment content
- Increases the allowable slot size and the effective well radius, improving the well's hydraulic characteristics
- Minimizes the quantity of native formation material entering the well following construction, reducing the time required for well development and the settlement of overlying annular materials
- Acts as a formation stabilizer during removal of the temporary steel casing[1].

Ideally, filter pack design is accomplished on a well-by-well basis, using grain-size distribution data for formation samples collected from the screened zone. However, this approach is rarely practicable because it requires that either a large selection of well screen and filter pack materials be maintained onsite, or that purchase of well screen and filter pack materials follow filter pack design. Because delaying purchase of well construction materials typically results in project delays, the filter pack material and well screen are often purchased prior to the startup of drilling activities. In many cases, these prepurchased materials are used for well construction, regardless of the aquifer material characteristics encountered. This approach often leads to sediment-laden groundwater samples (when the filter pack is too coarse) or wells that do not accurately reflect aquifer hydraulic properties (when the filter pack is too fine).

Two well screen/filter pack combinations were used for the Colbert Project as a practical compromise between well-specific design and a "one-size fits all" design approach. A 0.508-mm (0.020-in.) screen and 10-20 silica sand filter pack were used for relatively homogeneous sand and gravel zones that contained less than about 20 percent fines (material finer than a U.S. Standard No. 200 sieve). A 0.254-mm (0.010-in.) screen and 20-40 silica sand filter pack were used for stratified zones with fine-grained (silt or clay) units, and sand and gravel zones with greater than about 20 percent fines.

Construction materials for the two well screen/filter pack combinations used for the Colbert Project were delivered to the site prior to initiation of site activities. Grain-size analysis was accomplished in the field for a representative sample of formation material from each well boring. The most appropriate well screen/filter pack combination (of the two available) was used for well construction.

The primary performance criterion for the well screen/filter pack combinations used for the Colbert Project was low-sediment (less than 5 NTU) groundwater samples. Had well hydraulic performance (for slug tests or single-hole pumping tests) been an important consideration, a coarser well screen/filter pack combination would have been used in addition to the aforementioned combinations.

Annular Seal

The annular seal for a groundwater monitoring well is typically constructed using bentonite pellets, cement grout, or bentonite grout. Pure Gold Grout, a high percent-solids bentonite grout, was selected as the annular sealant for the Colbert Project monitoring wells. This selection was based on site hydrogeologic conditions, the advantages and disadvantages of available alternatives, and evaluation of the physical and chemical properties of Pure Gold Grout.

Bentonite pellets were considered impractical because of potential material placement difficulties. The annular sealant had to be placed through up to a 67-m (220-ft) water column. Bentonite pellets would swell significantly as they settled through the water column, and material bridging was an anticipated problem.

Cement grout was not considered for the annular sealant because the potential for elevated groundwater pH (due to the high alkalinity of cement)[1]. Elevated pH was a particular concern at the Colbert Project site because well clusters consisting of two to three monitoring wells located within 30 ft of each other were constructed, and the screened zone for shallow monitoring wells are adjacent to the annular sealant for deeper monitoring wells. The difficulty of maintaining a continuous seal when grout placement spans several days, the potential for permanent fluid migration pathways to form during cement curing, and the potential impact of the heat of hydration on PVC casing material strength characteristics also contributed to the decision not to use cement grout[1].

With the elimination of bentonite pellets and cement grout from consideration as annular sealants, bentonite grout emerged as the preferred alternative. However, bentonite grout also posed some potential performance problems, including filter pack intrusion, grout loss (in coarse formations), dehydration (and cracking) in the unsaturated zone, and settlement.

Many of these potential performance problems are directly or indirectly related to the low percent solids typically obtained with bentonite grout mixtures, which typically attain 20 percent solids. A recently introduced product, Pure Gold Grout, was considered due to the manufacturer's claim that it achieves 75 percent higher solids than other bentonite grouts.

Because Pure Gold Grout was a new product, its physical and chemical properties were evaluated prior to use. Evaluation included a limited number of laboratory tests to assess physical characteristics, and review of chemical analyses provided by the manufacturer. Laboratory tests verified that Pure Gold Grout achieved 35 percent solids[2] and maintained a workable consistency when mixed to the manufacturer's specifications. For comparison, wet densities were also determined for hydrated bentonite chips (raw sodium bentonite) and bentonite pellets. The wet density measured for Pure Gold Grout is about the same as that measured for bentonite chips, and only about 5 percent lower than the measured density of bentonite pellets. Although the high density of Pure Gold Grout reduced the concern of grout intrusion, a 0.65-m (2-ft) layer of fine (20-40) sand above the filter pack was incorporated into the monitoring well design to further reduce the potential for grout intrusion into the filter pack.

The rehydration characteristics of Pure Gold Grout were also evaluated. A sample of Pure Gold Grout was prepared in accordance with the manufacturer's recommendations and placed in two 6.1-cm (2.42-in.) diameter by 10.2-cm (4-in.) high brass tubes that were open at both ends. The samples were placed in an oven at 110°C for 24 hours, and then rehydrated by submerging in tap water for 48 hours. Each sample was subjected to three dehydration/rehydration cycles. Based on visual observation, both samples rehydrated to their initial volume (i.e., no cracks or space between the sample tube and sample were observed), except for a small amount of material lost from the top of each sample due to sample handling. Thus, dehydration cracks that may form in the vadose zone would tend to close when rehydrated.

The manufacturer provided chemical analysis results for EPA organic priority pollutants (volatile, semivolatile, pesticide, and PCB compounds), and EP Toxicity metals. All organic compounds were below laboratory detection limits. Although EP Toxicity (EPA Method 1310) results were below Resource Conservation and Recovery Act (RCRA) criteria for hazardous constituents, the cadmium concentration of 0.69 mg/L was relatively high (compared to the RCRA hazardous waste criteria of 1 mg/L) for the initial set of analyses provided. The manufacturer provided two additional laboratory analyses for EP Toxicity metals for Pure Gold Grout that indicated cadmium levels of 0.010 and 0.007 mg/L, respectively. Based on these data, and because cadmium was not a constituent of concern for the Colbert Project, the chemical characteristics of Pure Gold Grout were considered acceptable and no additional chemical analyses were performed. It should be noted that EPA replaced the EP Toxicity test with the Toxicity Characteristic Leaching Procedure (TCLP; EPA Method 1311) and the concentrations of leachable metals may be different using this new procedure.

Casing Centralizers

The primary purpose of casing centralizers is to maintain the well casing in the center of the boring, allowing even distribution of annular materials around the well casing. Casing centralizers also provide lateral casing support and can reduce the probability of small radius bends developing in the well casing. However, casing centralizers are common locations for material bridging and they tend to interfere with the insertion of a tremie pipe into the annular space.

---

[2] A wet density of 1,260 kg/m^3 (78.5 lb/ft^3) achieved.

The difficulties associated with casing centralizers tend to increase with well depth and the number of centralizers. As a result, the use of casing centralizers was limited to the screened section of monitoring wells constructed for the Colbert Project. As described in the following paragraphs, the use of grout for an annular sealant and the reduction of casing axial load eliminated the need for casing centralizers within the annular seal.

Because of its flow characteristics, a grout material does not require as large an annular space to create an effective annular seal as do bentonite pellets, which reduces the need for casing centralizers within the annular seal. The need for centering the casing within the annular seal is further reduced in coarse formations because grout intrudes into the formation and extends the effective radius of the annular seal beyond the borehole.

The use of casing centralizers to provide practical and effective lateral casing support for monitoring wells should be limited to shallow wells, where the axial load (due to the casing weight) is relatively small and lateral support can be provided with a small number of centralizers. As the number of centralizers increases with well depth, it becomes increasingly difficult to insert a tremie pipe into the annular space. This difficulty is exacerbated by the need to decrease centralizer spacing with depth, to account for the increased axial load on the casing due to its own weight.

An effective alternative to the use of casing centralizers for lateral casing support is reducing the axial load on the well casing by applying an upward (tensile) force during placement of annular materials. Applying a tensile force reduces the tendency of the casing to deflect under its own weight and form small radius bends, achieving the same goal as casing centralizers without the associated construction difficulties. The method used to maintain a tensile force on the casing for the Colbert Project is described in the Monitoring Well Installation section of this paper.

BORING ADVANCEMENT

Borings were advanced using air rotary and cable tool drilling techniques. While cable tool drilling provides excellent hydrogeologic information during boring advancement, it is a slow drilling method. Consequently, cable tool drilling was limited to the first (and deepest) boring at each well cluster location to obtain detailed hydrogeologic data. Air rotary drilling was used for subsequent borings at well cluster locations, where hydrogeologic conditions could be based largely on the data collected from the adjacent cable tool boring.

Most borings were advanced using threaded steel casing for boring sidewall support. Threaded steel casing is not commonly available in the northwestern U.S., and was obtained by the driller specifically for the Colbert Project. The casing was manufactured from 1.27-cm (0.5-in.) thick steel pipe commonly used in the petroleum exploration industry, and was threaded using a rope thread pattern. The primary advantages of using threaded casing are:

- A time savings of about one-half hour per casing section because welding is not required.
- The outside diameter of the 15.2-cm (6-in.) temporary casing is adequate (17.8-cm) to achieve the Washington State minimum annular space requirement of 5.1-cm (between boring and well casing) when installing 6.4-cm diameter PVC well casing. A 20.3-cm (8-in.) diameter temporary casing may be required to meet the annular space requirements when standard dimension welded temporary casing is used.
- Threaded casing has a flush drive shoe, which eliminates the need to cut and isolate the drive shoe following boring advancement.
- Drilling is safer in potentially explosive environments, such as near a solid waste landfill, because welding is not required.
- The threaded casing is thicker and more rigid than welded casing, which appears to result in fewer alignment difficulties during boring advancement.

There was some concern that the temporary casing might "unthread" when advanced to depths exceeding about 75 m, the maximum depth to which the driller had previously used threaded casing. However, threaded casing was advanced to 130 m without experiencing significant loosening at casing joints. When available, the advantages of threaded temporary steel casing far outweigh the additional costs (typically $6 to $10 per foot depending on diameter) commonly charged by drilling contractors for its use.

MONITORING WELL INSTALLATION

All monitoring wells were installed using cable tool drilling equipment, including wells installed in borings advanced by air rotary methods. For a large drilling project, such as the Colbert Project, it is significantly less expensive to use cable tool equipment for installation because of its lower hourly rate (typically about half that of air rotary). For the Colbert Project, cable tool equipment installed wells at about the same footage rate as air rotary equipment advanced borings, resulting in good balance between boring advancement and well installation.

As previously discussed, vertical alignment of the PVC well casing was maintained by applying a tensile force to the permanent PVC casing during placement of annular materials. The tensile force was applied to the top of the casing using the drill rig sand line, and was monitored using an in-line spring scale as shown on Figure 4. Ideally, a tensile force equivalent to the casing weight would be applied so the well casing "hangs" in the boring. However, a tensile force equivalent to about 75 percent of the casing weight was applied to minimize the potential of pulling out the permanent well casing when tension is reapplied following the removal of each temporary casing section.

The primary alignment criteria for a groundwater monitoring well is that the well be sufficiently straight to allow the insertion of groundwater sampling equipment. A PVC pump blank with the same dimensions as the groundwater sampling pump selected for the Colbert Project was inserted into all the project wells periodically (about every 20 m) during well installation to confirm adequate alignment.

Filter pack material was placed by gravity (free fall). To minimize the potential for material segregation, uniform filter pack material was used (uniformity coefficient less than 1.5)[2]. Filter pack material was installed at a slow, steady rate for the entire length of the filter pack to further minimize the occurrence of material segregation, and to avoid material bridging.

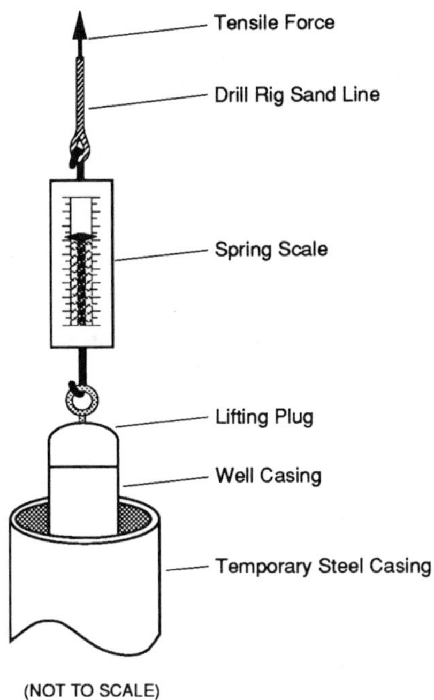

FIG. 4 -- Application of tensile force to well casing.

Pure Gold Grout was tremied into place following placement of the filter pack and the layer of fine (20-40) silica sand placed above the filter pack to minimize the potential for grout intrusion. Each monitoring well was evaluated for grout intrusion after the initial batch of grout was placed, by checking bailed water samples for the presence of grout. Grout intrusion was not observed in any of the monitoring wells constructed for the Colbert Project.

Grout was placed using a diaphragm pump. Although Pure Gold Grout is highly viscous, it appears to be somewhat thixotropic and remains workable when continuously mixed, and when placed in a timely manner.

The only significant problem encountered with the use of Pure Gold Grout was grout loss. Grout loss typically results from the filling of voids outside the boring created during drilling, and migration of grout into the formation through interstitial pore spaces and fractures.

The percentage of grout loss (GL) was estimated as follows based on the ratio of the calculated boring volume grouted (V_b) to the volume of grout used (V_g):

$$GL = (1 - \frac{V_b}{V_g})\ 100 \tag{1}$$

Based on previous experience, grout loss of up to 25 percent commonly occurs during monitoring well construction. However, somewhat higher grout loss was anticipated for the Colbert Project site due to the presence of coarse sand and gravel zones. Grout loss averaged about 39 percent for Pure Gold Grout, but approached 100 percent for some gravel zones. By comparison, Volclay[3] bentonite grout (20 percent solids) was used as the annular sealant for a limited number of Colbert monitoring wells, and averaged about 54 percent grout loss. Thus, although significant grout loss was experienced using Pure Gold Grout, it performed better than lower percent-solids grout.

Solid bentonite (chips or pellets) was used as the annular sealant for high-permeability zones where grout loss was excessive (approached 100 percent). Excessive grout loss was not considered a significant problem because it occurred infrequently, extended over a limited vertical distance (less than 10 m), and occurred primarily in the vadose zone where it could not impact groundwater sample quality or well hydraulic performance.

CONCLUSIONS

Thirty monitoring wells, including a total of about 2,400 linear meters (8,000 ft) of well casing, were installed at the Colbert Landfill site between August 1989 and July 1990. No significant well installation or performance difficulties were encountered during well construction. Additionally, all the monitoring wells constructed meet the overall performance criteria of producing groundwater samples with turbidity levels below 5 NTU. Conversely, two of the monitoring wells constructed during previous site activities required the use of small diameter sampling equipment due to short-radius bends in the well casing. Also, five of nine groundwater samples collected from previously constructed wells exceeded the 5 NTU turbidity standard, even after extensive purging.

The cost of implementing a monitoring well construction program is dependent on a number of factors, and the cost savings resulting from the monitoring well design and construction methods used for the Colbert Project are difficult to quantify. However, well construction cost savings were estimated based on the difference between boring

---

[3] Volclay is an American Colloid brand name.

advancement and well installation rates achieved for the Colbert Project versus other projects with similar hydrogeologic conditions (where other well construction methods were employed). The use of threaded temporary casing increased the average rate of boring advancement by about 20 percent. Well installation progressed about 30 percent faster because construction difficulties commonly associated with the installation of a bentonite pellet seal, frequent casing centralizers, and unacceptable boring alignment were avoided. This reduced construction time resulted in an estimated cost savings of about $100,000 for the Colbert Project.

ACKNOWLEDGMENTS

The author acknowledges with appreciation the assistance of Landau Associates technical staff members, particularly Brian Butler and Larry Eaton, in developing and implementing the well design and construction procedures described in this paper.

REFERENCES

[1] Aller, L., T.W. Bennett, et al., Handbook of Suggested Practices for the Design and Installation of Ground-Water Monitoring Wells, National Water Well Association, Dublin, Ohio, 1989.

[2] Driscoll, F.G., Groundwater and Wells, Johnson Division, St. Paul, Minnesota, 1986.

[3] United States Environmental Protection Agency, RCRA Ground-Water Monitoring Technical Enforcement Guidance Document, Office of Waste Programs Enforcement, Office of Solid Waste and Emergency Response, OSWER-99sd.1, United States Environmental Protection Agency, 1986.

[4] American Society for Testing Materials, "Standard Specification for Thermoplastic Well Casing Pipe and Couplings Made in Standard Dimension Ratios (SDR), SCH40 and SCH80," ASTM Designation: F480-89, ASTM, Philadelphia, Pennsylvania, May 1989.

Water-Level Measurement and Ground-Water Sampling

Gary N. Durham,[1] Tom G. Bumala[2]

USING VIBRATION STRIP TRANSDUCERS FOR MEASURING WATER LEVELS IN MONITORING WELLS

REFERENCE: Durham, G. N. and Bumala, T. G., "Using Vibration Strip Transducers for Measuring Water Levels in Monitoring Wells," Current Practices in Ground Water and Vadose Zone Investigations, ASTM STP 1118, David M. Nielsen and Martin N. Sara, Eds., American Society for Testing and Materials, Philadelphia, 1992.

ABSTRACT: A new system is presented that permits unattended measuring of water levels in monitoring wells. The system consists of a sensor connected by an umbilical cord to a data collector all of which can be installed in a monitoring well as small as 51 mm ID. The data collector stores up to 10,000 data points. Sampling frequency, start/stop times, and engineering unit selection are made by the user via portable PC DOS computer connected to the collector through the standard communication port of the computer. The described system provides the means to achieve the precision required in ASTM Standard D4750 "Standard Test Method for Determining Subsurface Liquid Levels in a Borehole or Monitoring Well (Observation Well)."

KEYWORDS: ground water, water level, monitoring wells, data acquisition

Water level measurements are probably the single most recorded information obtained from ground water monitoring wells. The large number of monitoring wells in place and planned implies that millions of water level measurements are made each year in the USA. For an overview discussion describing various techniques using water level data collection, the reader is referred to reference [1]. The cost associated with processing data secured over the life of a monitoring well is often greater than the installation. The use of a simple water level indicator or graduated tape is the tool most used for determining ground water levels within monitoring wells.

The sheer number of wells, high sampling rates, remoteness or inconvenience of monitoring wells, and the need for concurrent readings at multiple locations are some of the factors that add to the labor intensity of securing adequate ground water level data. Also, the requested accuracy of ground water level data has increased over the years. ASTM D4750 implies that water levels should be measured to the nearest 0.01 ft. (3 mm) [2].

[1]Vice President and General Manager, Brainard-Kilman Drill Company, 2175 West Park Court, Stone Mountain, GA 30086.

[2]Consultant and Past President, Slope Indicator Company, 3668 Albion Place, N., Seattle, WA 98103.

DEVICE DEFINITION

A device was conceived based upon the individual needs of a large and diverse user group. At the outset of product development, a generalized list of requirements was established.

1. Unattended Monitoring. The device would be self-contained and capable of being installed and left unattended to sense and record ground water levels at user defined intervals and duration.
2. Device would be able to be secured totally within the casing of a monitoring well with an ID of 2 in. (51 mm) and larger. We estimate that approximately 30% of the 1991 installed ground water monitoring wells will be 2 in. (51 mm) in diameter.
3. Device able to accurately and repeatedly measure changes in the ground water level of 0.01 ft. (3 mm) or better.
4. Device to have low electrical power consumption in order to allow weeks or even months between battery replacement.
5. Able to store sufficient data points for diverse applications.
6. MS-DOS and PC compatible.

FREQUENCY SENSOR

The system described here uses frequency sensors. Vibrating wire sensors have been used since the late 1800s [3] to measure mass, force, pressure and displacement. This century of use and refinement has provided confidence due to their long-term reliability, reasonably priced components, adaptability to wide ranges of measuring requirements and durability in harsh environments.

The use of vibrating wire (VW) sensors by hydrologists and geo-environmentalists is neither common in the geotechnical laboratory nor testing field. However, the specialists concerned with field instrumentation to measure geotechnical behavior such as water and soil mass pressures, land movement, or loads and forces are very familiar with frequency sensors such as the vibrating wire (VW). VW sensors were first used on a large scale in connection with geotechnical instrumentation nearly fifty years ago in France [4].

Operating Principle of Vibrating Wire Sensors

A vibrating wire sensor employs a tensioned wire and an electro-magnet. The electro-magnet is used to excite the wire which then vibrates at a frequency related to the tension in the wire. In the "pluck" method of excitation, voltage pulses passed through the electro-magnet cause it to draw and release the wire at selected frequencies. After the pluck, the wire is left to oscillate at its natural resonant frequency in the same way that a plucked guitar string does. The oscillation of the wire in the magnetic field now generates a series of voltage pulses in the electro-magnet. These pulses occur at the same frequency at which the wire is vibrating. The pulses are then amplified and transmitted to the readout or data logger.

The force transducer described in this paper uses a newly introduced "vibrating strip" [5] in place of the conventional "vibrating wire." This strip is welded into the system.

A fundamental difference between the vibrating wire and strip transducer is that the wire is pre-stressed during assembly such that an inverse relation exists between wire tension and increases of force or pressure. Whereas, stretch occurs in the strip configuration as force or pressure is applied to the transducer. Users of vibrating wire transducers generally agree that a disadvantage is the possibility of creep of the tensioned wire or slippage at the clamp. [6].

SYSTEM DESCRIPTION

The system was designed in two basic configurations: Where very high accuracy and relatively small changes in water levels occur, the preferred sensor is a vibrating strip force transducer coupled to a floating buoyancy tube (Fig. 1). For many normal applications, a VWP pressure transducer is preferred (Fig. 2).

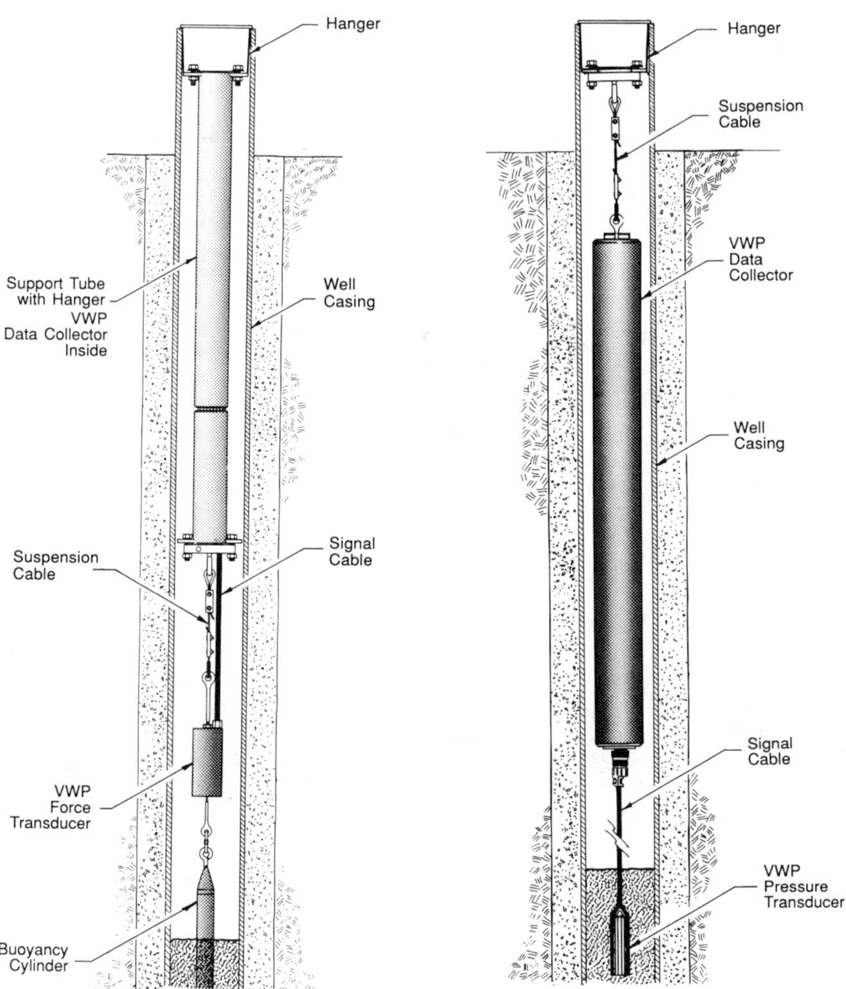

FIG. 1 -- VSP Force Transducer System

FIG. 2 -- VWP Pressure Transducer System

The entire system has three main components:

- ♦ Data Collector
- ♦ Frequency Sensor
- ♦ Data Manager Software

Data Collector

The VWP Data Collector consists of a microprocessor system and a power pack.

Microprocessor System: The microprocessor system is sealed within a 1.7 in. O.D. x 12 in. (43 x 305 mm) cylinder made of corrosion resistant stainless steel. One end of the cylinder provides a waterproof connector for the sensor signal cable. The other end is threaded to accept the power pack. Inside are circuit boards (Fig. 3) containing a Motorola 68H11 processor, programmable array logic (PAL), read-only memory (ROM), random access memory (RAM) with a capacity of 10,000 readings, and a number of other circuits. Critical components are assembled using surface-mount technology which enhances reliability.

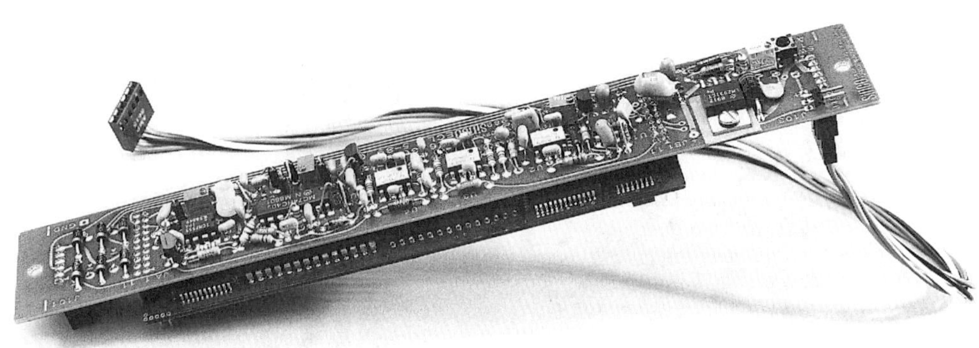

FIG. 3 -- Electronic Circuits of Data Collector

Power Pack: The power pack is 1.7 in. x 15 in. long (43 x 380 mm) and houses 4 alkaline D cells capable of powering the system for up to 12 months (frequent sampling or lower ambient temperatures can shorten battery life). One end of the power pack threads onto the microprocessor cylinder. The other provides an RS-232 port for programming and transferring data to a computer.

Connections: The upper end of the Data Collector is equipped with a RS-232 connection port. With a short cable coiled near the top of the mounting well, the operator need only remove the well cover and connect to a portable IBM PC or compatible computer for retrieving data or initiating a new or continuous data collection.

The lower end of the assembly has a standard six pin connector for terminating the sensor's signal cable and a connector eyelet for connecting a stainless steel suspension cable to support either the VWP force transducer or the pressure transducer if long cables are required.

Transducers

Vibrating Wire Pluck (VWP) Pressure Transducer: The VWP Pressure Transducer converts ground water pressure to a frequency signal which is sensed and recorded by the Data Collector. VWP transducers are available with various pressure ranges from 5 to 500 psi (34.5 to 3447 kPA), typically 5 to 25 psi (34.5 to 345 kPA) are used for monitoring ground water levels. The accuracy of pressure transducers is a function of the particular transducer's design and range selected. Transducers are generally rated in terms of their precision over the full range. Common precisions are $\pm$0.01% to $\pm$0.025% over the rated range. A 0-5 psi transducer will provide measurements twice as precise as a 0-10 psi transducer of equal precision.

Pressure transducers, whether of the strain gauge or vibrating wire type, are all affected by changes in the atmospheric pressure. For data collection periods lasting more than a few hours (assuming constant atmosphere) in duration, it is necessary to correct the recorded data for variations in atmospheric pressure and incorporate some means of compensation within the sensor. Traditionally, the latter has been accomplished by incorporating a small diameter vent tube in the sensor's cable. The vent tube terminates within the cavity of the sensor and is vented to atmosphere. Typically the ID of the vent tube is 1 to 2 mm diameter. Changes in atmospheric pressures at the vent opening will not instantaneously occur at the transducer end of the vent tube. Pressure changes occur with volume change of the enclosed gas according to Boyle's law. The resulting time lag can be pronounced when the capillary effect of long, small diameter vent tubes influence the rate at which the enclosed gas can equalize.

If condensation occurs within the vent tube line, capillary tension of the condensation will effectively lock the vent line rendering it useless. Special precautions are often taken to add desiccant at either end, i.e., within the sensor and at the outlet. However, the very humid conditions of most monitoring wells will result in quick contamination of the desiccant requiring continuous checking and replacing when necessary. The preferred method is to use a recording barometer so the time and magnitude of atmospheric pressure changes can be correlated. This information can also be obtained from a weather station near the site.

Vibrating Strip Pluck (VSP) Force Transducer: VSP force transducers coupled with buoyancy cylinders are recommended when precise measurement changes in water levels are required. This system consists of a force transducer and a buoyancy cylinder. The force transducer (Fig. 4) is a newly designed frequency sensor that uses vibrating strip technology. The force transducer produces a frequency that has a known relationship to the mass of the immersion cylinder. Part of the buoyancy cylinder is always in water, but the water must not rise above the top of the cylinder. As the water level rises or falls, the buoyancy of the cylinder decreases or increases, and the force transducer outputs a corresponding change in frequency.

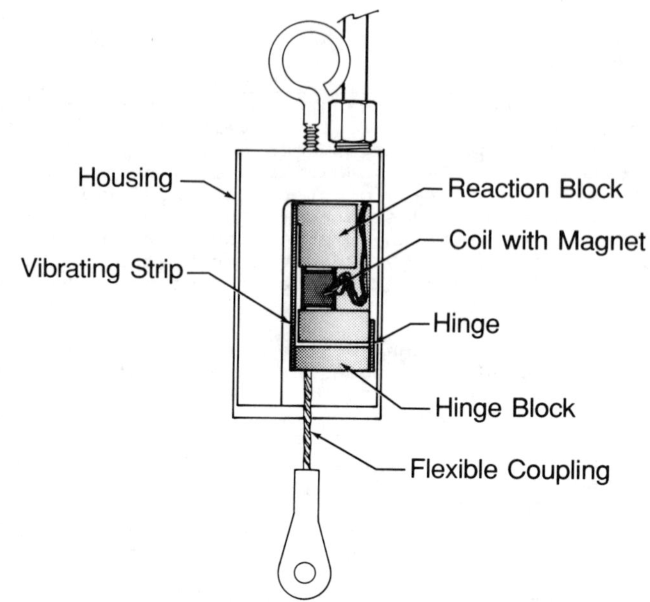

FIG. 4 -- VSP Force Transducer Schematic

The measuring range of the system can be extended by lengthening the buoyancy cylinder. Sensitivity of the instrument can be increased by employing a larger diameter cylinder. A centering mechanism can be employed above the force transducer to keep the suspension cable centered within the borehole as erroneous readings occur if the buoyancy cylinder contacts the walls of the well casing. The well and casing need to be vertical where this system is employed. Typical ranges of measurement and associated precision for various cylinder diameters are illustrated in Table 1.

TABLE 1 -- System Range and Precision

Cylinder Diameter	Maximum Range	System Precision
1.0" (25 mm)	240" (6096 mm)	±0.10" (2.45 mm)
1.5" (38 mm)	180" (4570 mm)	±.045" (1.14 mm)
2.0" (50 mm)	120" (3048 mm)	±.025" (0.62 mm)
4.0" (102 mm)	20" (508 mm)	±.005" (0.15 mm)

Data Collector Manager (DCM) Program

The DCM program operating in an MS-DOS microcomputer is used to program the data collector and to later retrieve the stored data. The DCM program runs under MS-DOS 3.0 or later and requires a minimum of 512 KB memory.

Typical functions of the DCM include:

1. Creating a monitoring program and sending the program to the data collector.
2. Initiating data logging.
3. Retrieving data and storing retrieved data in a database for viewing and editing.

BOREHOLE INSTALLATION

The selection of the buoyancy cylinder coupled with a VSP Force Transducer or a VWP Pressure Transducer depends largely upon the physical constraints and the desired accuracy.

Buoyancy Cylinder/Vibrating Strip Force Transducer (VSFT)

The buoyancy cylinder/VSFT system is recommended where high precise changes in the ground water level are required and the water level fluctuation is small to moderate over the sampling period. The system requires special above-ground provisions where the water level will rise within 3 ft. (1 m) of the top of the well casing as sufficient head room above the water level is not available for the VSFT and data collector. The buoyancy cylinder/VSFT system is felt to be most suitable for water level changes of 20 ft. (6 m) or less. The longer the installed buoyancy cylinder, the greater the ID of the well casing should be to safeguard against the cylinder coming in contact with the casing wall. In contrast, the greater the accuracy, the larger the diameter of the buoyancy cylinder.

The length of the buoyancy cylinder should be as short as practical and the diameter as small as possible to satisfy accuracy needs. If the anticipated water level variation cannot be reasonably estimated or the variation is considerable yet changes slowly, the user might consider periodically raising or lowering the buoyancy cylinder rather than using a longer cylinder to cover for possible contingencies.

The buoyancy cylinders are precision machined and calibrated for water at 4° C. If the specific gravity of the borehole fluid is significantly different, the buoyancy mass of the cylinder as measured by the VSP force transducer will be affected.

The data collector may be placed near the top of the casing or suspended a nominal distance down the casing. The latter is generally recommended for operations where the ambient temperature is below freezing over prolonged periods. Slack is left in the signal cable so that the data collector can be removed for battery changes.

VWP Pressure Transducer

In selecting the pressure transducer, the rated working pressure should be matched as nearly as possible to the anticipated variation of water levels in the borehole over the planned sampling interval. Because the precision of VWP transducers is a function of the full scale, it is important to select a transducer that most nearly matches the anticipated fluctuation of the ground water plus 50%. For example, ground water levels expected to vary by 20 ft. (6.1 m) equate to 8.7 psi (60 kPA) adding a 50% safety margin resulting in a sensor with a working pressure range of at least 13 psi (90 kPA). Most VWP pressure transducers can have their rated working pressure exceeded by 100% without damaging the unit; however, the device may not produce repeatable data when the working pressure is exceeded.

The transducer must be suspended below the minimum water level. Most transducers operate best in the mid 80% of their range which implies that the transducer should always have a water column not less than 10% of the rated pressure range, e.g. an equivalent head of 5 psi (34 kPA) for a 50 psi (345 kPA) pressure transducer.

CONCLUSIONS

This paper discusses a new system for unattended measuring of water levels in monitoring wells. The system consists of a sensor connected by an umbilical cord to a data collector all of which can be installed in a monitoring well as small as 2 in. (51 mm) ID. The data collector's 1.75 in. (89 mm) diameter is possible by using surface mount technology (SMT) that miniaturizes critical electronic circuits.

Two sensor types can be used to monitor water levels: buoyancy cylinder and pressure transducer. Both sensors use the vibrating wire or strip technology. The buoyancy cylinder water level sensor consists of a vibrating strip - pluck type (VSP) force transducer and a buoyancy cylinder. The force transducer produces a frequency signal proportioned to the mass of a buoyancy cylinder. The cylinder (2 to 20 ft. [.61 to 6.1 m] in length) is always partially immersed in the water. As the water level rises or falls, the buoyant mass of the cylinder changes, and the transducer outputs a corresponding change in frequency.

ACKNOWLEDGMENTS

The system described in this paper was designed and developed by the Engineering Department of the Slope Indicator Company, Seattle, Washington, under the direction of Dale Shoup, Engineering Manager. Numerous individuals contributed toward development of the force transducer sensor and data collector. The vibrating strip force transducer is based on the invention of David Clements [5]. Special recognition for the talent of David Clements, Jim Ernest, Tehmosp Khan and Dave Arnston who constituted the design team. The authors thank Dale Shoup and Rick Monroe who reviewed and offered appropriate suggestions of the paper's content. A special appreciation of those who typed and formatted the manuscript - Connie Johnson of Brainard-Kilman Drill Company and Sharon Donlan of Slope Indicator Company.

REFERENCES

[1] Nielsen, D.M. ed., Practical Handbook of Ground-Water Monitoring, Lewis Publishers, Inc., Chelsea, Michigan, 1991, pp 378-386.

[2] ASTM (1990), "Standard Test Method for Determining Subsurface Liquid Levels in a Borehole or Monitoring Well (Observation Well)," D4750-87, Annual Book of Standards, Vol. 04.08, ASTM, Philadelphia, PA.

[3] Patent #633,471 Issued to E. McGarvey, Weight & Pressure Indicator, September 1899.

[4] Bordes, J.L. and Debreuille, P.J. (1985), "Some Facts About Long-Term Reliability of Vibrating Wire Instruments," Reliability of Geotechnical Instrumentation, Trans. Res. Record, No. 1004, pp 20-27.

[5] Patent #4,914,962 Issued to D.J. Clements, Vibrating Strip Transducer, April 10, 1990.

[6] Dunnicliff, J., Geotechnical Instrumentation for Monitoring Field Performance, John Wiley & Sons, New York, 1988, p 103

Van Maltby[1] and Jay P. Unwin[1]

A FIELD INVESTIGATION OF GROUND-WATER MONITORING WELL PURGING TECHNIQUES

REFERENCE: Maltby, V. and Unwin, J. P., "A Field Investigation of Ground-Water Monitoring Well Purging Techniques," Current Practices in Ground Water and Vadose Zone Investigations, ASTM STP 1118, David M. Nielsen and Martin N. Sara, Eds., American Society for Testing and Materials, Philadelphia, 1992.

ABSTRACT: A field investigation of commonly-used monitoring well purging techniques was conducted under different conditions including type of pump, pump inlet location, and the use of packers. Tracers including deionized water, fluorescent dye, and lithium chloride, were used to define the amount of stagnant water at any given time in the pump discharge. Tests were conducted in shallow 5 cm (2 in) diameter wells. The effects of drawdown were examined.

Tests conducted in the absence of drawdown with the pump inlet in a fixed position at or above the screen showed a highly variable and unpredictable inclusion of stagnant water. The use of packers did not completely prevent the inclusion of stagnant water into the pump inlet. The inclusion of stagnant water into a sample was minimized by purging from some distance above the screen followed by relocation of the pump inlet into the screen for sample collection. In wells where drawdown occurred during purging, stagnant water inclusion was minimized by reduced pumping rates to allow for sample collection during periods of well recharge. Real time monitoring of indicator parameters such as pH, temperature and specific conductance was not generally successful in indicating when purging was complete.

KEY WORDS: ground water, monitoring well, sampling, purging, tracer study, packers

[1]Research Scientist and Regional Manager, respectively, National Council of the Paper Industry for Air and Stream Improvement, Inc., Central-Lake States Regional Center, (NCASI), Western Michigan University, Kalamazoo, MI 49008.

MONITORING WELL PURGING

The water standing in the well casing is commonly referred to as being stagnant; that is, the water has been isolated from the aquifer at least since the last time the well was sampled. It is generally recognized that the composition of the stagnant water within a monitoring well above the screened section is probably not representative of the overall ground-water quality at the sampling site. During that time, the chemical quality of the stagnant water may have changed by (a) direct introduction of foreign material into the well, (b) interactions with the well casing or at the interface with the atmosphere, or (c) biological activity. Even without such alterations, the stagnant water would not reflect any changes in the ground-water quality that may have occurred since the last time the well was sampled. Because the investigator cannot be certain which, if any, of these influences has occurred or whether inclusion of some of the stagnant water in a sample from the well would significantly change the conclusions that might be drawn from the data, the safe thing to do is to prevent or minimize such inclusion. Such an approach may be particularly important when sampling for trace level organics.

One presently-used purging technique presented as a cooperative agreement by the Illinois State Water Survey (ISWS) and Illinois State Geological Survey (ISGS) (1) is to pump the well and calculate the percentage of water at any given time in a pump discharge that can be attributed to drawdown. This approach is based on the knowledge of time-drawdown characteristics of a well and does not account for contributions of stagnant water from any source other than drawdown. Another commonly-used purging technique presented by the U.S. Geological Survey (USGS) (2) is to pump a well until indicator parameters such as pH, temperature, and conductivity stabilize. This approach ignores the possibility that a near-constant contribution of stagnant water into the sample may result in stabilized readings for the observed parameters. It also fails to account for contributions of stagnant water that are too small to notably affect the measured parameters, but which may significantly alter the outcome of an analysis. Probably, the most commonly-used purging practice is to purge an arbitrary number of casing volumes (well casing) with little or no regard to drawdown or indicator parameters.

This research reflects the need for documentary evidence regarding the hydraulic behavior of a monitoring well during purging. By examining truly trace concentrations, the fraction of stagnant water entering a pump inlet can be better defined as a function of casing volumes pumped (or time) and inlet position.

CRITICAL REVIEW OF EXISTING INFORMATION ON MONITORING WELL PURGING

Illinois State Water Survey/Geological Survey

The Illinois State Water Survey (ISWS) and Illinois State Geological Survey (ISGS) have published "Procedures for the Collection of Representative Water Quality Data From Monitoring Wells" (1) which describes in detail guidelines for monitoring well purging. The basic assumption made in this research was that during the initial pumping of a small diameter monitoring well, a significant fraction of the pump discharge comes from stagnant water within the well casing. This effect is due to drawdown. The procedure uses an equation which develops time-drawdown data based on individual monitoring well hydrologic data obtained during pump tests. The resulting theoretical drawdown curve is used to predict the time at which the effects of casing storage due to drawdown become negligible. This curve is intended to be used as a guideline, in conjunction with the observation of indicator parameters, for the selection of an appropriate pumping rate and number of casing volumes to be pumped prior to sample collection.

Six monitoring wells at different locations within the State of Illinois were used for the development of this purging protocol. The National Council of the Paper Industry for Air and Stream Improvement (NCASI) encountered several difficulties while reviewing the results of the study on these six wells. These difficulties included (a) reproduction of the theoretical drawdown curves presented, and (b) interpretation and conclusions drawn from the monitored indicator parameters. NCASI and others have been unable to reproduce any of the published theoretical drawdown curves for six wells examined by ISWS/ISGS. For each well, the ISWS/ISGS have provided theoretical drawdown curves derived from the Papadopulos and Cooper equations which generally show good agreement with the actual drawdown curves presented. NCASI has used these equations in the manner described by the cooperative agreement to produce theoretical drawdown curves which bear little resemblance to those published. A careful examination by NCASI has not revealed the reason for these discrepancies.

The ISWS/ISGS examined the effects of well purging on the chemical composition for six monitoring wells. Five of the six wells were described as having site specific limitations which hindered interpretation of the results. The single well (Site 5), in which a clearly indicated effect of purging on indicator parameters was noted, directly contradicted information from the pump test portion of the study. In spite of limitations described for each of the six sites examined, ISWS/ISGS concluded that "the chemical data from this portion of the study have verified the theoretical ratios of aquifer to stored water predicted during the pump tests". A subsequent publication by the ISWS "Practical Guide For Ground-Water Sampling (3) has endorsed the above mentioned purging protocol.

United States Geological Survey

The United States Geological Survey (USGS) (2) states that in order to obtain a representative sample from an aquifer at a given location, a well must be pumped until indicator parameters such as pH, temperature, and conductivity are constant. Guidelines for indicator parameter stability have been published by Gibs and Imbrigiotta (4) and Hardy et al. (5). Measurement of drawdown during the purging period is recommended because changes in the indicator parameters may reflect water from different zones of the aquifer being drawn into the well. This procedure is stated as the minimum required precaution for insuring that a sample adequately represents the water quality in the aquifer.

Research by Slawson et al. (6) examined the variability of indicator parameters and other constituents in well discharges during continuous pumping. Appreciable changes were observed in several of the parameters, most notably conductivity. These changes were attributed to naturally occurring vertical and horizontal variability within the aquifer from natural influences. Such a situation would not allow for the universal use of indicator parameters to determine when a ground water sample should be collected.

Consideration should also be given to the general nature of the specified indicator parameters. Conductivity and pH may be affected by changes in temperature and pressure during sampling. Pressure changes can cause rapid degassing of carbon dioxide and other gases that could affect sample pH and specific conductance.

NCASI Laboratory Purging Investigations

NCASI conducted a laboratory investigation to examine factors other than drawdown that could cause stagnant water to enter a pump inlet (7). Tests were conducted at constant head, thereby disregarding the effects of drawdown. In the research, stagnant water in a well column was spiked with a fluorescent dye. Care was taken to minimize density differences induced by either temperature gradients between stagnant and aquifer water or concentration-induced density gradients. The well was then sampled with the pump inlet in various positions while the tracer concentration in the pump effluent was continuously monitored. Results demonstrated that an average of about 2 to 4 percent of the water pumped from locations above the screen and an average of about 1 percent of the water pumped from within the screen of the monitoring well came from the stagnant water located above the pump inlet. The mechanism that caused the overlying stagnant water to reach the pump inlet was not investigated, though mixing caused by turbulence around the pump inlet was hypothesized.

University of Waterloo

Robin and Gillham (8) conducted a study using

non-reactive tracers to judge the effectiveness of various purging procedures. The results suggest a sharp interface and little mixing between fresh water in the screen or below a pump inlet and the stagnant water in the casing. For wells not completely evacuated, pumping from immediately beneath the air/water interface for 2 or 3 casing volumes was deemed sufficient to collect a representative sample. Three tracers were used for the study: deionized water, NaCl (conductivity), and bromide. Of these, NaCl was demonstrated to be an inappropriate tracer due to mixing caused by density differences between the tracer and the fresh water. While deionized water was determined to be an appropriate tracer (verified with bromide) an even greater density difference existed between deionized water and the natural ground water at the site. Although the deionized water was less dense than the ground water, the effect of this density difference may have acted to discourage mixing around the pump inlet.

EXPERIMENTAL PROCEDURE

Wells

Two shallow monitoring wells were selected for this research. Both wells were constructed of 5-cm (2-in) ID PVC with commercially available PVC screened sections (Timco 0.010 slot). Both wells are situated in a shallow glacial unconfined aquifer composed of sand and clay. Each well is approximately 9.2 m (30 ft) deep and has a standing water level within 1.5 m (5 ft) of the ground surface. Although the wells were only approximately 62 m (200 ft) apart, local variability within the aquifer accounted for marked difference in the hydraulic performance of each well. At the purging rates used in this research, one well experienced very minimal drawdown (less than 0.6 cm, 0.25 in), while the other well experienced extreme drawdown and could easily be pumped dry.

Equipment

Two pumps were used for purging in this research. The majority of the purging tests were conducted using an above-ground peristaltic pump (Masterflex, #70-15 head) with a maximum flow rate of approximately 800 ml/min. Several of the purging tests were conducted with a submersible pump (Keck #84) with a flow rate of approximately 4.5 L/min.

The fluorescent tracer concentration in the purging pump discharge was detected with a fluorometer (Turner #111) with a flow-through cell for continuous measurement. Conductivity, drawdown, temperature, and pH were monitored continuously using methods described elsewhere (9). Drawdown was monitored with a submersible pressure transducer. Data from the instruments with the exception of pH were recorded on a portable computer. Values for pH were recorded manually due to pH signal recording difficulties.

A portable electric generator provided electrical power where needed.

The amount of stagnant water in the pump discharge at any given time was measured directly by the use of one or more of the following tracers: Rhodamine WT, lithium (as lithium chloride), or deionized water. Rhodamine WT was used as the tracer of choice during this research. Rhodamine WT is a non-toxic fluorescent xanthene dye commonly used in percolation studies, potable water systems, and surface water systems. Additionally, Rhodamine WT exhibits both low reactivity and sorption tendencies (10) making it well suited for this research. The initial concentration after the dye had been added to the stagnant water was approximately 200 ppb. At this concentration in water, the dye imparts no color detectable with the naked eye. In several of the purging tests in the well in which drawdown occurred, the presence of turbidity interfered with the detection of the fluorescent dye. In this situation, lithium was used as the tracer. For several of the tests, deionized water was used as a tracer in conjunction with Rhodamine WT.

General Procedures

At the start of each purging test, the static water level was measured. Tracer was added to the stagnant water in the well in a manner that resulted in a homogeneous concentration within the stagnant water column, without migration into the screen area. To accomplish this, an inflatable packer was used to hydraulically isolate the screened portion of the well from the cased portion above. The packer was designed as a flow-through device so that water could be collected from the screen area during the time the packer was inflated and in position.

The stagnant water above the packer was pumped out and collected in a container at the surface. Aquifer water from the screen area was pumped to the surface at the planned purging rate in order to zero the fluorometer and obtain background readings for pH, conductivity and temperature. With the fluorometer zeroed for the aquifer water, the previously collected storage water was pumped through the fluorometer in a closed loop system. Tracer was added to the casing water until the fluorometer readout was 100 percent. In order to chemically differentiate the casing water from the aquifer water, the pH was lowered one pH unit by adding a few milliters of acid. This also caused a corresponding increase in conductance. This casing water containing the tracer was poured back into the well to a water level slightly below the static water level. This insured that the volume displaced by the packer would be replaced by aquifer water moving into the screened section rather than by spiked water moving down into the screened section when the packer was deflated and removed.

The stagnant water containing the tracer was kept isolated from the screened section with the inflated packer for a minimum of 12 hours to insure that the undisturbed temperature-depth profile of the stagnant water would be re-established. A preliminary investigation of the wells used for this research revealed that the temperature-depth profile determined after a month of non-pumping would be re-established approximately 8 hours after having been disturbed.

A test was initiated by slowly deflating the packer and carefully removing it from the well. The pump inlet was placed at a predetermined depth for the test. Pumping and automated data collection were started. Fluorometry readings were recorded every two seconds, other readings except pH were recorded every five seconds. Readings for pH were recorded every 105 seconds. The computer and software allowed for real-time observation of data throughout the purging test.

At the end of a test, the pump inlet was raised to the air/water interface while pumping continued until the fluorometer readings returned to background. At this point new tracer could be added for the next test.

For the tests where it was applicable, the number of casing volumes pumped was based on the volume between the pump inlet (bottom of sample line) and the top of the well screen. For the tests where the pump inlet was placed immediately above the screen top, the duration of a test was measured in the number of liters pumped, rather than casing volumes. Because the measured tracer concentration in the aquifer water was zero for all tests, the fraction of the pumped volume that came from the stagnant zone above the pump inlet is simply the ratio of the measured concentration of tracer in the pumped water to the initial concentration in the stagnant zone.

Two packers were investigated to determine their effect on stagnant water concentration during purging and sampling. Such packers form a seal against the inside of the well casing to isolate the stagnant water above from the pump inlet. In effect, the volume of standing water above the pump inlet is reduced. One packer was a commercially-available unit attached to the top of the Keck pump. The other packer was laboratory-made and designed as a flow-through device.

The effect of purging and sampling in a well experiencing a significant degree of drawdown was investigated. During drawdown, the balance of the water in the pump discharge not accounted by flow into the well from the screen comes from the stagnant casing water above the pump inlet. A sample was collected from such a well by purging the well at a rate great enough to produce drawdown and thus reducing the level of stagnant water. Sample collection occurred at a reduced flowrate during water level recovery. Purging and sampling in this manner allowed the stagnant water/aquifer water interface to move upward and away from the pump inlet, reducing the chance for stagnant water to become captured by the pump inlet. As noted previously, due to excessive turbidity in this well, lithium chloride at an initial concentration of 62 mg/L was used as the tracer.

The following purging configurations were investigated during this research: (a) fixed pump inlet positions at approximately 5 cm (2 in) below the air/water interface; mid-casing, and at the top of the well screen; (b) a comparison of Rhodamine WT and deionized water as tracers with the pump inlet repositioned between the screen and the air/water interface several times during the test; (c) purging from above the screen, followed by sampling within the screen; (d) packers; (e) drawdown and recover. Details specific to each test are given in the section below.

RESULTS AND DISCUSSION

With the exception of the last test in the following series, drawdown did not become significant (maximum 0.6 cm, 0.25 in) at the specified purging rates.

Peristaltic Pump, Inlet Near Static Water Level

Figure 1 graphically presents the results of a purging test with the pump inlet located approximately 5 cm (2 in) below the static water level in the monitoring well. The purging rate during this test was 1.01 L/min.

During most of the time required to remove the initial casing volume of water from the well (14.3 minutes), the concentration of stagnant water in the pump inlet was 100 percent. As the column of fresh water moving up the well casing approached the pump inlet, there was a corresponding rapid decrease in the stagnant water concentration detected in the pump discharge. During the removal of subsequent casing volumes the concentration of stagnant water continued to decrease but was still detectable in the pump discharge for a relatively long time (9.0 casing volumes, 128 minutes). This unpredictable inclusion of stagnant water in the pump discharge closely matches the results obtained in a laboratory test under similar conditions (7). At this point the majority of the stagnant water had been removed from below the pump inlet; however, it may not have been entirely removed from above the pump inlet. Earlier research (11) using visible dye concentrations in transparent well casings provided evidence that the intermittent inclusion of stagnant water detected in the pump discharge may have been related to turbulence and subsequent mixing around the pump inlet. The reason for the relatively small concentrations of stagnant water detected at the end of the test is probably related to the fairly small volume of stagnant water above the inlet, (0.1L). This volume would most likely have been diluted by mixing during the pumping of 9 casing volumes (130L). At 9 casing volumes, the pump inlet was lowered into the screen area. With the pump inlet located in this position, one casing volume of water (14.4 liters) was collected without the detection of stagnant water in the pump discharge. Figure 2 presents a five-minute expanded segment of Figure 1 between 4.0 to 4.3 casing volumes. It shows the intermittent nature of the intrusion of stagnant water into the pump inlet. The casing volume axis has been converted to time. While the contribution of stagnant water is relatively small at this point, it is nonetheless measurable, and for intervals of up to 30 seconds.

One implication of this is that approximately 500ml of water could be collected as a ground water sample over a 30 second time interval. Depending on the purpose for which the sample was being collected, inclusion of almost 1 percent stagnant water could be problematic.

As shown in Figure 3, temperature, pH, and conductivity during this test displayed an initial response during the removal of the first casing volume. However, they exhibited little or no change during the pumping of subsequent casing volumes. A visual inspection of these parameters between casing volumes 1 and 2 indicate little if any change, whereas the tracer concentration shown in Figure 1 indicates a stagnant water concentration decreasing between 11 and 2 percent. These parameters were too insensitive for the detection of trace concentrations of stagnant water in this test. An examination of these parameters for all of the other purging tests revealed similar results. Research by Gibs and Imbrigiotta (4) revealed that monitoring field water-quality measurements was not a reliable indicator of when to collect a representative sample for purgeable organic compounds.

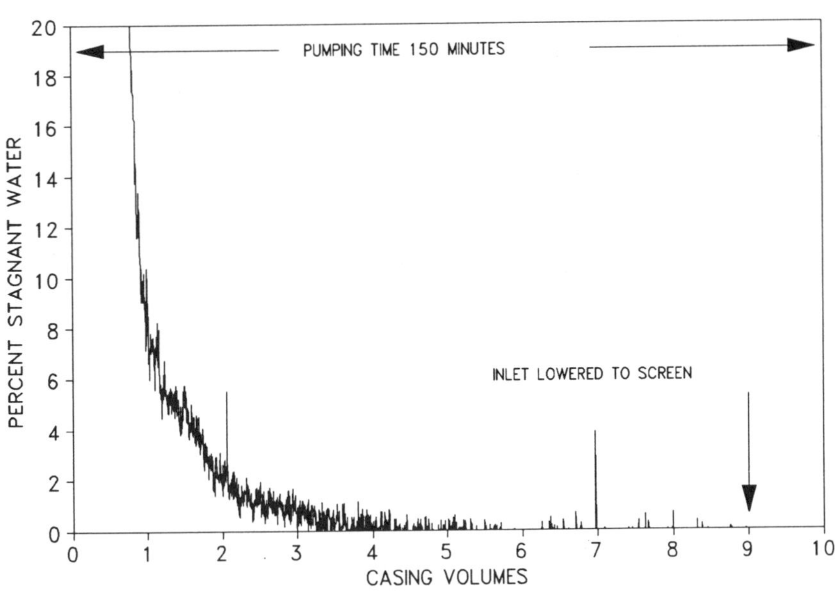

Fig. 1- Inlet near static water level

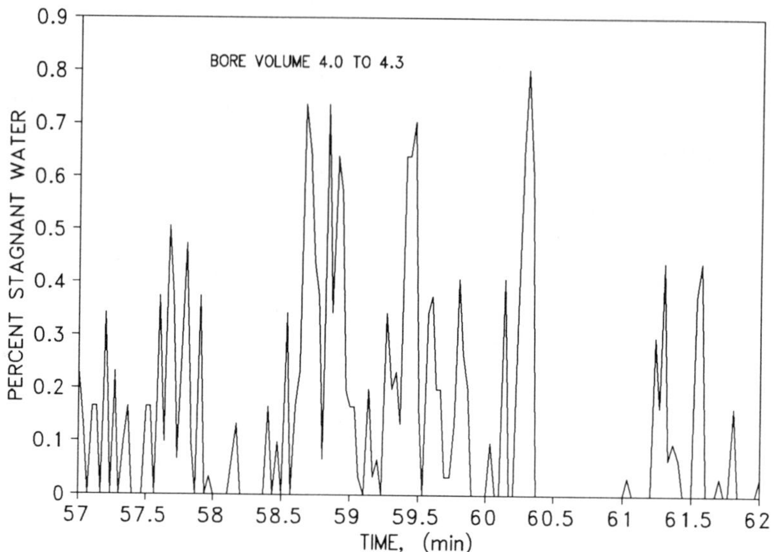

Fig. 2- Expanded view, bore volume 4.0 to 4.3

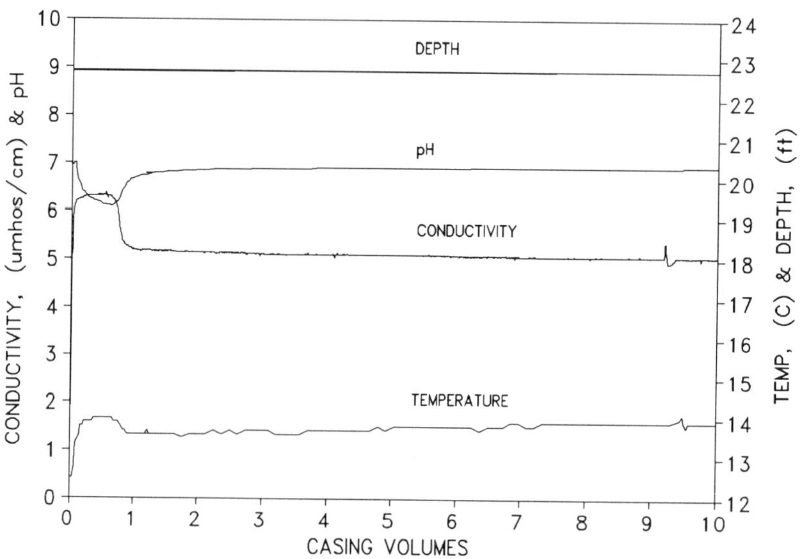

Fig. 3- Indicator parameters

Peristaltic Pump, Inlet at Mid-casing

Figure 4 displays the results of a purging test with the pump inlet located in the mid-casing position with equal amounts of water above and below the inlet. Examination of Figure 4 reveals results similar to the purging test displayed in Figure 1 in that the majority of the stagnant water below the pump inlet was removed with the initial casing volume of water. While the stagnant water concentration tapered off more slowly between casing volumes 2 and 4 with the inlet position located in the mid-casing position, the subsequent bore volumes pumped after the fourth casing volume appear to have elevated stagnant water concentrations when compared to the test displayed in Figure 1.

Peristaltic Pump, Inlet at Screen Top

Figure 5 displays the results of a purging test with the pump inlet situated immediately above the well screen.

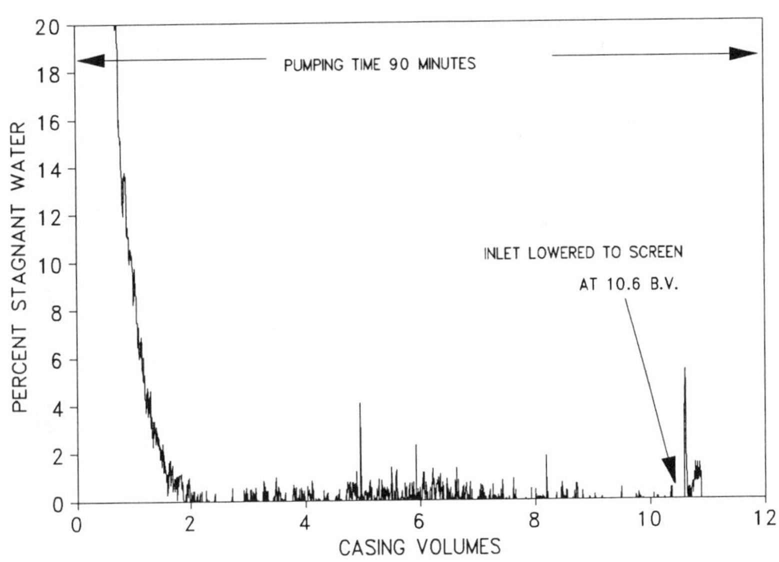

Fig. 4- Inlet at mid-casing

Because there was no volume of stagnant water below the pump inlet, the lower x-axis in Figure 5 refers to the number of liters pumped. The absence of stagnant water below the pump inlet is evidenced by the lack of an initial interval during which the pump discharge was 100 percent

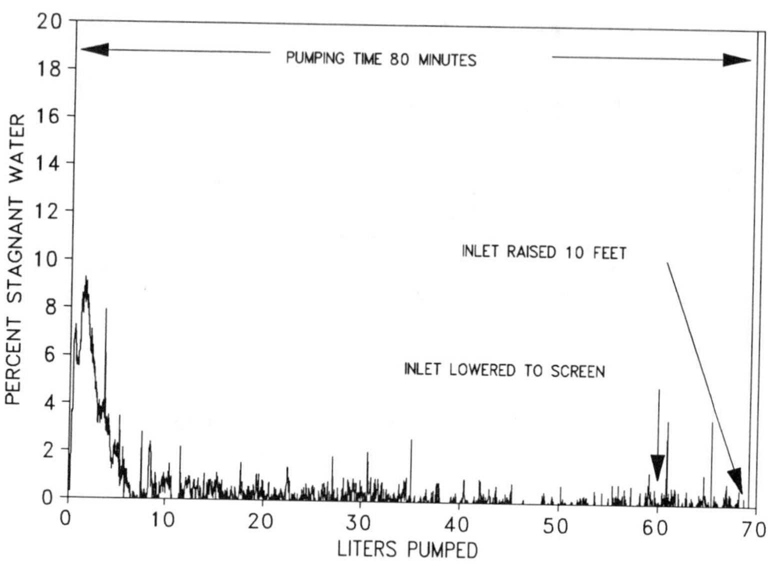

Fig. 5- Inlet at top of screen

stagnant water. This indicates that the pump inlet was positioned at the stagnant water/fresh water interface. The results suggest that the stagnant water detected in the pump discharge during the 80-minute pumping period was caused by the stagnant water overlying the inlet mixing with the fresh water from the screen. At 60 minutes, the pump inlet was lowered 5 cm (2 in) to the top of the screen with little effect on the stagnant water concentration. At 69 minutes, the inlet was raised 3 m (10 ft) which increased the stagnant water concentration to 100 percent.

Source of Stagnant Water

The results of the preceding three purging trials suggest that the stagnant water detected in the pump discharge was from the dyed water overlying the pump inlet. The reasons are twofold. First, the concentration of stagnant water detected in the pump discharge was apparently affected by the volume of dye overlying the inlet. A mathematical average of the stagnant water concentration detected between the fifth and the sixth casing volumes was determined to be 0.026 percent, 0.199 percent, and 0.103 percent for Figures 1, 4, and 5 respectively (Fig.5 concentration determined by examining the data between 38 and 45 minutes, an interval equal to that selected for the concentration determined in Figure 4). One explanation for the average stagnant water concentration in Figure 1 being approximately one order of magnitude smaller than that of Figures 4 and 5 is the smaller volume of stagnant water (0.1L -vs- 7.2L and 14.4L) available for mixing above the pump inlet. The much smaller volume would have been diluted to a greater extent over a given time period, accounting for the smaller percentage of stagnant water in the pump discharge. Second, the results of the test displayed in Figure 5 verify that the pump inlet was positioned at the stagnant water/fresh water interface. The detection of stagnant water for the duration of the 80-minute test indicates that the tracer above the pump inlet was being captured.

Multiple Inlet Positions, Dual Tracers

In an attempt to repeat a purging test conducted by Robin and Gillham (6), deionized water was used as a tracer in addition to Rhodamine WT. During this purging test, varying inlet positions were used starting from the screen top. The purging rate was 1.1 L/min. The water above the screen was replaced with deionized water containing Rhodamine WT using the inflatable packer in the manner described previously. The well was pumped with the inlet immediately above the screen while the fluorescence and conductivity were being monitored. The well was pumped in this manner until the fluorescence approached zero. The inlet was then raised 1.2 m (4 ft) and pumping was continued at the new level until the fluorescence approached zero. This process was repeated several times until the air/water interface was reached.

The results of this test are presented in Figure 6. In general, the shapes of the curves obtained with the fluorescent dye are very similar to the initial portions of curves shown in Figures 1, 4, and 5 in that removal of the majority of the stagnant water below the inlet occurs fairly quickly. The curves from the deionized water conductivity measurements, however, contradict those for the fluorescent dye. The deionized water tracer suggests background water quality was reached within a few minutes of each relocation of the pump inlet, whereas the curves from the fluorescent dye tracer still show stagnant water detectable in the percent range several minutes later. The pattern of response for each of the tracers was repeated each time the pump inlet was raised.

Deionized water may be too insensitive a tracer when used to differentiate water sources in a monitoring well. Conclusions as to required purging volumes based on deionized water tracer tests may, therefore, be inappropriately low for many purposes.

Peristaltic Pump, Sampling Below Purging Point

Figure 7 displays the results of a purging test during which the pump inlet was initially positioned five feet above the screen for approximately five casing volumes

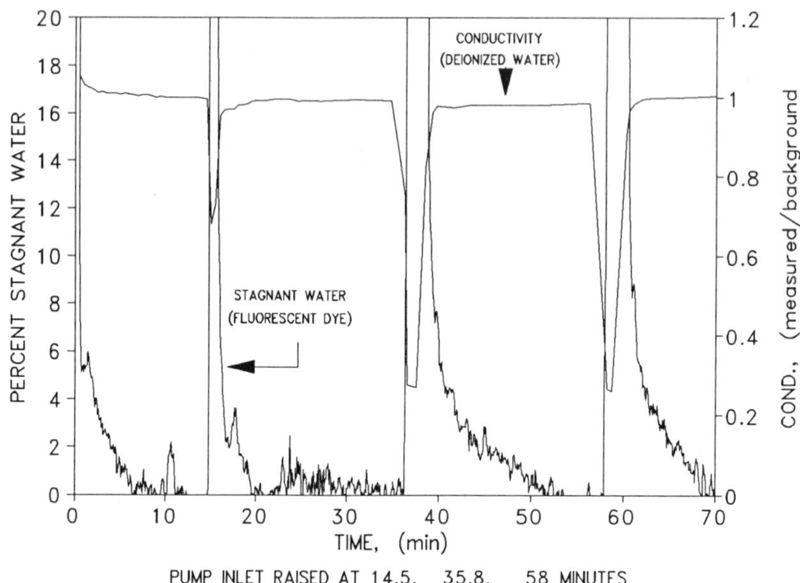

Fig. 6- Comparison of Rhodamine WT and deionized water as tracers

(18.4L) and then repositioned into the screen for sample collection. The advantage of purging and sampling in this fashion is that the entire column of stagnant water need not be purged in order to collect a representative ground-water sample. This test was conducted in triplicate, with similar curves observed for all three trials. During the removal of the initial five casing volumes of water, a clean zone generally free of stagnant water was developed below the pump inlet. After lowering the pump inlet, a minimum 40 liter pumping period followed, during which stagnant water was not detected in the discharge. Stagnant water was eventually detected in the pump discharge. This occurred while the pump inlet was still in the screen area, meaning that the stagnant water had migrated over the five-foot separation between the initial inlet position and the screen. The mechanism for this migration has not been investigated, but is likely due to a combination of diffusion and advection.

Packers

Figures 8 (Keck pump) and 9 (laboratory manufactured packer) present the results of purging tests using packers. Tracer was added to the well for each trial in the manner described earlier. At the start of each test the packer used for tracer addition was removed. The packer used for

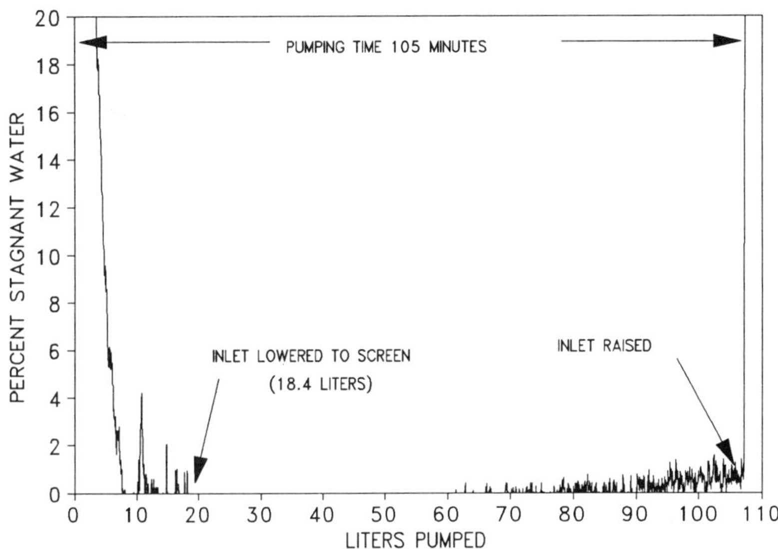

Fig. 7- Sampling below purging point

sampling was slowly lowered so that the pump inlet for each trial was positioned immediately above the screen. The packer was inflated and pumping was initiated.

Neither packer functioned as anticipated. This is evidenced by the unexpected contribution of stagnant water to the pump discharge. While stagnant water was detected in both tests, there are marked differences between Figures 8 and 9. Both curves exhibit the same general shape. However, the curve for the Keck pump (Figure 8) is dramatically shifted to the right. A portion of the curve from Figure 8 is superimposed onto Figure 9 for comparison. There are three possible causes for the higher than expected contribution of stagnant water in these tests. First, leakage past the packer induced by a decrease in head below the packer during pumping may have caused stagnant casing water to migrate past the packer into the pump inlet. Such a leak could have been caused by an irregularity or crack in the casing wall. Second, in the case of the Keck pump and packer, a significant volume of water is displaced (2.66 L) when the pump is lowered into position. This volume is equivalent to a water level rise of 1.31 m (4.3 ft) in a 5 cm (2 in) well casing. It is assumed therefore, that stagnant water was displaced downward through the screen into the aquifer. This may have caused the dramatic shift to the right of the curve for the Keck pump. Finally, stagnant water held between the packer and the pump inlet may have been drawn in by the same phenomenon

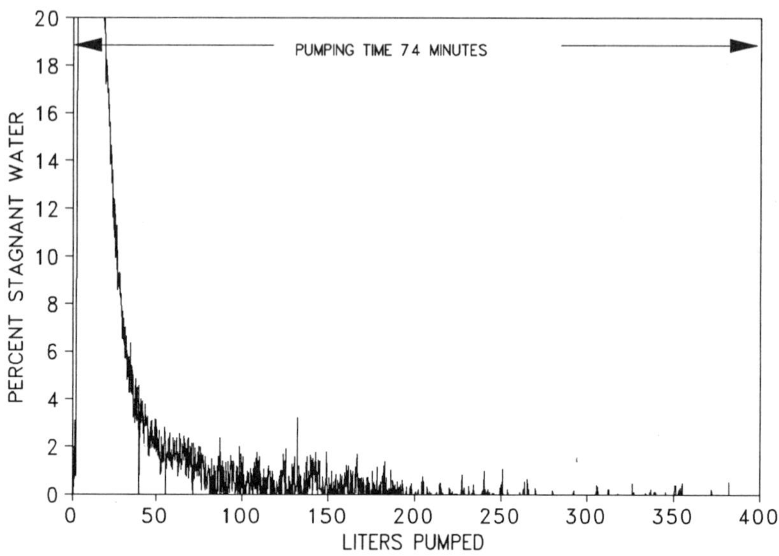

Fig. 8- Keck Pump with packer

observed in other tests were stagnant water was above the pump inlet. The inlet on the laboratory-manufactured flow-through packer was immediately below the packer. However the Keck pump inlet was situated approximately 30 cm (12 in) below the bottom of the packer. This may explain why the test with the Keck pump showed much more frequent and higher spikes of stagnant water than did the flow-through packer.

Purging/Sampling With Drawdown

Figure 10 displays stagnant water concentrations for a well sampled by the drawdown and recover method described earlier. The pump inlet was positioned at the screen top for the duration of the test. The initial pumping rate for the first 25 minutes was approximately 750 ml/min. During the initial portion of this time period when the rate of drawdown was the greatest (as indicated by the slope of the curve marked "depth") a relatively larger portion of the pump discharge would have been from stagnant water. This is confirmed by the higher lithium concentrations. As the rate of drawdown decreases (indicated by decreasing slope of the "depth" curve) the lithium concentration also decreases, indicating a smaller fraction of stagnant water in the pump discharge. At 25 minutes the pumping rate was decreased to 300 ml/min to allow the well to recharge. Evidence of the stagnant water/fresh water interface moving

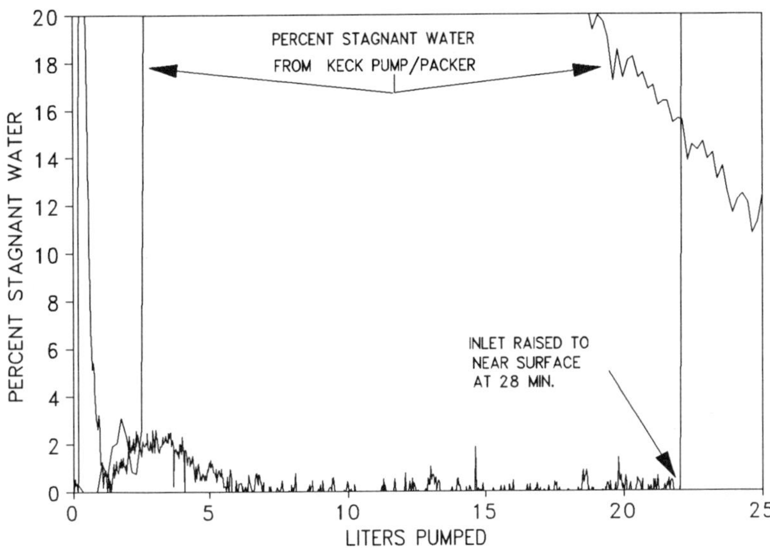

Fig. 9- Peristaltic pump with packer

upward away from the pump inlet is shown by the rapid decrease in lithium concentration. Sample collection during the period of recovery would minimize the inclusion of stagnant water. Further research is necessary to determine the optimum sample collection period using this method.

The principle illustrated here is related to that used in the ISWS/ISGS procedure (1). However, since the mathematical methods used there were unable to predict the drawdown for this well, it was necessary to measure drawdown directly in order to determine when a sample could be obtained. Direct measurement is not difficult in such a low yielding well, and it could be part of the sampling protocol for such a well.

SUMMARY AND CONCLUSIONS

The results of the research presented here show that the procedures used to sample a well can have an effect on the amount of stagnant water in a ground water sample. All tests conducted in the absence of drawdown with the pump inlet in a fixed position at or above the screen showed a highly variable and unpredictable inclusion of stagnant water. This inclusion of stagnant water may have been caused by turbulence around the pump inlet. Packers were not generally effective in preventing the inclusion of

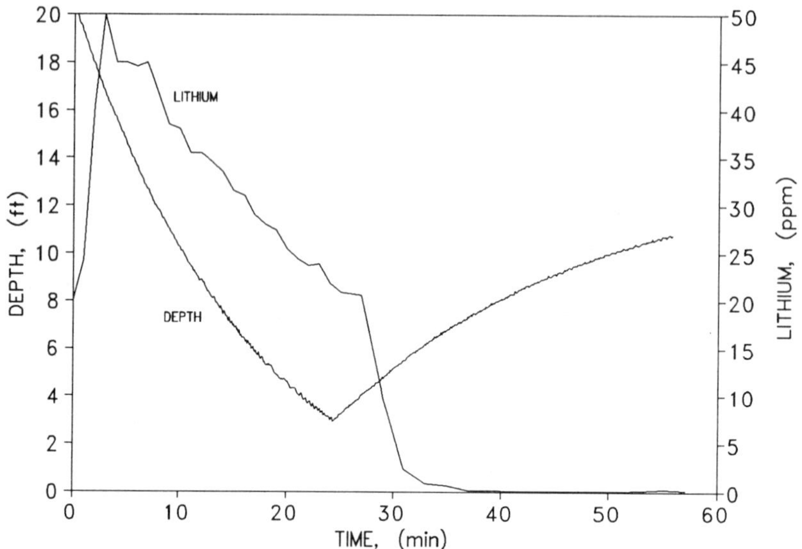

Fig. 10- Purging/sampling with drawdown

stagnant water into the pump inlet. Deionized water as a tracer in this study was generally ineffective and possibly misleading when compared to Rhodamine WT. The results of tests conducted with deionized water as a tracer (which suggest pumping 2 to 3 casing volumes from near the air/-water interface) may be inappropriately low. Real-time monitoring of indicator parameters such as pH, temperature and conductivity was not successful in indicating when purging was complete.

The inclusion of stagnant water into a sample was minimized by purging from some distance above the screen followed by relocation of the pump inlet into the screen for sample collection. In wells where drawdown occurred during purging, stagnant water inclusion was minimized by reduced pumping rates to allow for sample collection during well recharge. The ISWS/ISGS procedure for calculating the effect of drawdown was found not to be usable for the wells in this research.

REFERENCES

(1) Gibb, J.P., Schuller, R.M., and Griffin, R.A., "Procedures for the Collection of Representative Water Quality Data from Monitoring Wells", Illinois State Water Survey and Illinois State Geological Survey, Cooperative Groundwater Report 7, State of Illinois Depart-

ment of Energy and Natural Resources, 1981.

(2) Wood, W.W., Guidelines for Collection and Field Analysis of Ground-Water Samples for Selected Unstable Constituents, U.S. Geological Survey Techniques of Water Resources Investigations, Book 1, Chapter D2, 1981.

(3) Barcelona, M.J., Gibb, J.P., Helfrich, J.A., and Garske, E.E., Practical Guide for Ground-Water Sampling, Illinois State Water Survey, November 1985.

(4) Gibs, J., and Imbrigiotta, T.E., Well Purging Criteria for Sampling Purgeable Organic Compounds, Ground Water, Volume 28, Number 1, January-February 1990.

(5) Hardy, M.A., Leahy, P.P., Alley, W.M., "Well Installation and Documentation, and Ground-Water Sampling Protocols for the Pilot National Water-Quality Assessment Program", U.S. Geological Survey, Open-File Report 89-396, 1989.

(6) Slawson, G.C., Jr, Kelly, K.E., and Everett, L.G., "Evaluation of Ground-Water Pumping and Bailing Methods-Application in the Oil Shale Industry", Ground Water Monitoring Review, Summer 1982.

(7) Unwin, J.P., and Maltby, C.V., "Investigations and Techniques for Purging Ground-Water Monitoring Wells and Sampling Ground Water for Volatile Organic Compounds", Ground-Water Contamination: Field Methods, Collins, A.G., and Johnson, A.I., Eds., ASTM STP 963, 1988.

(8) Robin, M.J.L., and Gillham, R.W., "Field Evaluation of Well Purging Procedures", Ground Water Monitoring Review, Fall 1987.

(9) "Guide to Groundwater Sampling", NCASI Technical Bulletin No. 362, January 1982.

(10) Replogle, J.A., Myers, L.E., and Brust, K.J., "Flow Measurements with Fluorescent Tracers", Journal of Hydraulics Division, September 1966.

(11) Unwin, J.P., and Huis, D., "A Laboratory Investigation of the Purging Behavior of Small-Diameter Monitoring Wells", Proceedings of the Third National Symposium on Aquifer Restoration and Ground-Water Monitoring, Nielsen, D.M., Eds., National Water Well Association, May, 1983.

Ronald Schalla[1]

DEPLOYMENT AND PURGING FOR IN SITU SENSORS IN MONITORING WELLS

REFERENCE: Schalla, R., "Deployment and Purging for In Situ Sensors in Monitoring Wells," Current Practices in Ground Water and Vadose Zone Investigations, ASTM STP 1118, David M. Nielsen and Martin N. Sara, Eds., American Society for Testing and Materials, Philadelphia, 1992.

ABSTRACT: Ongoing development of in situ sensing systems will enable accurate, rapid measurement of many chemical and physical parameters at monitoring wells before the year 2000. The potential reduction of sampling and analysis costs, which are often a major portion of costs for monitoring at CERCLA and RCRA facilities, is the primary reason for the emphasis toward lower-cost in situ methods for obtaining ground-water quality data. The costs are lowered by reducing field labor, reducing or eliminating purge-water-disposal costs, and reducing the number of laboratory analyses. Relatively little has been done either to adapt new in situ technologies to the configuration of existing monitoring wells or to make deployment adaptations for minimizing purge water. Proper deployment of in situ sensor systems in monitoring wells will require an understanding of the monitoring well design with respect to the stratigraphy, lithology, hydraulics, size and configuration of in situ chemical sensor system(s), and purging requirements relative to accuracy of contaminant concentrations.

KEY WORDS: in situ sensors, sampling, deployment, measurement, ground water, monitoring well, purging, purge-water.

INTRODUCTION

Devices installed in monitoring wells to obtain chemical data by direct measurement at the location of interest are in situ chemical sensor systems. Four methods are available for in situ sensing: optical, electrochemical, mass, and thermal [1]. In situ sensor systems are available or under development for measuring the concentration of a wide variety of physical parameters, specific chemical species, or groups of chemicals in ground water [2-6]. Regardless of the sensing method, these devices require contact with the contaminated ground water or the vapor phase above water to determine the concentration of specific contaminants or groups of contaminants.

When the contaminants of concern at a site have been identified by initial characterization for each monitoring well, sensors can be very useful. Initial characterization enables selection of sensors needed to monitor and characterize the site for identified contaminants of concern.

---

[1] Senior Research Engineer, Pacific Northwest Laboratory, P.O. Box 999, Richland, WA 99352

The alternative to in situ technology is to use conventional sampling technology and analytical laboratory methods, but the enormous sample collection and analytical costs require funds that could be used for other monitoring, characterization, or remediation activities at hazardous waste sites. The potential cost savings for automated, in situ sensor systems for measuring physical parameters has been demonstrated [5,6]. Similar savings may be possible with in situ chemical sensors.

In addition to cost savings, there are two other major advantages to measuring the concentration of ground-water contaminants or physical parameters in situ: measuring the concentration in place eliminates the alteration of the sample, which may occur during the sample collection process, and in situ systems provide results immediately in the field. To achieve accuracy equal to standard sampling methods, stagnant water in the well must be purged before in situ measurement if such purging is necessary for standard sampling. This need for purging to obtain representative samples or measurements is perhaps the most significant disadvantage to in situ sensor technology because of the cost and complexity of proper purging and instrument deployment. Without purging and proper deployment of in situ sensors, obtaining "representative" measurements of chemical constituents in the ground water is often difficult or impossible because constituents are often affected by monitoring well design. The alternative to purging prior to measurement is to accept an additional, often significant amount of error that is introduced when the stagnant water is measured directly. The same conditions would apply if the sensors were in the air above the water surface to detect volatile organic compounds in the vapor phase, and Henry's law constants were used to calculate the concentration of constituents in ground water. This disadvantage offsets some of the sample integrity advantage gained by avoiding the error introduced by conventional sample collection methods.

This paper deals primarily with the sensor deployment and purge water issues related to obtaining "representative" measurements in conventional monitoring wells using conventional and in situ sensor technology.

PURGING AND DEPLOYMENT OF PUMPS FOR SAMPLING

The importance and necessity for well purging to obtain "representative" samples or measurements is based on the assumption that the quality of the water in the monitoring well casing and screened interval is not "representative" of ground water in the formation. Purge-water disposal is time consuming and costly, therefore, any way to reduce purge water without sacrificing accuracy of sample measurement is an important step forward. This purging necessity will also affect the accuracy and cost-effectiveness of in situ sensor systems. Empirical data collected and evaluated by others for cations, anions and volatile organic compounds suggest there are errors that vary from the "representative" or "stable" value by a significant percentage [7,8]. The author's own experience and those of his colleagues indicate a significant variation of measured parameters and chemical constituents with purging time, that is, casing volumes removed. For example, in Fig. 1 the vertical axis is the measured values for hexavalent chromium by both inductively coupled argon plasma emission spectroscopy (ICAP) and adsorptive stripping voltammetry method plotted against the time since pumping (i.e., purging) started in minutes. The concentration of chromium in the water pumped from the well increased dramatically from stagnant conditions indicating less than 10 ppb to "stabilization" of concentration values at approximately 75 ppb for ICAP method and approximately 58 ppb for voltammetry after 22 minutes of purging, which is equal to 2 casing volumes. Each casing volume was equivalent to 60 gallons. When electrical conductivity (i.e., specific conductance) rather than purge volume is used as a measure of purge efficiency,

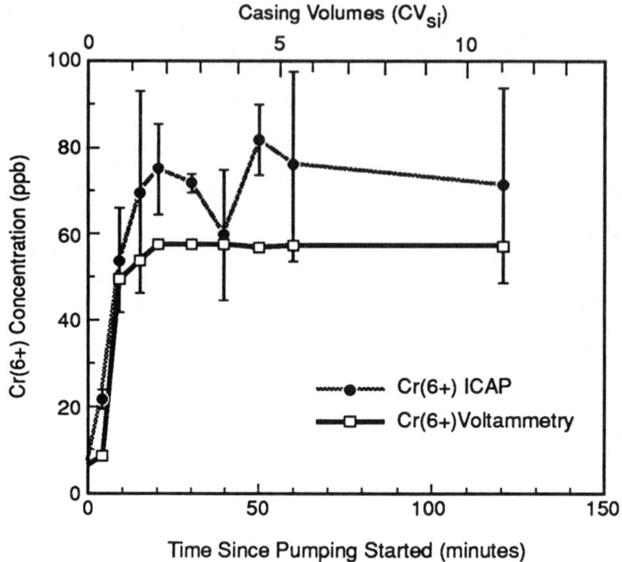

Fig. 1--Chromium concentration measured by inductively coupled argon plasma (ICAP) emission spectroscopy and adsorptive stripping voltammetry.

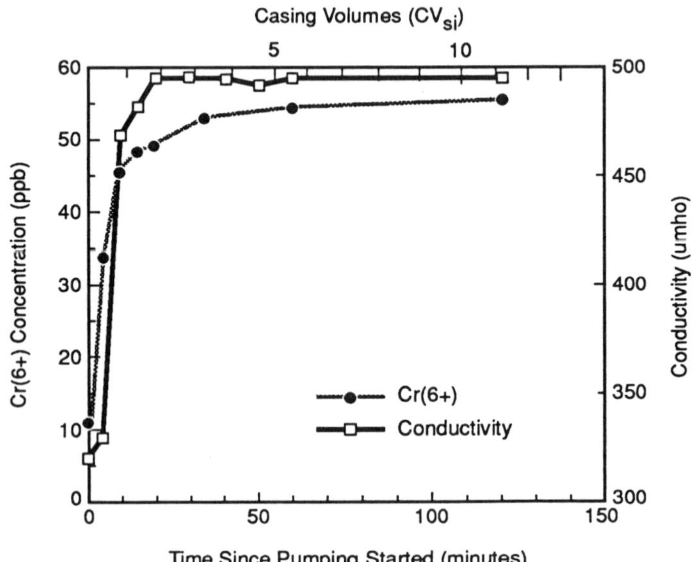

Fig. 2--Chromium concentrations with electrical conductivity.

there is a close correlation between the stabilization of specific conductance and the stabilization of chromium concentration as shown in Fig. 2. Other authors have noted similar results for other chemical parameters including volatile organic compounds [7,8].

Purging requirements have been based not only on variation in parameters, but often on removal of a number of casing volumes (CV) [9,10]. The amount of water that is to be removed from a well casing to achieve a "representative" ground-water sample that is representative of the formation is usually expressed in the number of casing volumes to be removed or purged.

The precise interpretation of the definition of CV varies, as shown in Fig. 3. The most commonly used CV includes not only the water in the screened interval, but also the water in the well casing above the screened interval if any is present. More recent studies suggest that with proper pump placement (just above the screened interval), only partial purging of the stagnant water above the screened interval is necessary to obtain a "representative" sample or measurement [11]. Also, the amount of purging required for the screened interval is commonly less than the conventional 3 equivalent volumes [7,8,11].

When water is present above the screened interval, which occurs in a well screened in a confined aquifer or screened only in the lower part of an unconfined aquifer, conventional practice has been to purge the well at a rate and for a sufficient duration to remove virtually all of the water from above the screened interval, as shown in Fig. 4.

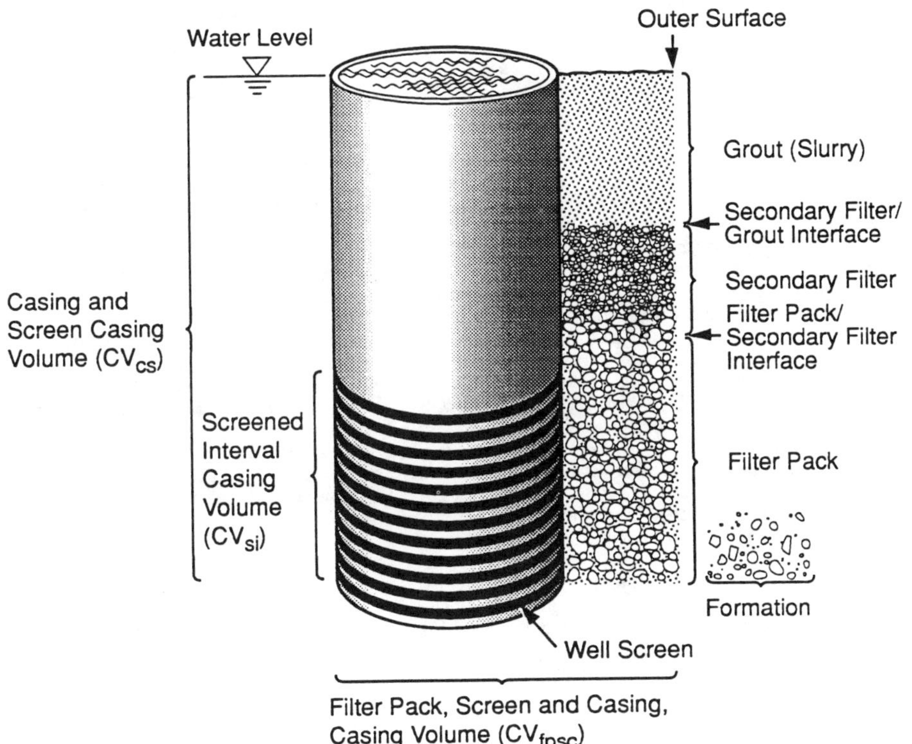

Fig. 3--Monitoring well and casing volume terminology.

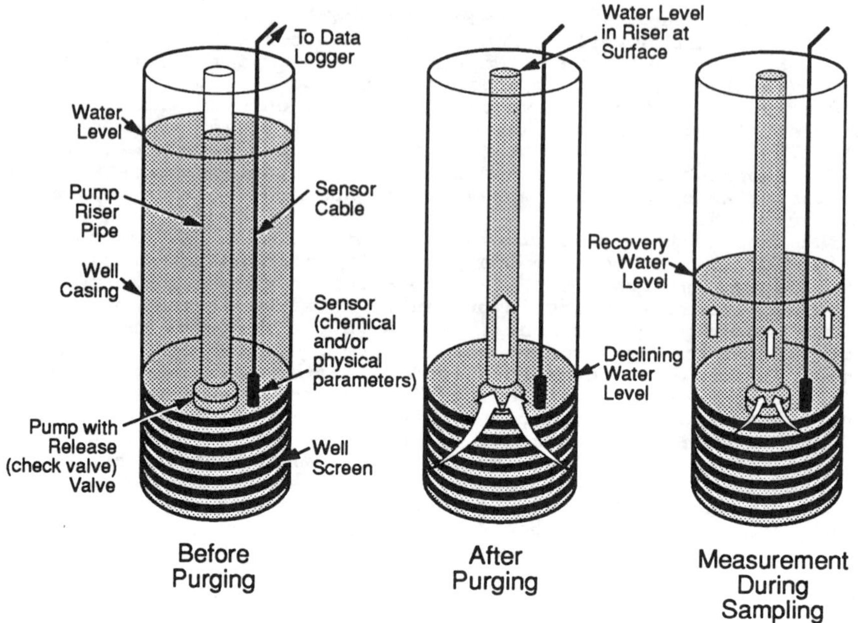

Fig. 4--Conventional purging and sampling without a packer.

This approach is not efficient because it needlessly produces large amounts of purge water that has to be disposed of, and it is not practical because with existing pump technology it is impossible to remove all of the stagnant water in the casing above the well screen. This approach also runs the risk of "breaking suction" and introducing air into the discharge line. Also, recent research demonstrates that this excessive purging is unnecessary [11].

A more effective method shown in Fig. 5 is to set the pump intake a few well casing diameters (not shown at scale because of horizontal exaggeration of well diameter) above the top of the well screen to avoid turbulence, then to purge the well at a rate that minimizes turbulent flow in the screened interval and casing. Following removal of at least 1 casing volume from the screened interval plus the amount of drawdown in the casing during purging, the purge rate is reduced to allow at least partial recovery of water level in the well. This rise in water level at the lower discharge rate will allow water entering from the formation to drive the stagnant column of water in the casing above the pump intake and allow sampling of ground water from the formation without mixing with the stagnant casing water.

When water is present only in the screened interval, such as in a well completed at the top of an unconfined aquifer, the pump must be placed near either the top or the middle of the aquifer, as shown in Fig. 6. It is generally not advisable to place the pump near the bottom of the screened interval because the proximity to the sediment trap and stagnant water contained in the trap may lead to increased turbidity in samples and possible mixing with stagnant water.

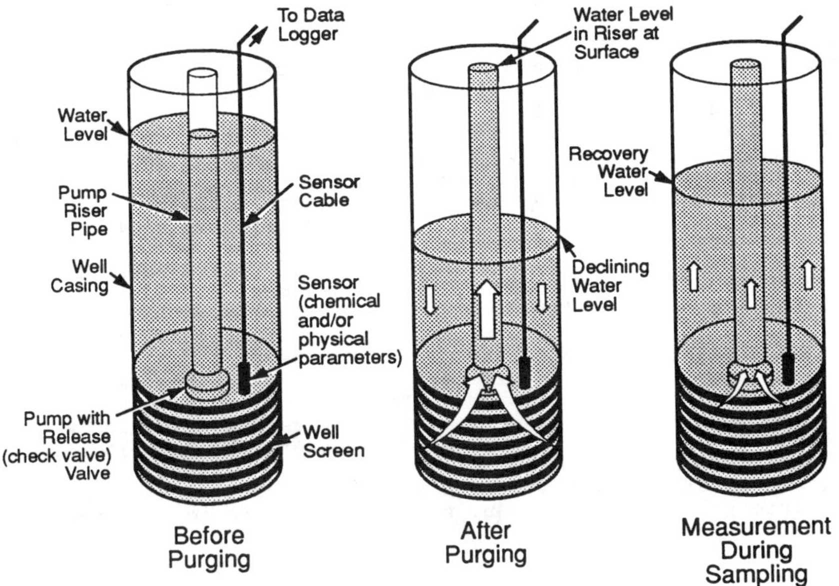

Fig. 5--More effective purging and sampling without a packer.

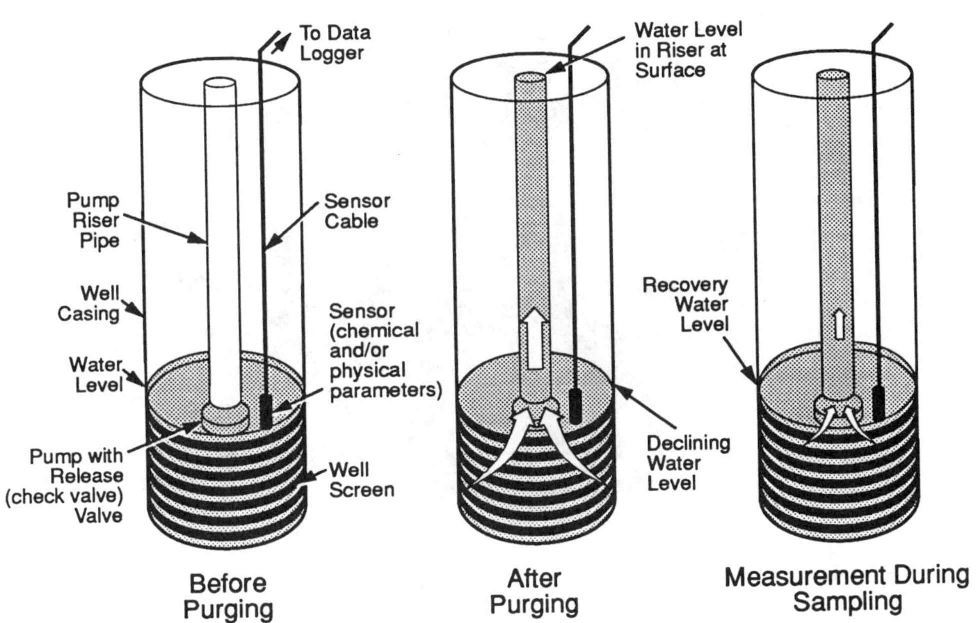

Fig. 6--Purging and sampling without a packer (unconfined).

PURGING AND DEPLOYMENT WITH IN SITU SENSORS

Sensor systems for chemical characterization and monitoring must be integrated with other technologies for effective subsurface deployment. For these integrated systems, we need to develop associated field deployment and utilization protocols and instrumentation that minimize the amount of purge water that may require expensive handling and disposal action. Use or integration with other systems and proper deployment will be necessary for in situ sensor systems to obtain accurate measurements and gain regulatory acceptance. Without purging, the accuracy of in situ sensor systems will be compromised, in some instances to an unacceptable degree.

In Figures 4, 5, and 6, the sensor is shown placed near the pump because the formation is assumed to be homogeneous, and near to the pump intake assures close correlation with actual samples collected from the pump riser. In real-world situations, this is usually the most practical approach for sensor placement. However, if the flow into the well screen is known in three dimensions, sensors can be placed adjacent to the screen (Fig. 7). This can be accomplished in two ways. First, the flow into the well can be defined by temporarily inserting a three-dimensional sensor array into the well to measure the distribution and movement of a tracer in a borehole dilution test. Once the most permeable zones of interest are identified by stratigraphy within the screened interval and confirmed by the test with the sensor array,

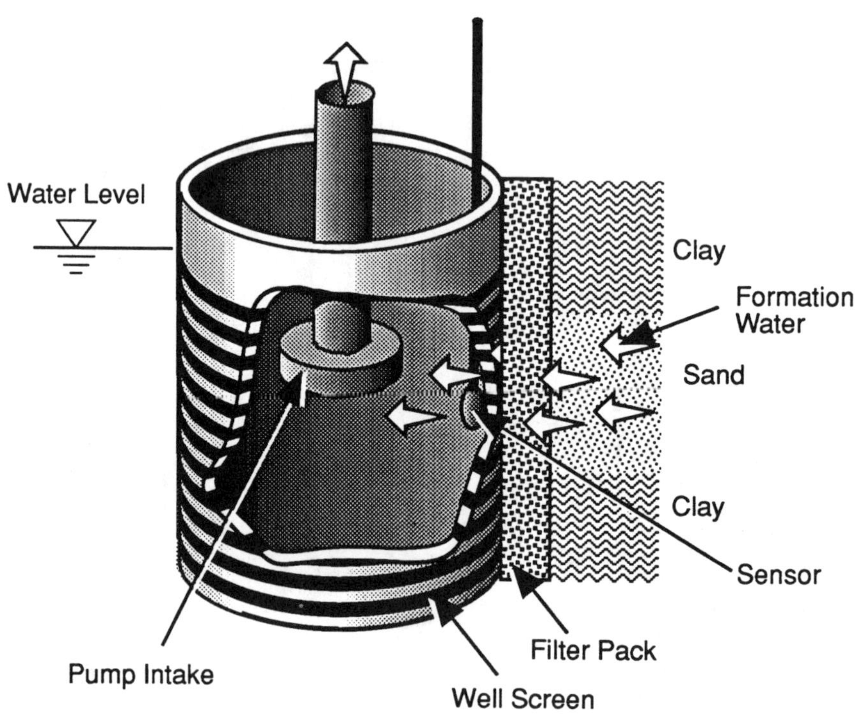

Fig. 7--Placement of sensors next to the well screen.

the permanent sensor can be placed where formation water will first enter the screened interval (Fig. 7). The nearness of the sensor to the inside diameter of the well screen reduces the amount of water needed for purging. A second alternative is to limit the amount of stagnant water in the screened interval. This can be accomplished by displacing the stagnant water with a displacement device that limits the amount of purge water to be removed (Fig. 8). The sensors can be placed outside this displacement system but within the well screen, as shown in Fig. 8. Such devices can reduce purge water by as much as 80%, as can the first alternative, but these devices offer a greater degree of certainty.

Even with these alternatives there still is a small amount of purge water. In most monitoring wells, purging volumes may be small compared to the total length of the monitoring well; therefore, the actual purge volume may be rather small compared to the storage capacity in the monitoring well casing if a packer is used. The usefulness of this storage capacity is strongly influenced by the local hydraulic gradient and hydraulic conductivity of the formation [7].

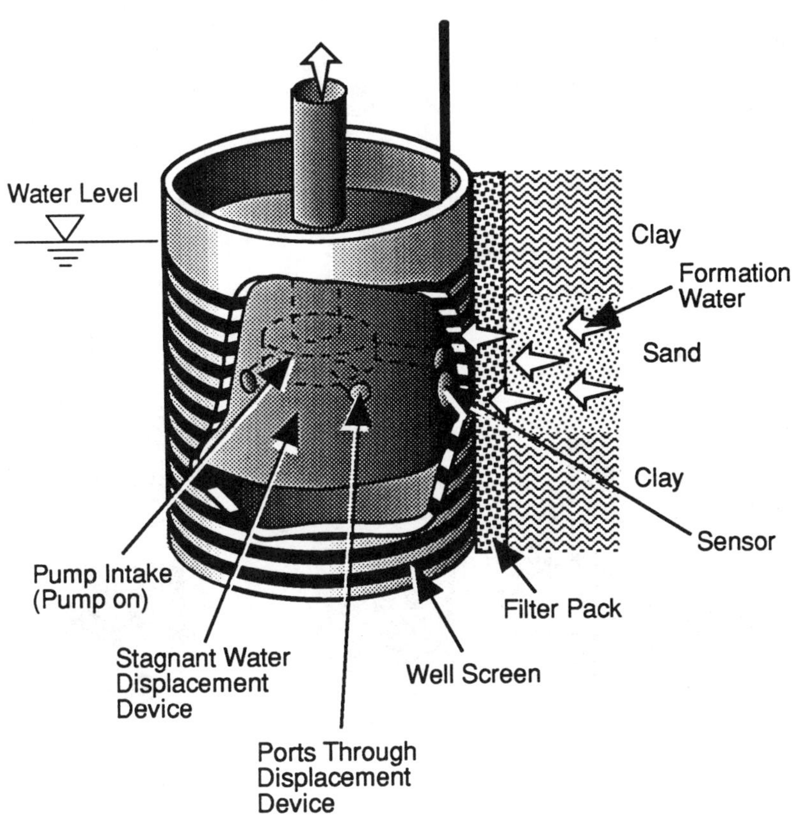

Fig. 8--Placement of sensors with water displacement device.

If the hydraulic gradient and hydraulic conductivity are not large enough, purge water released into the formation may have an impact on sample quality and frequency of sampling, because the purge water may still be adjacent to the screened interval if the sampling cycles are short; thus some of the stagnant purge water released to the formation during the previous sampling cycle may be recaptured. Adequate subsurface storage capacity is essential to avoid costs for disposal and treatment of the purge water when it is normally brought to the surface. If surface disposal is required and subsurface storage is only a temporary measure before later removal after several samplings, then using these methods to reduce purge-water volumes per sampling is essential, because, after all, the subsurface storage capacity for any well is finite.

The subsurface purging system (Fig. 9) for sensor probes would have a packer permanently inflated in the monitoring well casing. Water is purged from the screened interval and is stored above the packer. In lieu of a packer, purge water could be stored in a large-diameter riser; however, if an expandable riser was used for increased storage a packer would be needed for support (Fig. 9); then a measurement is taken and the stored purge water is released back into the formation. Again, if releasing purge water back into the formation is unacceptable from either a technical (e.g., where the screened interval has very low permeability) or a regulatory perspective (e.g., in the case of highly contaminated water), the water can simply be stored above the packer and removed later for disposal at a more convenient time. Also, not having to handle the purge water each time the well is purged is an important cost-saving and safety advantage.

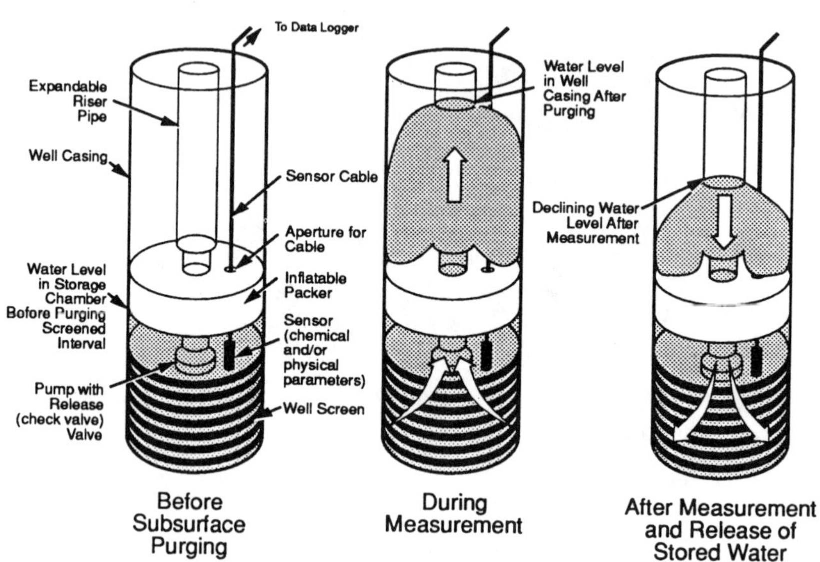

Fig. 9--Subsurface storage using a packer and expandable riser.

The displacement system shown in Fig. 8 can be added to this subsurface purging system. The primary purpose of the displacement device (Fig. 8) in this arrangement is to minimize the amount of stagnant water (i.e., water that is not representative of formation conditions) that must be removed (purged) from the screened interval in the well to obtain accurate chemical measurements from the sensor(s). Site-specific data will be needed to determine whether purging must be done after each sampling if a packer is used with this deployment configuration.

Storage in an enlarged riser rather than in the well casing above the packer eliminates contact of contaminants with the well casing. This change can be particularly important if high-concentration pulses move through the ground water and contaminate the well casing. If necessary, the storage riser can be flushed with clean water, or even removed from the well. This may be particularly important if radioactive materials are encountered.

SUMMARY

In situ sensors afford the greatest advantage for the acquisition of large quantities of water chemistry data, in real time and in a convenient manner; however, a decision must be made on the quality of the data to be obtained. If compromises in accuracy caused by not purging the monitoring well are unacceptable, then purging is required. Choosing the amount and method of purging and proper deployment of in situ sensor systems in ground-water monitoring wells requires an understanding of the design of the well with respect to the stratigraphy, lithology, hydraulics, specific in situ sampling system(s), and deployment systems, and an understanding of the water chemistry in the well relative to the formation. Purge-water minimization and subsurface storage technologies eliminate or reduce the amount of contaminated purge water that must usually be stored, treated, and disposed of and so reduces cost.

ACKNOWLEDGMENTS

This research was supported by the U.S. Department of Energy under Contract DE-AC06-76RLO1830. Pacific Northwest Laboratory is operated for the U.S. Department of Energy by Battelle Memorial Institute.

REFERENCES

[1] Janata, J. and Bezegh, A., "Chemical Sensors," Analytical Chemistry, Vol. 60, No. 12, 1988, pp 62-74R.

[2] Eccles, L. A., Simon, S. J., and Klainer, S. M., "In Situ Monitoring at Superfund Sites with Fiber Optics, I. Rationale," EPA/600/X-87/156, Las Vegas, NV, 1987.

[3] Eccles, L. A. and Simon, S. J., "In Situ Monitoring at Superfund Sites with Fiber Optics, II. Plan for Development," EPA/600/X-87/415, Las Vegas, NV, 1987.

[4] Murphy, E. M. and Hostetler, D. D., "Evaluation of Chemical Sensors for In Situ Ground-Water Monitoring at the Hanford Site," PNL-6854, Pacific Northwest Laboratory, Richland, WA, 1989.

[5] Campbell, M. D. and Schalla R., Field Tests of Automatic Water-Level Monitoring Systems: Technology Development Program - Site Investigation Technology Project," PNL-7497, Pacific Northwest Laboratory, Richland, WA, 1990.

[6] Campbell, M. D., Schalla R., and Newcomer, D. R., Accuracy and Cost Effectiveness of Manual and Automated Water-Level Monitoring Technology," PNL-7566, Pacific Northwest Laboratory, Richland, WA, 1991.

[7] Robin, M. J. L. and Gilliam R. W., "Field Evaluation of Well Purging Procedures," Ground Water Monitoring Review, Vol. 7, No. 4, 1987, pp 85-93.

[8] Gibs, J. and Imbrigiotta, T. E., "Well-Purging Criteria for Sampling Purgeable Organic Compounds," Ground Water , Vol. 28, No. 1, 1990, pp 68-78.

[9] Unwin, J. R. and Maltby, V., "Investigations of Techniques for Purging Ground-Water Monitoring Wells and Sampling Ground Water for Volatile Organic Compounds," A. G. Collins and A. I. Johnson, Eds., ASTM STP 963, 1988, pp 240-252.

[10] U.S. Environmental Protection Agency., RCRA Ground-Water Monitoring Technical Enforcement Guidance Document., U.S.EPA/OSWER-9950.1, 1986.

[11] Maltby, V. and Unwin, J. R., "A Field Investigation of Ground-Water Monitoring Well Purging Techniques," D. M. Nielsen and M. N. Sara, Eds., ASTM STP 1118, 1992.

Jeffrey W. Griffin, Khris B. Olsen

A REVIEW OF FIBER OPTIC AND RELATED TECHNOLOGIES FOR ENVIRONMENTAL SENSING APPLICATIONS

REFERENCE: Griffin, J. W. and Olsen, K. B., "A Review of Fiber Optic and Related Technologies for Environmental Sensing Applications," Current Practices in Ground Water and Vadose Zone Investigations, ASTM STP 1118, David M. Nielsen and Martin N. Sara, Eds., American Society for Testing and Materials, Philadelphia, 1992.

ABSTRACT: Realization of most fiber optic chemical sensor (FOCS) concepts is the result of numerous recent developments in fiber optic, optoelectronic, and computer technologies. For the most part, these developments have been motivated and funded by the communications industry. This paper provides a review of current technologies and an overview of FOCS concepts currently under development for environmental sensing and monitoring applications.

KEYWORDS: fiber optics, chemical sensors, biomedical sensors, environmental sensors, groundwater, vadose zone, optoelectronics, computer technology

INTRODUCTION

Motivation

During the 1970s and 1980s, there was a heightened concern about environmental issues resulting from the contamination of the environment from anthropogenic activities. One particularly notorious problem surfaced in 1976 with the disclosure that hazardous materials, improperly disposed at Love Canal, were adversely affecting the health of individuals living in that immediate area. This public awareness of environmental issues resulted in the legislative enactment of the Resource Conservation and Recovery Act (RCRA) in 1976; the Comprehensive Environmental Response, Compensation and Liability Act (CERCLA) in 1980; and renewal of CERCLA by the Superfund Amendments and Reauthorization Act (SARA) in 1986. RCRA was established to protect

J. W. Griffin and K. B. Olsen are Research Scientists at the Pacific Northwest Laboratory, which is operated for the U.S. Department of Energy by Battelle Memorial Institute under Contract DE-AC06-76RLO 1830.

human health, the environment, and to conserve environmental and energy resources. CERCLA and SARA were established to provide liability, compensation, cleanup, and emergency response for hazardous materials released into the environment and to clean up inactive hazardous waste disposal sites. One direct result of the aforementioned and other environmental legislation was a significant increase in the number of solid, liquid, and gas samples submitted to laboratories for analysis of hazardous chemicals. The sheer number of environmental samples submitted for analysis has overwhelmed the analytical laboratory system. Furthermore, these analyses cost the government and private industry hundreds of millions of dollars each year. Clearly, a reduction in the number of samples submitted for laboratory analysis would proportionately reduce costs. A reduction in the volume of samples collected during a field investigation or monitoring project could be achieved by using field screening and monitoring methods. Since the contaminants of concern for a specific investigation are usually limited to a few species or constituents, chemical sensors would provide a potential pathway to limit analytical costs. In addition, chemical sensors, either compound or class specific, could have further advantages. They could be used to measure analyte species in situ, provide real-time measurement capability, and/or be used to measure labile chemical species in the field. Furthermore, sensors have extended applications for real-time measurement of process-related parameters. The intent of this paper is to provide the environmental monitoring community with a review of the current fiber-optic-based optical sensor technology and an overview of fiber optic chemical sensor (FOCS) concepts presently under development for environmental sensing and monitoring applications.

Sensor Concepts

The term "fiber optic chemical sensor" is used to describe a broad class of chemical sensors that use an optical fiber as the communication conduit between a sensing probe and an electro-optic data acquisition system. A schematic of the concept appears in Figure 1. The end of the fiber at which the sensing probe is located is customarily called the "distal end." The "proximal end" of the fiber generally interfaces to an optoelectronic system comprised of optical sources, filters, and detectors, plus some type of (micro or mini) computer-based control and data-acquisition system. The umbilical containing the optical fiber(s) may, in some cases, also contain electrical conductors and/or tubing to supply gases, liquids, or a vacuum to the sensing probe.

The implementation of fiber optic chemical sensor concepts for environmental sensing applications offers several advantages. First, these devices present the potential for in situ, real-time monitoring of the groundwater, the vadose zone, and the atmosphere at a single site. The word "potential" is important in the previous statement because 1) there have been very few successful demonstrations of FOCS in a field environment, and 2) in some cases conventional sampling/sensing methods are entirely adequate (e.g., the use of filter infrared methods to detect and quantify atmospheric trace gases). A second latent asset of the FOCS concept is the capability for sensing a wide variety of atomic and molecular analytes. This is because a FOCS can either exploit the optical properties of the target analyte

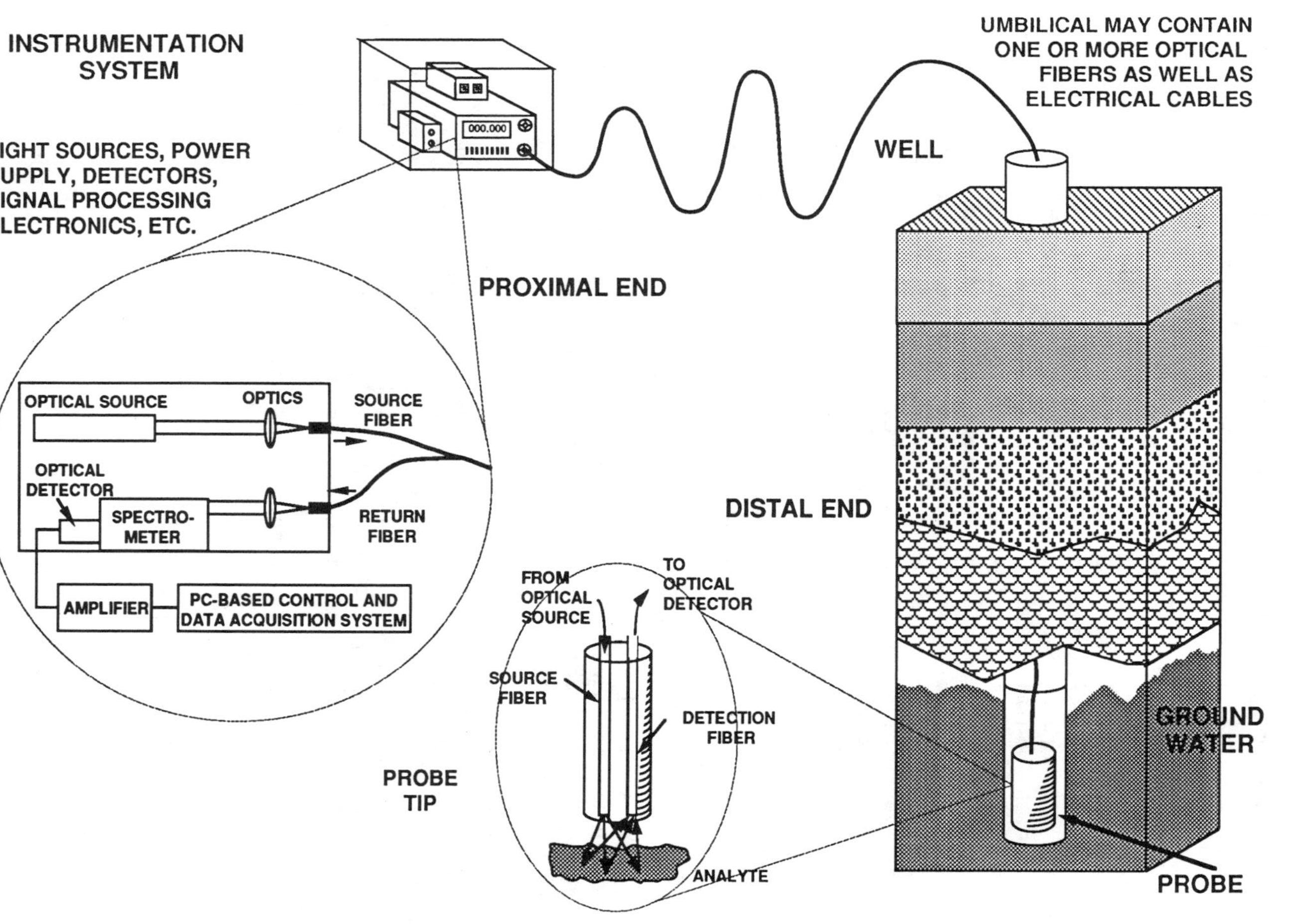

FIG. 1 -- Typical fiber optic sensor system.

directly or, through the use of a colorimetric or fluorometric reagent, the analyte can be identified and quantified indirectly through a chemical reaction at the probe tip. These attributes of the FOCS open the door for detection and quantification of a wide range of analytes, i.e., if the analyte lacks optical distinction, react it with the proper reagent and monitor the optical properties of the perturbed reagent or a reaction product. A third valuable characteristic of fiber optical chemical sensors is the demonstrated capability for multiplexing multiple sensors over a single fiber optic link. This characteristic results from the fiber optic transmission medium, not the optical or chemical properties of the probe. This feature makes possible (in principle) extended area environmental monitoring systems of much less complexity than those comprised of multiple, discrete sensor systems with their accompanying power and signal cables. Multiplexing schemes have been demonstrated for a wide variety of fiber optic physical sensors (e.g., sensors for magnetic or electric fields and temperature) but not, to the authors' knowledge, for fieldable chemical sensor systems. A final, much-touted attribute of FOCS systems is their low cost, but this is a dubious claim. This claim derives from the fact that the materials used to manufacture the sensor probe itself (optical fiber, small amounts of chemical coatings, optical components, etc.) are generally insignificant. However, if one includes the costs associated with the optical excitation/detection systems and the computer-based control and data-acquisition systems (and the associated customized software), FOCS systems typically do not appear nearly as attractive cost-wise as drawing a sample and sending it to the nearest analytical laboratory. Another important consideration is that, to date, nearly all reported fiber optic chemical sensor systems proposed for environmental monitoring require significantly more development if they are to be considered for extended field deployment. If these development costs are considered as a component of the sensor cost, it is clear that the "cost advantage" of FOCS may be a myth.

Before continuing with a review of recent FOCS research and development activities, it is important to review the desirable traits of a fieldable chemical sensor system suitable for environmental monitoring applications. These traits include portability (preferably battery-powered), ruggedness, good chemical sensitivity (typically the lowest detection limit possible), wide dynamic range (ideally the sensor should be capable of measuring the entire range of analyte concentrations anticipated), a high degree of chemical specificity, excellent measurement accuracy and precision, low cost, and simple operation and maintenance procedures. These traits should be kept in mind during the following discussions of the present state of FOCS development and the associated fiber optic, optoelectronic, and data processing technology building blocks.

SUMMARY OF RECENT WORK

Over the past five years there has been significant technical activity in the development of fiber optic chemical sensors. An appreciation for the extent of this recent work can be obtained from Reference List A [1 through 26]. However, despite the large number of research publications, there are only a handful of commercially available FOCS. The field of greatest commercial activity appears to be medical diagnostics, Reference List B [27 through 61]. There are several reasons for this. First, since fiber optic probes are fabricated from relatively inert, non-conducting materials (glasses and plastics), they allow the medical diagnostician to probe the human body without electrical or chemical hazards. Second, the payback (or profit) associated with medical instrumentation appears to be more than sufficient to attract sensor development funds (especially private sector venture capital). On the other hand, there appears to be some reluctance on the part of the government to invest in new methods and technologies for environmental chemical sensors, accompanied by an insufficiency of research and development monies. Once the private sector takes an interest, however, this situation may very well change. A third major reason for the advancement of medical FOCS devices is largely technical and is related to the environment in which the sensor must operate. For the human body, the chemical and physical parameters of the operating environment are well known (e.g., dissolved oxygen, pH, conductivity, temperature, trace chemical species, etc.) and the range of anticipated values (for a living person) is generally narrow. Consequently, the probability of the sensor being "fooled" by some extraneous chemical interferant is very low. Such is not the case for environmental sensors. Each field site may offer an entirely different array of chemical constituents present in widely varying concentration levels. In this situation the probability of sensor failure from chemical interferences can be very high. Consequently, extensive FOCS characterizations <u>at each field site</u> (or within each sample matrix) are necessary before confidence will be placed in the sensor output.

Currently, the major research activities in FOCS development for environmental monitoring are concentrated in only a few university, national laboratory, and industrial research groups. These groups, for the most part, are focusing their activities on detection of hazardous chemical species in the groundwater and the vadose zone. Species of interest include trichloroethylene (TCE), chloroform, carbon tetrachloride, petroleum products (e.g., jet fuel and gasoline) and metal ions. A review of these research activities can be obtained from the Reference List C [62 through 74]. The balance of this paper will address the optical fiber, optoelectronic, and computer technologies that have made all of these research efforts possible.

REVIEW OF CURRENT FIBER OPTIC, OPTOELECTRONIC, AND MINI-COMPUTER TECHNOLOGIES

There are a number of relatively new technologies that have contributed to the development of fiber optic chemical sensors. For the most part, these technologies have been developed at the expense of (and for the benefit of) the communications industry. In particular, current FOCS systems take advantage of 1) the wide variety of commercial optical fiber materials and designs, 2) availability of miniature optical components (especially gradient-index optics), 3) inexpensive optical sources (especially gas and solid-state lasers), 4) improved optical detectors, and 5) inexpensive portable control and data-acquisition equipment.

Optical fiber technology has seen some very significant advances in the past 10 years [75]. Silica glass fibers are now available at low cost and in a wide range of diameters. These fibers are used when the FOCS is to be operated in the near-ultraviolet to near-infrared spectral region (250 nm to 1200 nm). This type of fiber is essential for implementation of many fluorescence-based chemical sensor designs where significant fluxes of ultraviolet photons are necessary for optical excitation at the sensor tip. For less demanding applications, plastic fibers are finding wide application, their major advantages deriving from their low cost. While the optical losses are higher for plastic fibers, their chemically inert nature and their resistance to water are particularly well-suited for measurements in aqueous solutions over extended durations. Both silica glass and plastic fiber technologies take advantage of a relatively new development in miniature optical focusing components. These components, called gradient-index (or GRIN) optics, have a cylindrical geometry identical to optical fibers and can be fabricated at dimensions comparable to the fiber diameter. This is a very important feature when it becomes necessary to minimize the FOCS sensing probe volume (e.g., in biomedical applications). The recent introduction of fluoride-glass fibers has opened a new horizon for fiber optic chemical sensors. At this writing, zirconium-fluoride fibers are now commercially available that exhibit good optical transmission at wavelengths spanning the visible to the infrared (beyond 4 microns). Fluoride-glass fibers allow exploitation of fundamental and overtone molecular vibration-rotation bands for chemical sensing schemes. However, these fibers are not particularly well-suited for aqueous measurements since the fiber material is soluble in water. Another important class of optical fibers currently under development is fabricated from chalcogenide glasses (e.g., germanium-arsenide-selenide). Although commercial fibers are not yet available (short lengths of prototype materials are available for evaluation), these materials will eventually enable the spectral band from 2 to 12 microns to be exploited for FOCS development. This spectral region allows access to a number of fundamental molecular vibration-rotation bands and will find wide application, especially for detection and monitoring of gaseous molecular species.

Recent developments in optoelectronic technology have also been crucial to the development of today's fiber optic chemical sensors [76, 77]. Developments in this category have been driven by the communication and defense industries led largely by a collection of small entrepreneurial research and development firms. To understand the importance of these optoelectronic developments, it must be understood that the starting point for virtually all fiber optic chemical sensing systems consists of "stuffing light into the fiber." If the source is an extended blackbody (e.g., a tungsten lamp), energy transfer considerations dictate that the fiber have a large diameter in order to get sufficient optical energy to the sensing probe (the distal end). Such a requirement is often not acceptable or feasible. New solid-state sources such as light-emitting diodes (LEDs), superluminescent diodes (SLDs), and diode lasers are ideally suited to fiber optic sensor applications because of their small size and the ease with which they may be interfaced with optical fibers. The LEDs and SLDs have relatively broadband optical outputs (tens of nanometers) and are available in a range of wavelengths spanning the visible and near infrared. Improved spectral purity and higher spectral radiances are obtainable with diode (semiconductor) lasers. Heretofore these devices were only available in near infrared wavelengths (700 to 1600 nm), but recently visible devices operating in the red spectral region (670 nm) have become available. In addition to solid-state optical sources, a variety of low-power, low-cost gas lasers suitable for FOCS applications are now available. An example is the helium-neon laser that can be obtained at a number of discrete visible and near-infrared wavelengths. The optical power output of these devices is generally limited to less than 1 milliwatt; however, this is often sufficient for chemical sensing applications. The small size of these lasers and the capability for battery operation are major assets when designing fieldable FOCS systems.

Recent developments in optical detectors have also had an impact on FOCS development. Currently, the most popular detectors for fiber-optic applications are silicon devices. Depending on the required frequency bandwidth and detector area, they can be obtained in conventional photodiode, positive-intrinsic-negative (PIN), or avalanche photodiode (APD) configurations. The wavelength response of these devices (400 to 1100 nm) is well-matched to the spectral outputs of many LEDs, SLDs, and diode lasers. Low-cost linear and two-dimensional silicon detector arrays have greatly simplified spectral discrimination in FOCS optical detection schemes. These devices allow nearly instantaneous acquisition of fiber optical output over a predetermined spectral range, thereby eliminating the need for scanning diffraction gratings, prisms, or optical filters. These arrays also expedite the processes of co-adding spectral scans and background subtraction resulting in much-improved signal/noise ratios. If wavelength response beyond 1100 nm is required (e.g., when using zirconium-fluoride fibers), there are also now available a variety of single- and multi-element solid-state photoconductive/photovoltaic and pyroelectric detectors. Many of the currently available optical detectors can be obtained as hybrid units with an integral pre-amplifier, optical bandpass filter, and thermoelectric cooler (for temperature stabilization) in the same package, resulting in less output drift, lower electronic noise, and improved reliability.

Computer technology has also enhanced the attractiveness of fieldable FOCS systems. It is now possible to purchase inexpensive computer systems (e.g., "laptops") with low power consumption, fast data processing speeds, and real-time displays. The flexibility of these computer systems is enhanced by a wide variety of available data acquisition and data processing software as well as a plethora of compatible data acquisition and control add-on boards. This new computer technology is especially valuable when it is necessary to process FOCS spectra, for example, from a linear silicon array. Not only can the spectra be acquired rapidly, stored, co-added, and manipulated, but processing of the spectra with partial least squares algorithms for pattern recognition is also facilitated [78].

CHEMICAL SENSING MECHANISMS

To provide a justification for the current activity in FOCS development, it is appropriate to review the optical sensing schemes incorporated in these devices. This review should also point out the importance of the technology developments described above. Note that very few of these sensing schemes are really new. For the most part they have been practiced in the laboratory for many years. The resurgence in interest in these techniques derives from the potential for implementing these schemes remotely, i.e., at the distal end of an optical fiber. This important capability allows in situ measurement, whereas, in the past, the sample had to be collected and transferred to the laboratory.

Current FOCS concepts generally implement one of eight chemical sensing mechanisms. These include optical fluorescence, colorimetry, optical scattering (elastic and inelastic), spectrochemical emission, infrared absorption, attenuated total internal reflection (evanescent wave absorption), refractometry, and chemically induced mechanical perturbations. Brief descriptions of these methodologies follow with appropriate references.

Optical Fluorescence

Fluorescence-based FOCS utilize the optical emission from a fluorescent compound to perform their sensing function. The fluorescent compound may be the analyte of interest, a compound formed via reaction between the analyte and a probe-borne reagent, or an inert compound whose fluorescence properties are altered by the presence of analyte. Fluorescence excitation is generally achieved by transmitting ultraviolet or visible optical radiation to the sensing volume via an optical fiber. Fluorescent emissions are then transmitted back to the detection system through a second fiber. Alternatively, a single fiber can be used to transmit both the excitation and fluorescence radiation in counter-propagating directions. An example of this method is provided by the work of Zung, et al. [79]. In their paper the authors report the use of a fluorescence-based FOCS to monitor emissions from blue-green algae when perturbed by small quantities of analyte.

Colorimetry

Colorimetric fiber sensors generally exploit color changes induced in a probe-borne reagent by the target analyte. Once again, the source radiation to the sensor probe is usually provided through one optical fiber while sample-transmitted light is transported back to the detection system via another fiber. In some cases illumination and return optical signals are conveyed in opposite directions over the same fiber. An example of this type of sensor is provided by the work of Angel and Ridley [80], who have developed a sensor for TCE and chloroform. Their sensor implements two fibers--a liquid (colorimetric) reagent and a semi-permeable Teflon membrane that allows diffusion of analyte into the reagent volume. One shortcoming of sealed colorimetric sensors is that the reagent is eventually depleted, necessitating frequent replacement of the probe tip. This problem has been remedied by Berman and Burgess [82], who have designed a probe incorporating a reagent delivery capillary and a drain tube. Prior to initiating a measurement, the probe is remotely flushed and filled with fresh colorimetric reagent. This FOCS scheme was demonstrated for the detection of ammonia.

Optical Scattering

There are two types of fiber optic scattering sensors. The first is based on modulation of probe-backscattered light caused by interaction with an analyte (elastic scattering). In this situation the scattered light is of the same wavelength as the illumination light. Sensors of this type are called "intensity sensors" and are typically very simple to construct. For example, one might monitor the diffuse reflectance of a paper target to detect the presence of liquid water. The other type of optical scattering sensor is based on the detection of inelastic photon scattering by molecules. Inelastic scattering processes result in a spectral shift of the scattered photons from their initial wavelength. This method is more commonly known as Raman spectroscopy. Implementation of this method always requires the use of a laser source and generally necessitates the use of a high-resolution spectrometer and photon-counting detection instrumentation because of the low scattered photon fluxes. This method has been used by Myrick and Angel [82] for the detection of benzene. In some cases the Raman scattering of molecular species is significantly enhanced if they are deposited onto metal surfaces (e.g., copper, silver, or gold) prior to excitation. This measurement methodology is termed surface enhanced Raman spectroscopy (SERS). Angel and Myrick [83] have demonstrated this technique over an optical fiber using a near-infrared diode laser source. The analyte was tris(2,2'-bipyridine)ruthenium(II), commonly designated as RB3, in bulk solution. The SERS method may be extendable to environmentally important analytes in the future.

Spectrochemical Emission

This unique method implements spectrochemical emission analysis over an optical fiber. In this technique an electrically generated plasma serves as the excitation source for the analyte. This plasma may be generated either via electrical spark (suitable for solid and liquid aerosols, vapors, and liquids) or by radio-frequency induction using helium as the working gas (suitable for vapors only). The optical fiber is simply used as a light conduit to convey the analyte emissions back to the detection system. The umbilical carrying the fiber also carries electrical power, working gas, vacuum, etc., to the probe tip. Griffin et al. have demonstrated this method for detection of chlorine (from carbon tetrachloride) in air [84, 85].

Infrared Absorption

The recent introduction of zirconium fluoride optical fibers has made it possible to implement FOCS at near-infrared wavelengths. An example of this implementation is provided by the recent work of Matson and Griffin [86]. In their paper they describe and demonstrate an infrared fiber optical chemical sensor concept for the detection of explosive hydrocarbons via the measurement of optical absorption attributed to fundamental carbon-hydrogen stretching vibrations in the 3.3 micron region. The optical source used was a helium-neon laser operating at 3.39 microns. The report describes hydrocarbon detection using three FOCS configurations. The first method uses a photoacoustic cell; the second, a conventional dual-pass absorption cell; and the third uses a commercially available fourier transform infrared (FTIR) spectrometer.

Attenuated Total Internal Reflection

The attenuated total internal reflection method relies on the detection of optical absorption caused by analyte that has diffused into the fiber cladding material. The absorption is attributed to the fact that, while photons in a fiber optic are constrained to propagate in the fiber core, the electric field caused by these photons does extend for some distance into the cladding material ("the evanescent tail") where it may be partially attenuated if the appropriate chemical species are present. This method has become a standard laboratory technique for the characterization of optically dense liquids and slurries using a planar waveguide geometry but has only recently been demonstrated as a FOCS detection scheme. A sensor that implements this scheme has been demonstrated by DeGrandpre and Burgess [87], who have used a silica fiber with a polysiloxane cladding to detect hexane, ethanol, chloroform, and carbon tetrachloride. The measurements were performed in the near-infrared using a FTIR spectrometer system.

Refractometry

Chemical detection via refractometry relies on removing light propagating in the optical fiber core by modifying the optical properties of the cladding material when it comes into contact with a target analyte. Light propagation through an optical fiber requires that the refractive index of the cladding material be (slightly) lower than that of the core material. When this relationship is not obeyed, light will leave the fiber core, propagate into the cladding, and eventually be lost from the fiber entirely because of scattering at the cladding/air interface. A sensor that uses this scheme for the detection of gasoline and gasoline constituents (liquid and/or vapor phases) has been demonstrated by Klainer et al. [88]. Their sensor probe is comprised of a conventional optical fiber clad with a proprietary coating, which changes its index of refraction when gasoline constituents are allowed to diffuse into it.

Chemically Induced Mechanical Perturbations

Butler et al. [89] report a sensor concept for volatile organic compounds that exploits the swelling property of a plasma-polymerized fluorocarbon film deposited onto the end of a mirrored optical fiber. This film, in conjunction with an adjacent chromium coating, forms a Fabry-Perot etalon whose reflectance (at a single wavelength) is a function of the coating thickness. As organic compounds diffuse into the coating, its thickness changes (i.e., it swells) resulting in a change of reflected light intensity at the FOCS system proximal end. The sensor is reversible because the fluorocarbon film returns to its original thickness once the absorbed compound is allowed to diffuse out.

CONCLUSIONS

From the above discussions, it should be obvious that the viability of fiber optic chemical sensors for environmental monitoring applications has largely been and will continue to be technology-driven. Quantitative, fieldable fiber optic chemical sensors are not well developed. The sensor types closest to commercial implementation are targeted to the detection of chlorinated organics and gasoline constituents in groundwater. The main limitation of FOCS technology today is the dependence upon chemical transduction schemes, which are all subject to chemical and physical interferences, thus lacking compound or compound class specificity . For this reason, fiber optic chemical sensor development should, in the future, focus on the development of more efficient, chemically specific transduction methods (i.e., chemical reactions). This needs to be accomplished in the laboratory before considering implementation via fiber optic technology.

ACKNOWLEDGMENTS

The authors would like to express their appreciation to J. A. Nolting for producing the drawings used in the oral presentation and the text of this paper.

REFERENCES

Reference List A

[1] Guthrie, A. J., Narayanaswamy, R., and Russell, D. A., "Optical Fibres in Chemical Sensing - A Review," Transactions of the Institute of Measurement and Control, Vol. 9, No. 2, April-June 1987, pp. 71-80.

[2] Goswami, K., Kennedy, J. A., Dandge, D. K., Klainer, S. M., and Tokar, J. M., "A Fiber Optic Chemical Sensor for Carbon Dioxide Dissolved in Sea Water," Chemical, Biochemical and Environmental Fiber Sensors, Proc. SPIE 1172, Bellingham, Washington, 1990, pp. 225-232.

[3] Schnitzer, I., Katzir, A., Schiessl, U., Riedel, W., and Tacke, M., "Fiber-Optic-Based Evanescent Field Infrared Spectroscopy Using Tunable Diode Lasers," Infrared Fiber Optics, Proc. SPIE 1048, Bellingham, Washington, 1989, pp. 133-140.

[4] Zhang, Y., Grant, C. L., Russell, R., Zhujun, Z., Ma, W., Seitz, W. R., and Sundberg, D., "Polyvinyl Alcohol as a Matrix for Indicators for Fiber Optic Chemical Sensors," Proceedings of the Ninth Annual Conference of the IEEE Engineering in Medicine and Biology Society; available from IEEE Service Center (Catalog No. 87CH2513-0), Piscataway, New Jersey, 1987, pp. 800-801.

[5] Shahriari, M. R., Zhou, Q., and Sigel, G. H., Jr., "Porous Optical Fibers for High-Sensitivity Ammonia-Vapor Sensors," Optics Letters, Vol. 13, No. 5, May 1988, pp. 407-409.

[6] Guiliani, J. F., and Bey, P. P., Jr., "Computer Modelling for Optical Waveguide Sensors," Applied Optics, Vol. 27, No. 7, 1 April 1988, pp. 1353-1357.

[7] Schaffar, B. P. H., and Wolfbeis, O. S., "Fibre Optic Chemical Sensors," DOPS-NYT, Vol. 2, No. 4, October 1987, pp. 12-15.

[8] Wlodarczyk, M. T., Vickers, D. J., and Kozaitis, S. P., "Evanescent Field Spectroscopy with Optical Fibers for Chemical Sensing," Fiber Optic and Laser Sensors IV, Proc. SPIE 718, Bellingham, Washington 1987, pp. 192-196.

[9] Miller, H. H., and Hirschfeld, T. B., "Fiber Optic Chemical Sensors for Industrial and Process Control," Fiber Optic and Laser Sensors IV, Proc. SPIE 718, Bellingham, Washington, 1987, pp. 39-45.

[10] Yuan, P., and Walt, D. R., "Calculation for Fluorescence Modulation by Absorbing Species and Its Application to Measurements Using Optical Fibers," Analytical Chemistry, Vol. 59, No. 19, 1 October 1987, pp. 2391-2394.

[11] Scheggi, A. M., and Baldini, F,, "pH Sensing by Fibre Optics," Optica Acta, Vol. 33, No. 12, December 1986, pp. 1587-1597.

[12] Peterson, J. I., "Fiber Optic Chemical Sensors - A View from the Past to the Future," Proceedings of the Symposium on Biosensors; available from IEEE Service Center (Catalog No. 84CH2068-5), Piscataway, New Jersey, 1984, pp. 35-39.

[13] Druy, M. A., Elandjian, L., and Stevenson, W. A., "Autoclave Monitoring of Composite Resin Chemistry in Laminates with an In-Situ Fiber Optic Polymer Reaction Monitor," National SAMPE Symposium and Exhibition (Proceedings), Vol. 35, No. Book 2, SAMPE, Covina, California, 1990, pp. 1517-1522.

[14] Tokar, J. M., Woodward, W. E., and Goswami, K., "Measurement of Oceanic Biogeochemical Compounds Using Fiber Optic Chemical Sensors," Oceans '89. Part 5: Diving Safety & Physiology; Ocean Engineering/Technology; available from IEEE Service Center, Piscataway, New Jersey (Catalog No. 89CH2780-5), Piscataway, New Jersey, 1989, pp. 1360-1365.

[15] DeGrandpre, M. D., and Burgess, L. W., "Fiber-Optic FT-NIR Evanescent Field Absorbance Sensor," Applied Spectroscopy, Vol. 44, No. 2, February 1990, pp. 273-279.

[16] Badini, G. E., Grattan, K. T. V., Palmer, A. W., and Tseung, A. C. C., "Fiber Optic Chemical Sensor Based on Immobilized Fluorescent Dye Techniques," Summaries of Papers Presented at the Conference on Lasers and Electro-Optics; available from IEEE Service Center (Catalog No. 89CH2638-5), Piscataway, New Jersey, 1989, pp. 106-107.

[17] Blyler, L. L., Jr. (Ed.), "ANTEC '88 - Plastics in Lightwave Technology," Polymer Engineering and Science, Vol. 29, No. 17, Society of Plastics Engineers, Brookfield Center, Connecticut, Mid-September 1989, pp. 1159-1218.

[18] Williams, M. C., and Cicero, D. C., "Department of Energy's Program for Characterizing PFBC Exhausts with the METC Fiber-Optic Alkali Monitor," American Society of Mechanical Engineers, ASME, New York, JPGC/EC11 10, 1988.

[19] Zhang, Z., Zhang, Y., Ma, W., Russell, R., Shakhsher, Z. M., Grant, C. L., Seitz, W. R., and Sundberg, D. C., "Poly(vinyl alcohol) as a Substrate for Indicator Immobilization for Fiber-Optic Chemicals," Analytical Chemistry, Vol. 61, No. 3, February 1, 1989, pp. 202-205.

[20] Bacci, M., Baldini, F., Brenci, M., Conforti, G., Falciai, R., Mignani, A. G., and Scheggi, A. M., "Model for an Optical Fiber pH Sensor," Optical Fibers in Medicine II, Proc. SPIE 713, Bellingham, Washington, 1987, pp. 88-92.

[21] Grattan, K. T. V., Mouaziz, A., and Palmer, A. W., "Dual Wavelength Optical Fibre Sensor for pH Measurement," Biosensors, Vol. 3, No. 1, 1987, pp. 17-25.

[22] Tenge, B., Grunke, A. Q., and Honigs, D. E., "Fiber Optic 'Chemfuse' Sensors," InTech, Vol. 34, No. 10, October 1987, pp. 29-33.

[23] Harris, J. M., Hunnicutt, M. L., Carr, J. W., and Weaver, M. R., "Spectroscopic Studies of the Silica/Solution Interface," Electrochemical Society Extended Abstracts, Vol. 85-2, Electrochemical Society, Pennington, New Jersey, 1985, p. 488.

[24] Saari, L. A., "Trends in Fiber Optic Sensor Development," TrAC, Trends in Analytical Chemistry (Personal Edition), Vol. 6, No. 4, April 1987, pp. 85-90.

[25] Munkholm, C., Walt, D. R., Milanovich, F. P., and Klainer, S. M., "Polymer Modification of Fiber Optic Chemical Sensors as a Method of Enhancing Fluorescence Signal for pH Measurement," Analytical Chemistry, Vol. 58, No. 7, June 1986, pp. 1427-1430.

[26] Arnold, M. A., and Ostler, T. J., "Fiber Optic Ammonia Gas Sensing Probe," Analytical Chemistry, Vol. 58, No. 6, May 1986, pp. 1137-1140.

Reference List B

[27] Munkholm, C., Yuan, P., Luo, S., and Walt, D. R., "Fiber Optic Biosensors for In Vivo and In Vitro Measurements of Clinically Important Species," Proceedings of the 9th Annual Conference of the IEEE Engineering in Medicine and Biology Conference; available from IEEE Service Center (Catalog No. 87CH2513-0), Piscataway, New Jersey, 1987, pp. 802-803.

[28] Walt, D. R., Munkholm, C., Jordan, D., Milanovich, F. P., and Daley, P. F., "Preparation of Small Diameter Sensors for Continuous Clinical Monitoring," Optical Fibers in Medicine II, Proc. SPIE 713, Bellingham, Washington, 1987, pp. 144-149.

[29] Narayanaswany, R., "Design of Optical Sensors for Biomedical Applications," Colloquium on Electrochemical Sensors for Biomedical Applications, IEEE, London, 1986, p. 10.

[30] Junker, B. H., Wang, D. I. C., and Hatton, T. A., "Fluorescence Sensing of Fermentation Parameters Using Fiber Optics," Biotechnology and Bioengineering, Vol. 32, No. 1, June 1988, pp. 55-63.

[31] Takatani, S., Noda, H., Kohno, H., Takano, H., and Akutsu, T., "Optical Sensor for HB Content and O//2 Saturation: Design, In Vitro and In Vivo Evaluation," Proceedings of the Ninth Annual Conference of the IEEE Engineering in Medicine and Biology Conference; available from IEEE Service Center (Catalog No. 87CH2513-0), Piscataway, New Jersey, 1987, pp. 810-812.

[32] Martin, M. J., Wickramasinghe, Y. A. B. D., Newson, T. P., and Crowe, J. A., "Fibre-Optics and Optical Sensors in Medicine," Medical & Biological Engineering & Computing, Vol. 25, No. 6, November 1987, pp. 597-604.

[33] Vo-Dinh, T., Tromberg, B. J., Griffin, G. D., Ambrose, K. R., Sepaniak, M. J., and Gardenhire, E. M., "Antibody-Based Fiber-optics Biosensor for the Carcinogen Benzo(a)pyrene," Applied Spectroscopy, Vol. 41, No. 5, July 1987, pp. 735-738.

[34] Regnault, W. F., and Picciolo, G. L., "Review of Medical Biosensors and Associated Materials Problems," Journal of Biomedical Materials Research, Vol. 21, No. A2, Suppl., August 1987, pp. 163-180.

[35] Klainer, S. M., and Harris, J. M., "The Use of Fiber Optic Chemical Sensors (FOCS) in Medical Applications: Enzyme-Based Systems," Optical Fibers in Medicine III, Proc. SPIE 906, Bellingham, Washington, 1988, pp. 139-147.

[36] Kulp, T. J., Camins, I., and Angel, S. M., "Enzyme-Based Fiber Optic Sensors," Optical Fibers in Medicine III, Proc. SPIE 906, Bellingham, Washington, 1988, pp. 134-138.

[37] Walt, D. R., Shufang, L., and Munkholm, C., "Fiber Optic Chemical Sensors Using Immobilized Bioreceptors," Optical Fibers in Medicine III, Proc. SPIE 906, Bellingham, Washington, 1988, pp. 60-64.

[38] Ping, Y., Jordan, D., Milanovich, F. P., and Walt, D. R., "Fibre-Optic Chemical Sensors Based on Energy Transfer," Optical Fibers in Medicine III, Proc. SPIE 906, Bellingham, Washington, 1988. pp. 28-29.

[39] Villarruel, C. A., Dominguez, D. D., and Dandridge, A., "Evanescent Wave Fiber Optic Chemical Sensor," Fiber Optic Sensors II, Proc. SPIE 798, 1987, pp. 225-229.

[40] Tarassenko, L., and Stange, U., "Review of Optical Sensors for Biomedical Applications," IEEE Colloquium on 'Electrochemical Sensors for Biomedical Applications' (Digest No. 51), London, 10 April 1986, pp. 9/1-9/5.

[41] Bluestein, B. I., Walczak, I. M., and Chen, S-Y., "Fiber Optic Evanescent Wave Immunosensors for Medical Diagnostics," Trends in Biotechnology, Vol. 8, No 6, June 1990, pp. 161-168.

[42] Merlo, S., Yager, P., and Burgess, L. W., "Optical Method for Detecting Anesthetics and Other Lipid-Soluble Compounds," Sensors and Actuators, A: Physical, Vol. 23, No. 1-3, April 1990, pp. 1150-1154.

[43] Poscio, P., Depeursinge, C., Vorin, C., Sheja, B., and Parriaux, O., "Realization of a Miniaturized Optical Sensor for Biomedical Applications," Sensors and Actuators, A: Physical, Vol. 23, No. 1-3, April 1990, pp. 1092-1096.

[44] Merlo, S., Burgess, L., and Yager, P., "Development of a Fluorescence-Based Fiber Optic Sensor for Rapid Detection of General Anesthetics and Other Lipid-Soluble Chemicals," Proceedings of the Annual Conference on Engineering in Medicine and Biology, Vol. 11, Pt. 4, Alliance for Engineering in Medicine & Biology, Bethesda, Maryland; available from IEEE Service Center (Catalog No. 89CH2770-6), Piscataway, New Jersey, 1989, pp. 1104-1105.

[45] Thompson, R. B., and Kondracki, L., "Waveguide Parameter for Waveguide-Binding Fiber Optic Biosensors," Proceedings of the Annual Conference on Engineering in Medicine and Biology, Vol. 11, Pt. 4, Alliance for Engineering in Medicine & Biology, Bethesda, Maryland; available from IEEE Service Center (Catalog No. 89CH2770-6), Piscataway, New Jersey, 1989, pp. 1102-1103.

[46] Brazy, J. E., and Vander-Vliet, F. J., "Clinical Application of Near Infrared Spectroscopy to Neonatal Intensive Care," Proceedings of the Annual Conference on Engineering in Medicine and Biology, Vol. 11, Pt. 1, Alliance for Engineering in Medicine & Biology, Bethesda, Maryland; available from IEEE Service Center (Catalog No. 89CH2770-6), Piscataway, New Jersey, 1989, pp. 337-338.

[47] Hara, T., Tsukagoshi, K., Arai, A., and Imashiro, Y., "Highly Sensitive Fiber-Optic Immunosensor Using a Metal-Complex Compound as a Chemiluminescent Catalyst," Bulletin of the Chemical Society of Japan, Vol. 62, No. 9, September 1989, pp. 2844-2848.

[48] Payne, P. A., "Transducers, Sensors, and Instrumentation in Clinical Biomechanics," Journal of Biomedical Engineering, Vol. 11, No. 3, May 1989, pp. 180-184.

[49] Besar, S. S. A., Kelley, S. W., and Greenhalgh, P. A., "Simple Fibre Optic Spectrophotometric Cell for pH Determination," Journal of Biomedical Engineering, Vol. 11, No. 2, March 1989, pp. 151-156.

[50] Krull, U. J., Brown. R. S., Vandenberg, E. T., and Hougham, B. D., "Optical Biosensors from Chemical Transduction by Lectins and Neural Receptors," IEEE/Engineering in Medicine and Biology Society Annual Conference Part 2 (of 4); available from IEEE Service Center (Catalog No. 88CH2566-8), Piscataway, New Jersey, 1988, pp. 793-794.

[51] Baldini, F., Falciai, R., Scheggi, A. M., and Bechi, P., "New Fiber-Optic Sensor for Bile Reflux," Optical Fiber Sensors. 1988 Technical Digest Series, Vol. 2, Parts 1-2; available from IEEE Service Center (Catalog No. 88CH2524-7), Piscataway, New Jersey, 1988, pp. II/353-356.

[52] Katzir, A., "Biomedical Fiber Optic Sensors," Optical Fiber Sensors. 1988 Technical Digest Series, Vol. 2, Parts 1-2; available from IEEE Service Center (Catalog No. 88CH2524-7), Piscataway, New Jersey, 1988, pp. I/4-6.

[53] Schultz, J. S., "Sensitivity and Dynamics of Bioreceptor-Based Biosensors," Annals of the New York Academy of Sciences, Vol. 506, New York Academy of Sciences, New York, 1987, pp. 406-414.

[54] Wolfbeis, O. S., "Fibre-Optic Sensors in Biomedical Sciences," Pure and Applied Chemistry, Vol. 59, No. 5, 1987, pp. 663-672.

[55] Tromberg, B. J., Sepaniak, M. J., Vo-Dinh, T., and Griffin, G. D., "Fiber-Optic Chemical Sensors for Competitive Binding Fluoroimmunoassay," Analytical Chemistry, Vol. 59, No. 8, April 15, 1987, pp. 1226-1230.

[56] Kulp, T. J., and Angel, S. M., "Biomedical Applications of Fiber-Optic Chemical Sensors," Conference on Lasers and Electro-Optics - Digest of Technical Papers, Optical Society of America, Washington, D.C.; available from IEEE Service Center (Catalog No. 87CH2390-3), Piscataway, New Jersey, 1987, p. 76.

[57] Angel, S. M., "Fiber-Optic Sensors for Fluoroimmunoassay," Conference on Lasers and Electro-Optics - Digest of Technical Papers, Optical Society of America, Washington, D.C.; available from IEEE Service Center (Catalog No. 87CH2390-3), Piscataway, New Jersey, 1987, p. 48.

[58] MacDonald, A. M. G. (Ed.), "Plenary Lectures Presented at the Second International Symposium on Quantitative Luminescence Spectrometry in Biomedical Sciences," Pure and Applied Chemistry, Vol. 59, No. 5, May 1987, pp. 625-672.

[59] Parker, J. W., Cox, M. E., and Dunn, B. S., "Chemical Sensors Based on Oxygen Detection by Optical Methods," Fiber Optic Sensors, Proc. SPIE 586, Bellingham, Washington, 1986, pp. 156-162.

[60] Opie, R., "Fibre Optics Research Shines Through," Control and Instrumentation, Vol. 19, No. 1, January 1987, pp. 72-73.

[61] Smith, A. M., "Optical-Fibre Sensors - New Opportunities for Chemical Sensing?," Electronics & Power, Vol. 32, No. 11, November-December 1986, pp. 811-813.

Reference List C

[62] Dandge, D. K., Salinas, T., Klainer, S. M., Goswami, K., and Butler, M., "Fiber Optic Chemical Sensor for Jet Fuel," Chemical, Biochemical and Environmental Fiber Sensors, Proc. SPIE 1172, Bellingham, Washington, 1990, pp. 132-139.

[63] Oxenford, J. L., Klainer, S. M., Salinas, T. M., Todechiney, L., Kennedy, J. A., Dandge, D. K., and Goswami, K., "Development of a Fiber Optic Chemical Sensor for the Monitoring of Trichloroethylene in Drinking Water," Chemical, Biochemical and Environmental Fiber Sensors, Proc. SPIE 1172, Bellingham, Washington, 1990, pp. 108-114.

[64] Griffin, J. W., Matson, B. S., Olsen, K. B., Kiefer, T. C., and Flynn, C. J., "Fiber Optic Spectrochemical Emission Sensors: A Detector for Chlorinated and Fluorinated Compounds," Chemical, Biochemical and Environmental Fiber Sensors, Proc. SPIE 1172, Bellingham, Washington, 1990, pp. 99-107.

[65] Milanovich, F. P., Daley, P. F., Klainer, S. M., and Eccles, L., "Remote Detection of Organochlorides with a Fiber Optic Based Sensor. II. A Dedicated Portable Fluorimeter," Analytical Instrumentation, Vol. 15, No. 4, 1986, pp. 347-358.

[66] Milanovich, F. P., Garvis, D. G., Angel, S. M., Klainer, S. M., and Eccles, L., "Remote Detection of Organochlorides with a Fiber Optic Based Sensor," Analytical Instrumentation, Vol. 15, No. 2, 1986, pp. 137-147.

[67] Bach, D. T., and Lin, C., "Remote Fiber Optic Reflectometer," Reflective Optics II, Proc. SPIE 1113, Bellingham, Washington, 1989, pp. 257-264.

[68] Zhu, C., Bright, F. V., and Hieftje, G. M., "Simultaneous Determination of Br** Minus and I** Minus with a Multiple Fiber-Optic Fluorescence Sensor," Applied Spectroscopy, Vol. 44, No. 1, January 1990, pp. 59-63.

[69] Herron, N. R., Simon, S. J., and Eccles, L., "Remote Detection of Organochlorides with a Fiber Optic Based Sensor III: Calibration and Field Evaluation of an Improved Chloroform Fiber Optic Chemical Sensor," Analytical Instrumentation, Vol. 18, No. 2, June 1989, pp. 107-126.

[70] Seitz, W. R., Zhujun, Z., and Mullin, J. L., "Reversible Indicators for Alkali Metal Ion Optical Sensors," Optical Fibers in Medicine II, Proc. SPIE 713, Bellingham, Washington, 1987, pp. 126-131.

[71] Van Ryswyk, H., Carpenter, M. K., Blom, N., and Ellis, A. B., "Luminescent Fiber Optic Sensors Based on Schottky Diodes," Electrochemical Society Extended Abstracts, Vol. 85-2, Electrochemical Society, Pennington, New Jersey, 1985, p. 487.

[72] Klainer, S. M., Schwarz, W. J., and Milanovich, F. P., "Use of Fiber Optic Chemical Sensors for Monitoring Specific Parameters and Species in Aqueous Systems," Oceans '86; available from IEEE Services Center (Catalog No. 86CH2363-0), Piscataway, New Jersey, 1986, pp. 828-833.

[73] Milanovich, F. P., Daley, P. F., Klainer, S. M., and Eccles, L., "Remote Detection of Organochloride with a Fiber Optic Based Sensor - QII. A Dedicated Portable Fluorimeter," Analytical Instrumentation, Vol. 15, No. 4, December 1986, pp. 347-358.

[74] Milanovich, F. P., "Detecting Chloroorganics in Groundwater," Environmental Science and Technology, Vol. 20, No. 5, May 1986, pp. 441-442.

General Reference List

[75] Geckeler, S., Optical Fiber Transmission Systems, Artech House, Inc., Norwood, Massachusetts, 1987.

[76] Seippel, R. G., Optoelectronics, Reston Publishing Company, Inc., Reston, Virginia, 1981.

[77] Wilson, J., and Hawkes, J. F. B., Optoelectronics - An Introduction, Prentice Hall International (UK) Ltd., 1989.

[78] Sharaf, M. A., Illman, D. L., and Kowalski, B. R., Chemometrics, John Wiley & Sons, Inc., 1986.

[79] Zung, J. B., Woodlee, R. L., Fuh, M-R. S., and Warner, I. M., "Recent Developments and Applications of Multidimensional Fluorescence Spectroscopy," Fluorescence Detection III, Proc. SPIE 1054, Bellingham, Washington, 1989, pp. 69-76.

[80] Angel, S. M., and Ridley, M. N., "Dual-Wavelength Absorption Optrode for Trace Level Measurements of Trichloroethylene and Chloroform," Chemical, Biochemical and Environmental Sensors, Proc. SPIE 1172, Bellingham, Washington, 1989, pp. 115-122.

[81] Berman, R. J., and Burgess, L. W., "Renewable Reagent Fiber Optic Based Ammonia Sensor," Chemical, Biochemical and Environmental Sensors, Proc. SPIE 1172, Bellingham, Washington, 1989, pp. 206-214.

[82] Myrick, M. L., and Angel, S. M., "Normal and Surface-Enhanced Raman Scattering with Optical Fibers," Chemical, Biochemical and Environmental Sensors, Proc. SPIE 1172, Bellingham, Washington, 1989, pp. 38-48.

[83] Angel, S. M., and Myrick, M. L., "Near-Infrared Surface-Enhanced Raman Spectroscopy Using Fiber Optics," Chemical, Biochemical and Environmental Sensors, Proc. SPIE 1172, Bellingham, Washington, 1989, pp. 38-48.

[84] Griffin, J. W., Olsen, K. B., Matson, B. S., and Nelson, D. A., "Fiber Optic Spectrochemical Emission Sensors," Chemical, Biochemical and Environmental Applications of Fibers, Proc. SPIE 990, Bellingham, Washington, 1988, pp. 55-68.

[85] Griffin, J. W., Olsen, K. B., Matson, B. S., and Kiefer, T. C., "Fiber Optic Spectrochemical Emission Sensors: A Detector for Chlorinated and Fluorinated Compounds," Chemical, Biochemical and Environmental Sensors, Proc. SPIE 1172, Bellingham, Washington, 1989, p. 99-107.

[86] Matson, B. S., and Griffin, J. W., "Infrared Fiber Optic Sensors for the Remote Detection of Hydrocarbons Operating in the 3.3 to 3.36 Micron Range," Chemical, Biochemical and Environmental Sensors, Proc. SPIE 1172, Bellingham, Washington, 1989, pp. 13-26.

[87] DeGrandpre, M. D., and Burgess, L. W., "A Fiber Optic FT-NIR Evanescent Field Absorbance Sensor," Applied Spectroscopy, Vol. 44, No. 2, 1990, pp. 273-279.

[88] Klainer, S. M., Dandge, D. K., Goswami, K., Eccles, L. A., and Simon, S. J., "A Fiber Optic Chemical Sensor (FOCS) for Monitoring Gasoline," In Situ Monitoring with Fiber Optics, EPA/600/X-88/259 U. S. Environmental Protection Agency, Las Vegas, Nevada, June 1988.

[89] Butler, M. A., Ricco, A. J., and Buss, R., "Fiber Optic Micromirror Sensor for Volatile Organic Compounds," J. Electrochem. Soc., Vol. 137, No. 4, April 1990, pp. 1325-1326.

Determination of Hydraulic Parameters

Mohammad T. Izadi[1] and Richard W. Stephenson[2]

MEASUREMENT OF GAS PERMEABILITY THROUGH CLAY SOILS

REFERENCE: Izadi, M. T. and Stephenson, R. W., "Measurement of Gas Permeability Through Clay Soils," Current Practices in Ground Water and Vadose Zone Investigations, ASTM STP 1118, David M. Nielsen and Martin N. Sara, Eds., American Society for Testing and Materials, Philadelphia, 1992.

ABSTRACT: The engineering profession currently has a standard for measuring the coefficient of water permeability of saturated fine grained soils. However, there has not been any attempt to standardized the procedure of measuring the coefficient of gas permeability of unsaturated fine grained soils. Low permeability fine grained soils are often used in landfills as barriers to retard the movement of hazardous gases.

Petroleum engineers were primarily responsible for development of gas flow theories and measurements. Hence, the available test methods and equipment were developed to suit their demands. On the other hand, the increasing involvement of geotechnical and environmental engineers in design and construction of waste landfills necessitate the development of new techniques suitable to their need.

This investigation developed new apparatus and simple and reliable testing procedures for measuring gas permeability of unsaturated fine grained soils. A series of tests were performed on compacted clay samples to verify the technique.

KEY WORDS: Gas permeability, unsaturated soil, compacted soils, waste landfill, liners, permeability.

---

1 Graduate student, Department of Civil Engineering, University of Missouri-Rolla, Rolla, MO 65401

2 Professor of Civil Engineering, University of Missouri-Rolla, Rolla, MO 65401

INTRODUCTION

Soil lined facilities are used extensively for the containment and disposal of waste. In these facilities, low permeability natural clay-rich deposits, with hydraulic conductivities of 10^{-7} cm/sec or less, are often relied upon to retard the movement of liquids.

Traditional waste management strategies have almost exclusively addressed water pollution problems. Federal and state regulatory agencies have established standards for secure landfill design and construction principally aiming at controlling water run-on, run-off and collection, removal and monitoring of underground leachates.

On the other hand, there is not sufficient information available for the effective estimation of the movement and emission of gases from landfill sites (1). Estimation and modeling of gas movement can provide a useful tool in the management of hazardous wastes so a realistic assessment can be made on the environmental impact of existing or proposed landfills (2). Hence, controlling the movement of gases produced by decomposition of waste is an important part of landfill design.

A variety of gases are generated in landfills, among which methane and carbon dioxide are the major gases produced. Where methane is not controlled, fire and explosions have occurred. Concentrations above the lower explosive limit of methane have been reported at distance up to 1,000 feet off site. Loss of life and injuries have occurred in several cases(3). Excessive carbon dioxide reduces the oxygen content of the soil which can cause severe damage to vegetation(4).

THEORY

In 1856, Henry Darcy investigated the flow of water through homogenous sand filters. He concluded that the rate of flow (volume per unit time), Q, through porous media is directly proportional to the pressure gradient, ΔP, cross-sectional area, A, and inversely proportional to the length, h:

$$Q = -KA\frac{(h_2-h_1)}{h} \qquad (1)$$

where K is a constant characteristic of the porous media.

Darcy's law, in its original form, is rather restricted in its usefulness. Nutting (5) proposed a coefficient, k, as "specific permeability" of porous media such that $k=K\mu/\rho g$, where μ is the kinematic viscosity of the fluid and g is the acceleration of gravity:

$$Q = \frac{kA\rho g}{\mu}\left(\frac{\Delta P}{h}\right) \qquad (2)$$

Equations 1 and 2 apply to incompressible fluids only. For compressible fluids (i.e., gases), the rate of flow changes from point to point as the pressure decreases along the flow path, but the mass rate of flow, ρq, remains constant(6).

$$G=\rho q=-\frac{k\rho}{\mu}\frac{\delta P}{\delta x} \tag{3}$$

q=Q/At is the apparent linear rate of flow. Since μ is independent of P, the integration of equation 3 over the length of the sample, h, results in:

$$G=\rho q=\frac{k}{\mu}\rho_m\frac{\Delta P}{h} \tag{4}$$

where ρ_m is the density of the gas at the mean pressure, P_m.

Gas flow in porous media is a complex phenomenon since gas may flow through porous material by four different mechanisms: viscous flow; slip flow; molecular and diffusional flow; and adsorbed flow. There is no sharp transition between the various kinds of flow and flow changes gradually from one to the other. Variables such as pore size diameter, temperature, pressure, and gas type will determine the extent to which each mechanism plays a part in the flow. In some cases, because of the differences in the pore sizes, various types of flow may occur simultaneously.

In laminar flow of gases, when the mean free path of the gas is small compared to the pore diameters, the flow is governed by viscosity and obeys Poiseuille's type of flow. Resistance to the gas flow will depend mainly upon collision between gas molecules.

Slip flow occurs when the gas pressure is reduced until its mean free path is an appreciable fraction of the capillary radius. In this case the flow will be intermediate between molecular flow and viscous flow. In many fine pored media, the capillary diameters are less than or comparable with the mean free path even at atmospheric pressure. Jones et al. (7) concluded that gas slippages are large in low permeability tight gas sands.

Klinkenberg (8) developed a relationship in which he considered the effect of slip flow on gas permeability. He postulated that the effect of slip flow reduces with increasing pressure and at infinite pressure gas and liquid permeabilities are equal. At reduced pressure, gas permeability is the sum of viscous flow and slip flow and can be computed from

$$k_a=k\left(1+\frac{b}{P_m}\right) \tag{5}$$

k_a is the apparent permeability, k is the true permeability and b is the Klinkenberg parameter. However, Klinkenberg theory is only applicable to high permeability porous media and cannot be applied to gas flow through low permeability, fine grained soils.

VISCOUS FLOW THROUGH UNSATURATED SOILS

Rose (9) studied the gas permeability of unsaturated samples. He concluded that the effect of gas slippage on the measured effective gas permeabilities at various liquid saturations decreased with an increase in liquid saturation. Rose explained that when a condition of gas equilibrium saturation exists within the sample, gas will begin to flow through the larger pores. When the liquid saturation decreases, an increasing number of the smaller pores will be occupied by gas. Therefore, the average effective radius decreases with a decrease in liquid saturation.

Later Estes and Fulton (10) carried out permeability tests on sandstone and limestone samples. They also found that gas slippage decreased with increase in degree of saturation. Yashimi et al. (11) studied the air permeability of unsaturated compacted silty clay samples. They concluded that the relationship between air permeability and air void ratio was linear.

Langfelder et al. (12) studied the air permeability of compacted cohesive soils. Soil samples were compacted by static, dynamic or kneading compaction and were tested at the as compacted water content. They concluded that air permeability of the compacted soils was affected by water content, dry density and method of compaction. Kneading compaction resulted in samples having lower air permeabilities than samples compacted by dynamic compaction.

Jones and Owens (7) examined the effect of partial water saturation on gas permeability of low permeability, tight gas sands. They found that those samples whose specific water permeabilities are a great deal less than the Klinkenberg permeabilities of the dry samples also had corresponding low gas permeability in the presence of simulated connate water saturation.

Jones and Owens further reported that water greatly reduced permeability of gas tight gas sands in a manner different from its effect on higher permeability sands. The water adjacent to high-energy surfaces becomes ordered to result in viscosity increase or even solidification sufficient to reduce effective pore diameter.

In practice, the water content of soil changes with the amount of precipitation and seasonal change of weather, even though the soil is initially compacted at a given water content. Hence, it is likely that the soil liquid saturation changes with time. As the result, gas permeability of the soil will also change.

In investigations of the effect of liquid saturation on gas permeability, soil samples are usually compacted at various water contents and gas permeabilities are measured at as-compacted water contents. However, the effect of compaction moisture content on soil structure is significant in fine grained soils. Tests on liquid permeability of such soils have shown that coefficient of permeability changes by as much as 2 to 3 orders of magnitude depending on whether the soil is compacted on the dry or wet side of optimum(13). Likewise, the change in gas permeability is due to both changes in the liquid saturation and soil structure. Previous studies on gas permeability of saturated soils have not addressed this problem.

VISCOUS FLOW TESTING METHODS

Test methods for investigation of viscous flow of gases through porous media usually fall into two categories; non-steady state and steady state methods.

In the non-steady state methods, the soil sample is subjected to a pressure gradient which is allowed to dissipate. Gas permeability is calculated by monitoring the change in gradient with time. Three different techniques are usually employed: a) constant downstream method, b) constant upstream method and c) non-steady upstream and downstream method.

In the constant downstream method, a vessel is attached to the upstream end of the sample. The vessel is filled with gas under pressure and the pressure drop is recorded with time as gas flows through the sample. The rate of change of pressure depends on the permeability of the sample. The permeability can be calculated if the relationship between pressure and viscosity is known.

Heatherington et al. (14) proposed the following approximate empirical solution for calculating constant downstream gas flow. The rate of flow per unit time was calculated from:

$$Q=a(P_1^2-P_2^2)^N t \tag{6}$$

where a and N are experimental constants.

Weaver (15) used constant downstream method and calculated the gas permeability from:

$$k=\frac{V\mu h}{AP_a t}\ln\left[\frac{(P_1+2P_a)(P_2)}{(P_2+2P_a)(P_1)}\right] \tag{7}$$

Davies and Booth (16) developed the following relationship for calculating gas permeability:

$$k=\frac{MhV}{AP_o t}\log\frac{P_1}{P_t} \quad (8)$$

where P_t is the upstream pressure at time t.

The constant upstream method is similar to the constant downstream method except that gas pressure changes at downstream end of the sample.

Calhoun (17) used the non-steady upstream and downstream method to study the gas permeability of dry porous plugs. He considered a permeable body placed between two vessels under pressure P_1 and P_2 respectively. Both vessels were closed so all gas leaving the first vessel flowed to the second vessel. Flow will continue until P_1 and P_2 are equalized. The equation that he developed is:

$$k=\frac{\mu hV}{2AP_m t}\ln\frac{(P_1-P_2)}{(P_1-P_2)} \quad (9)$$

TEST PROCEDURES AND MATERIALS

An air floated, intermediate particle size, pulverized kaolinite (Albion S-75) from Albion Kaolin Company was used in this study. The properties of the soil are given in Table 1. Because of concerns about the effect of soil moisture and permeant on permeability, a standardized permeant (0.005 N calcium sulfate) was used for all mixing water and all water permeability testing. All samples were compacted by kneading compaction with a method similar to Harvard Miniature Compaction method. Soil samples were 2.5" (6.35 cm) in diameter and 0.75" (1.9 cm) in height.

Table 1: Specimen Properties

Specific Gravity	2.64
Liquid Limit	55.6
Plastic Limit	32.3
Optimum Water Content (%)	27.5
Maximum Dry Density (kg/m^3)	1454

CHANGING THE DEGREE OF SATURATION

The main concern in changing the moisture content of samples was the possibility of developing micro cracks. Cracks usually form due to uneven shrinkage and moisture distribution in the samples.

To overcome this problem, a new method for changing the degree of saturation of samples was developed. Samples were placed in a triaxial permeability apparatus in pairs with filter paper placed between them. The reason for treating two specimens at the same time was to be able to check the reproducibility of test results and compare gas permeability for different types of gas at given degrees of saturation.

First, all samples were consolidated to 5 psi (34.5 kN/m^2). If it was desired to reduce the moisture content of the samples, the top and bottom access lines were opened and air was allowed to flow to the top and bottom of the samples. The circulation of dry air facilitated the evaporation of moisture. A constant effective confining pressure of 5 psi (34.5 kN/m^2) reduced the possibility of crack formation during the drying period. Air flow continued for the desired period of time to bring the samples' moisture content to the desired value.

When it was desired to increase the samples' moisture content, water was allowed to flow through the access lines and porous stones to the top and bottom of the samples. This process required only a short period of time and water was drained quickly from the lines.

After cutting the supply of air or water, the samples were kept inside the apparatus for a minimum of 24 hours. This helped to reduce uneven moisture distribution and suction pressure inside the samples. The samples were then taken out of the apparatus and kept in a moist curing room for at least 3 days.

GAS PERMEABILITY APPARATUS

A new gas permeability apparatus was designed and constructed. The apparatus mainly consists of a downstream vessel and base, an upstream vessel and a cylindrical wall which serves as a confining chamber for the cell fluid. Access lines are provided through the upstream and downstream bases to the vessels. The access lines are used for gas flow into or out of the vessels and for measuring gas pressure in the vessels. The vessels are 2.5 inches (6.35 cm) outside diameter, 2 inches (5.08 cm) inside diameter and were 2.5 inches (6.35 cm) tall. They are constructed of Plexiglass.

An access line is provided to the base for filling the cell with water and applying confining pressure. The cell fluid is isolated from the soil sample by a flexible rubber membrane wrapped around the sample.

In existing gas permeability of soil test equipment, a soil sample is confined in a rigid ring or cylinder. This method of confinement results in gas flow along the interface between the ring and the sample. In the new apparatus, however, a flexible membrane replaces the rigid confining ring. The flexible membrane is kept tightly against the sides of the sample by applying air pressure to the fluid in the cell. This insures against short circuiting of gas around the soil sample. Furthermore the sample can be confined to its in-situ or any other desired confining pressure more correctly modeling the stress state of the soil in the field. The apparatus is capable of measuring gas permeability of soil samples by three different non-steady state methods.

TEST SET UP AND PROCEDURE

A schematic diagram of the test setup used in this study is shown in Figure 1. The setup consists mainly of: a) gas cylinder, b) gas valves, c) gas humidifier tube, d) pressure transducers, e) upstream and downstream gas vessels, f) biaxial chamber, g) sample holder and rubber membrane, h) water tank, i) air supply line, j) O rings, k) 1/8 inch (3.2 mm) outside

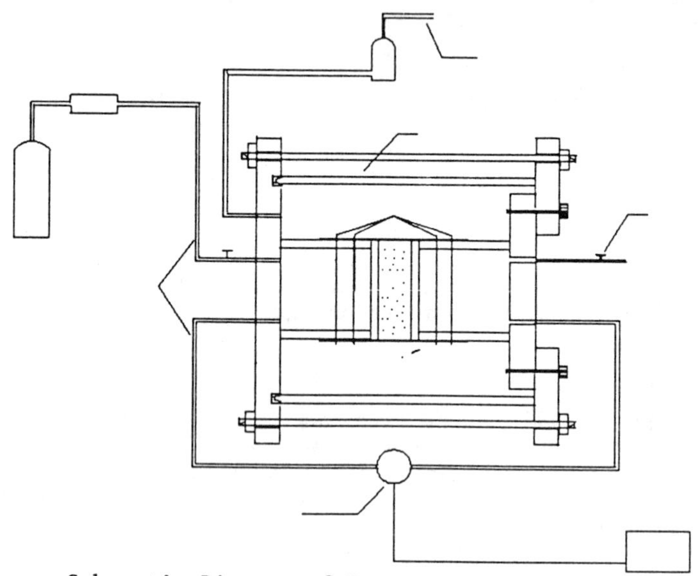

Figure 1: Schematic Diagram of Test Apparatus

diameter stainless steel tubing, l) digital pressure meter, and m) soil specimen.

The soil sample was placed between the upstream and downstream gas vessels. Porous stones or porous plastic rings were placed between the vessels and the sample to facilitate the free flow of gas through the entire cross-sectional area of the sample. Porous stones were used when the degree of saturation of the samples was less than 65 percent. Porous plastic rings were used for higher degree of saturation as it was found that when saturation was high, water migrated from the sample to the porous stones.

The cell was assembled and filled with water. The desired confining pressure was applied to the sample and the test was performed at a constant temperature (20° C).

For the non-steady upstream and downstream pressure method, the upstream chamber was filled with gas to the desired pressure. Both upstream and downstream valves were closed and gas was allowed to flow through the soil sample to the downstream vessel. The changes in the upstream and downstream pressure were recorded at the desired time intervals, using a data acquisition system. The time intervals were chosen according to the permeability of the samples.

The procedure for constant downstream method was similar to the above method except that the downstream pressure was kept at atmospheric pressure by leaving the downstream valve open.

In the case of constant upstream method, the upstream pressure was kept at atmospheric pressure by leaving the upstream valve open and applying vacuum to downstream vessel before closing the downstream valve.

The effect of carbon dioxide gas on the permeability of the clay was studied by allowing carbon dioxide to flow through the sample for a period of time depending on the permeability of the sample, while vacuum was applied to the downstream vessel. This facilitated the removal of air in the pores.

TEST RESULTS AND DISCUSSION

In order to check for leakage diffusion and possible flow of gas along the sample edges, a few tests were performed on samples compacted 2.3 percent wet of optimum (29.8 percent water content). Figure 2 shows the test results on two such samples. A small drop in the upstream pressure occurred at the beginning of the tests after which the pressure remained almost constant. The initial drop in pressure is believed to be due to increase in volume of upstream vessel due to the pressure increase. Figure 2 also shows that there was no appreciable change in pressure during the test indicating there was no flow of air along the boundaries. Therefore the specimen had zero air permeability as the upstream and downstream pressures remained constant during the testing period.

Comparison of gas permeability obtained by non-steady upstream and downstream method and constant down stream method are shown in Figure 3. Test results shows a good agreement between the two methods.

Test results on samples compacted at dry side of optimum (24.8 percent) are shown on Figure 4. Gas permeability increased by more than four orders of magnitude when the liquid saturation decreased from 94.4 percent to 0 percent. Surprisingly the soil was permeable to air at 94.4 percent liquid saturation ($k = 1.5 \times 10^{-14}$ cm^2). Samples compacted wet of optimum were not permeable to air at liquid saturation as low as 86 percent. This shows that soil structure and hence, particle orientation has a pronounced effect on the gas permeability of unsaturated clays.

Test results indicate that gas permeability is dependent on both the soil structure and the degree of saturation. Flocculated soil was much higher in permeability. Gas permeability increased more than four orders of magnitude when the liquid saturation decreased from 94.4 to 0.

The relationships between gas permeability and air voids for air and carbon dioxide are shown in Figure 5. Permeability of the soil to both

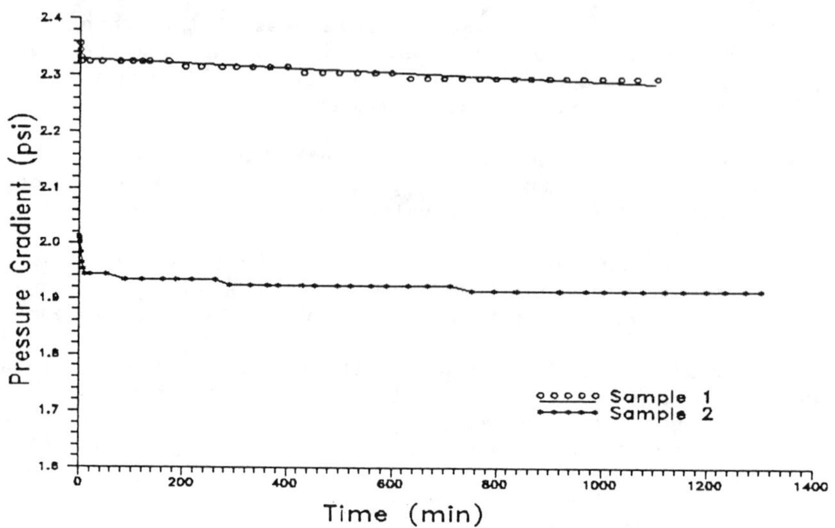

Figure 2: Pressure Drop Versus Time for Fully Saturated Specimen (1 psi - 7.0 kPa)

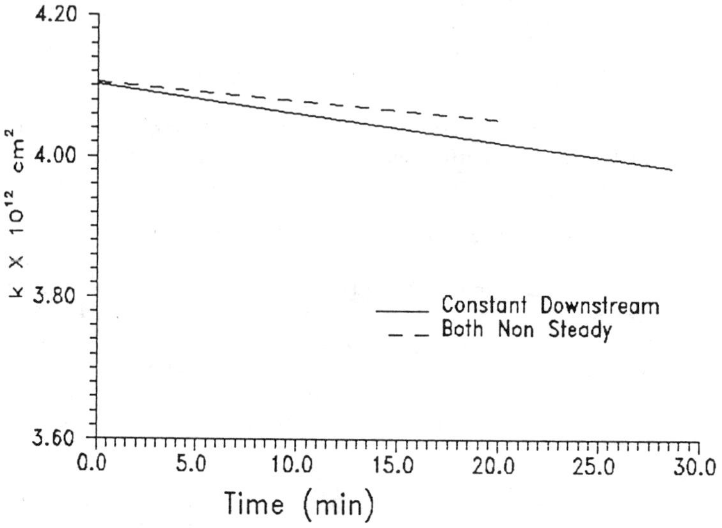

Figure 3: Comparison of Two Test Methods

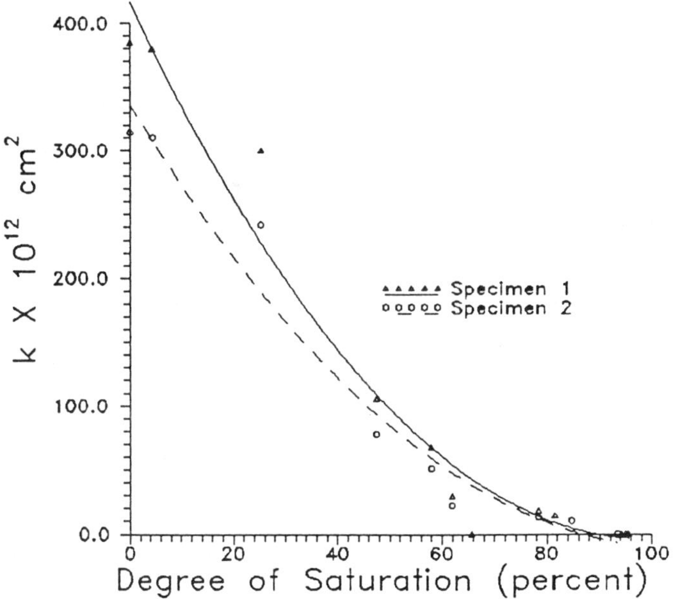

Figure 4: Gas Permeability For Specimen Compacted Dry of Optimum

gases increased with increasing air void volume. The rate of increase was slower for air voids up to about 0.4. As the volume of air increased further permeability increased at a faster rate. It is believed that as the water evaporates from the boundaries of the sample the water in the larger internal voids drains first. However, most of the flow channels are occluded at the narrow portion of the capillary. As more water evaporates water drains from the narrow portion of the channel. As the volume of narrow voids is smaller a larger number of flow channels will be open to flow for a given increase in air void and hence, gas permeability increases at a faster rate. There was little significant difference in the permeability of the two gases.

Figure 6 shows the ratio of air permeability to carbon dioxide permeability verses liquid saturations. Permeability ratio decreased by decreasing liquid saturation. This indicates that the effect of gas slippage on the measured gas permeability decreased with a decrease in liquid saturation. Test results are contrary to results obtained by previous studies on coarse grained soils and rock formations. This phenomenon is believed to be due to the nature of clay particles and effect of water molecules on them. In clays, unlike coarse grained soils,

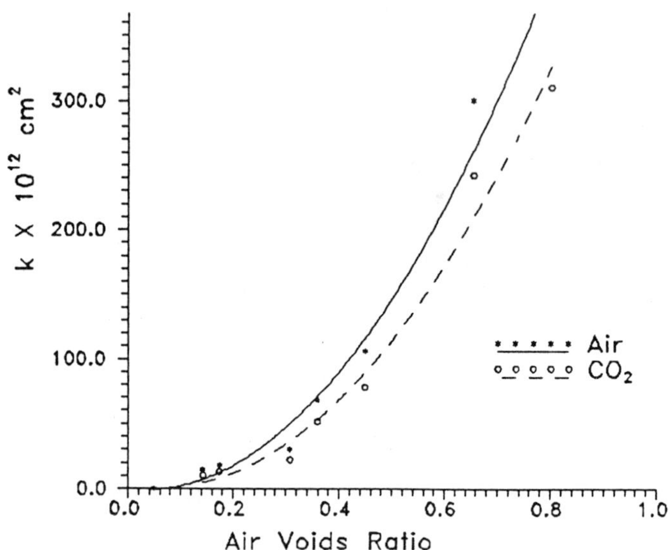

Figure 5: Permeability of Air and CO_2 Versus Degree of Saturation

the larger pores are not completely drained at partial saturation before the smaller pores are drained. As the liquid saturation decreases, thickness of the water layer adjacent to the clay surface decreases and the effective pore diameter increases. Another factor which might be responsible for this behavior is the shape of clay particles. The platey clay particles result in highly irregular channels and even though larger voids are drained first, gas has to pass through the narrow parts of the flow channels which in turn controls the gas flow.

Geotechnical engineers generally use K as the "coefficient of permeability" in cm/sec. Figure 7 shows the ratio of the coefficient of permeability of air to water. Air permeability of dry soil was about 28 times the water permeability. Air and water permeability were equal at 82 percent liquid saturation. For comparison, water permeability of soil compacted at dry side of optimum is also given.

CONCLUSION

A new gas permeability apparatus was designed and constructed. Test results showed that the apparatus provides a simple and reliable method of measuring gas permeability of unsaturated fine grained soils. The

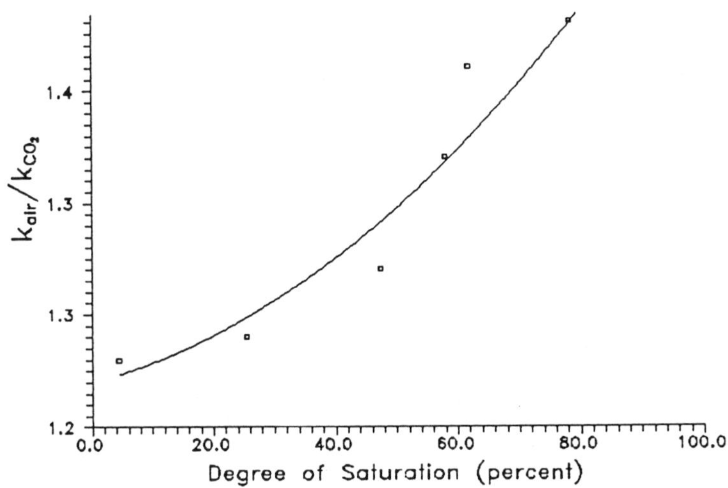

Figure 6: k_{air}/k_{co2} Versus Degree of Saturation

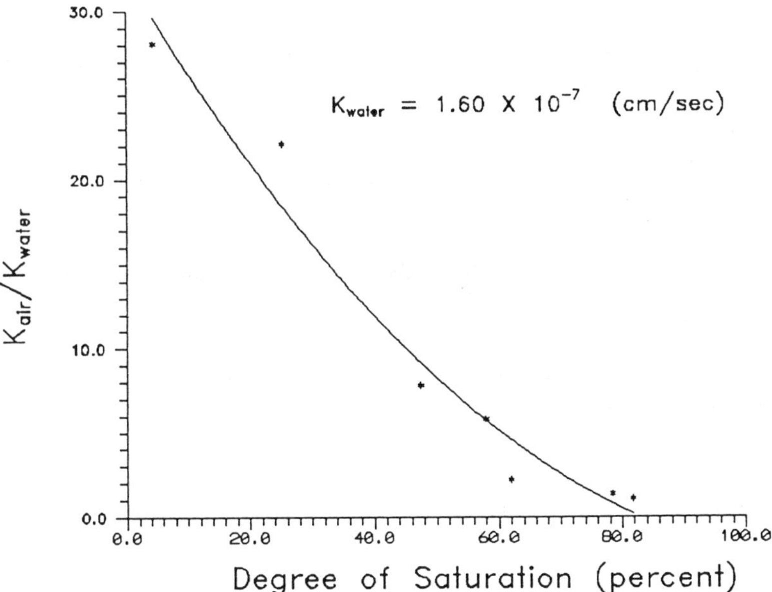

Figure 7: Permeability Ratio (cm/sec) Versus Degree of Saturation

apparatus is simple to build and tests are performed in a relatively short period of time. Testing time varied from one minute to several hours depending on the permeability of the sample. The apparatus is capable of measuring gas permeability by three different non-steady state methods. Non-steady state upstream and downstream method is mostly suited for low permeability samples with high liquid saturation. This is because in this method the system is isolated from the outside environment and evaporation of soil moisture is minimal.

Gas permeability increased by more than four orders of magnitude as the liquid saturation decreased from 94.4 to zero percent.

Contrary to the previous test results on coarse grained soils, the effect of gas slippage decreased with decreasing liquid saturation.

REFERENCES

1. Brown, K.W. (1988), "Review and Evaluation of the Influence of Chemicals on the Conductivity of Soil Clays," USEPA/600/S2-88/016, Apr. 1988.

2. Karimi, A., Ravindran, V. and Pirbazari, M. (1986), "Emission and Control of Volatile Organics in Hazardous Waste Landfills," Proceedings of the 1986 Specialty Conference in Environmental Engr., Env. Engr., 42, 1986, pp. 85-91.

3. Environmental Protection Agency (1988), "The Agency's Landfill Proposals," Waste Age, Dec. 1988, pp. 52-60.

4. HOEKS, J. (1972) "Changes in Composition of Soil Air Near Leaks in Natural Gas Mains", Soil Science, Vol. 113, No. 1, p. 46-54, 1972.

5. Nutting, P.G. (1930) "Physical Analysis of Oil Sands," Bulletin, Am. Assn. Petroleum Geologists, Vol.14, No. 10, 1930, pp. 1337-1349.

6. Carman, P.C. (1956), Flow of Gases Through Porous Media, Academic Press Inc., N. Y. 1956.

7. Jones, E. and Owens, W. (1980) "A Laboratory Study of Low-Permeability Gas Sands", Journal of Petroleum Technology, Sep. 1980, p. 1631-1640.

8. Klinkenberg, L.J. (1941), " The Permeability of Porous Media to Liquid and Gases," API Drilling and Production Practice, 1941, pp. 200-213.

9. Rose, W.D. (1948), "Permeability and Gas Slippage Phenomena," API Drill. and Prod. Prac., (1948), pp. 209.

10. Estes, R.K. and Fulton, P. F. (1956) "Gas Slippage and Permeability Measurements", Petroleum Transactions, AIME, Vol. 207, p. 338-342, 1956.

11. Yashimi, Y. and Osterberg, J.O. (1956) "Compression of Partially Saturated Cohessive Soils," J. Soil Mech. and Found. Engr., Vol. 89, July, 1963, pp. 1-24.

12. Langfelder, L.J., Chen, C.F. and Justice, J.A. (1968) "Air Permeability of Compacted Cohesive Soils", Journal of Soil Mechanics and Foundations Division, ASCE, SM 4, 1968, pp. 981-1001.

13. Lambe, T.W. (1958) "The Engineering Behavior of Compacted Clay," J. Soil Mech. and Found. Engr., Vol. 84, No. SM2, May, 1958, pp. 1655,1-35.

14. Heatherington, C.R., MacRoberts, D.T., and Huntington, R.L. (1942), "Unsteady Flow of Gas Through Porous Media," Transactions AIME, Vol. 146, 1942, p. 166.

15. Weaver, A.S. (1954), "Determination of Permeability of Granular Soil by Air Subjected to a Decreasing Pressure Differential," Special Technical Publication, ASTM, No. 163, 1954, pp. 123-130.

16. Davies, L.J. and Booth, H. (1961) "An Unsteady Flow Method for Air Permeability Measurement Of Refractory Materials," Ceramic Bulletin, Vol. 40, No. 12, 1961, pp. 744-747.

17. Calhoun, J.C. (1946) "New Method to Determine Permeability," Petroleum Engr., Feb., 1947, pp. 103-108.

Miguel Picornell, and Antonio Guerra

A COMPARISON OF FIELD AND LABORATORY MEASUREMENTS OF HYDRAULIC CONDUCTIVITY

REFERENCE: Picornell, M. and Guerra, A., "A Comparison of Field and Laboratory Measurements of Hydraulic Conductivity," Current Practices in Ground Water and Vadose Zone Investigations, ASTM STP 1118, David M. Nielsen and Martin N. Sara, Eds., American Society for Testing and Materials, Philadelphia, 1992.

ABSTRACT: The hydraulic conductivity of a soil was measured in the laboratory and in the field. Although each technique used yielded consistent results, the hydraulic conductivity obtained differed over several orders of magnitude. These differences in hydraulic conductivity can only be attributed to macroscopic features that weigh differently depending on the mass of soil affected by the different tests.

KEYWORDS: Hydraulic Conductivity, field tests, silts, halloysite.

INTRODUCTION

A natural basin within the explosion crater of an inactive volcano was proposed to be used as a water supply reservoir. The site is located near Barlovento, in northern La Palma, Canary Islands, Spain. The proposed reservoir capacity was of 5.45 x 10^6 m^3 (7.2 x 10^6 cubic yards) with a water depth of 35 m (115 ft). Outcrops of bedrocks were clearly visible on the slopes of the basin, while the lower part was covered with thick soil deposits. The expected differences in compressibility of the slope deposit and the bottom of the reservoir, coupled with the high stresses imposed by the expected water depth

Dr. M. Picornell is an Assistant Professor, Department of Civil Engineering, University of Texas at El Paso, El Paso, Texas 79968; Mr. Guerra is a Consulting Geotechnical Engineer, Agustin Betencourt 25, Madrid, Spain.

in the reservoir, weighed heavily against expensive alternatives of reservoir impermeabilization with synthetic liners.

The evaluation of the need to construct a liner justified a careful study of the hydraulic conductivity of the existing soils in the basin. This paper describes the site conditions and the results of field and laboratory tests used to assess the hydraulic conductivity of the existing soils at the proposed reservoir site.

SITE CONDITIONS

Surface Conditions

The proposed reservoir site makes use of a naturally occurring basin, created by an explosion crater of an inactive volcano. The plan shape of the basin is nearly elliptic with axes 615 m (2,020 ft) and 510 m (1670 ft) in the crest and axes of 250 m (820 ft) and 145 m (476 ft) along the lower level of the basin. The walls of the basin slope at about 30%. **An index map and a site plan showing the reservoir and the location of all site investigation are presented in Figure 1.**

Pyroclastic rock outcrops were evident along the western edge of the crater. A soil zone including large boulders was evident in the south-western wall, and outcrops of basaltic flows were present in the south-east wall of the crater. The bottom of the depression was covered with a deposit of fine-grained soils. Upon clearing the vegetation, the northern half of the basin appeared covered with a reddish-brown fine grained soil. In the southern half, the soils were lighter in color, exhibiting a yellowish-brown tint.

Subsurface Conditions

The subsurface conditions were investigated by drilling eleven borings up to 25 m (82 ft) deep and excavating twelve exploratory pits up to depths of more than 6 m (20 ft). The majority of the basin is covered by a layer of clayey silt resulting from in-place weathering of the volcanic rock. The thickness of this layer in the center of the basin exceeds the 25 m (82 ft) depth investigated with borings. This layer thins out towards the edges of the depression to thicknesses of at least 10 m (33 ft).

This surface layer is underlain by predominantly granular soils such as gravelly and sandy soils with variable amounts of silt. These granular materials were encountered in borings drilled near the edges of the crater. Bedrock, consisting of gray basalt, underlies the granular deposits along the eastern and southern edges of the basin. In the rest of the edge of the basin bedrock was not reached.

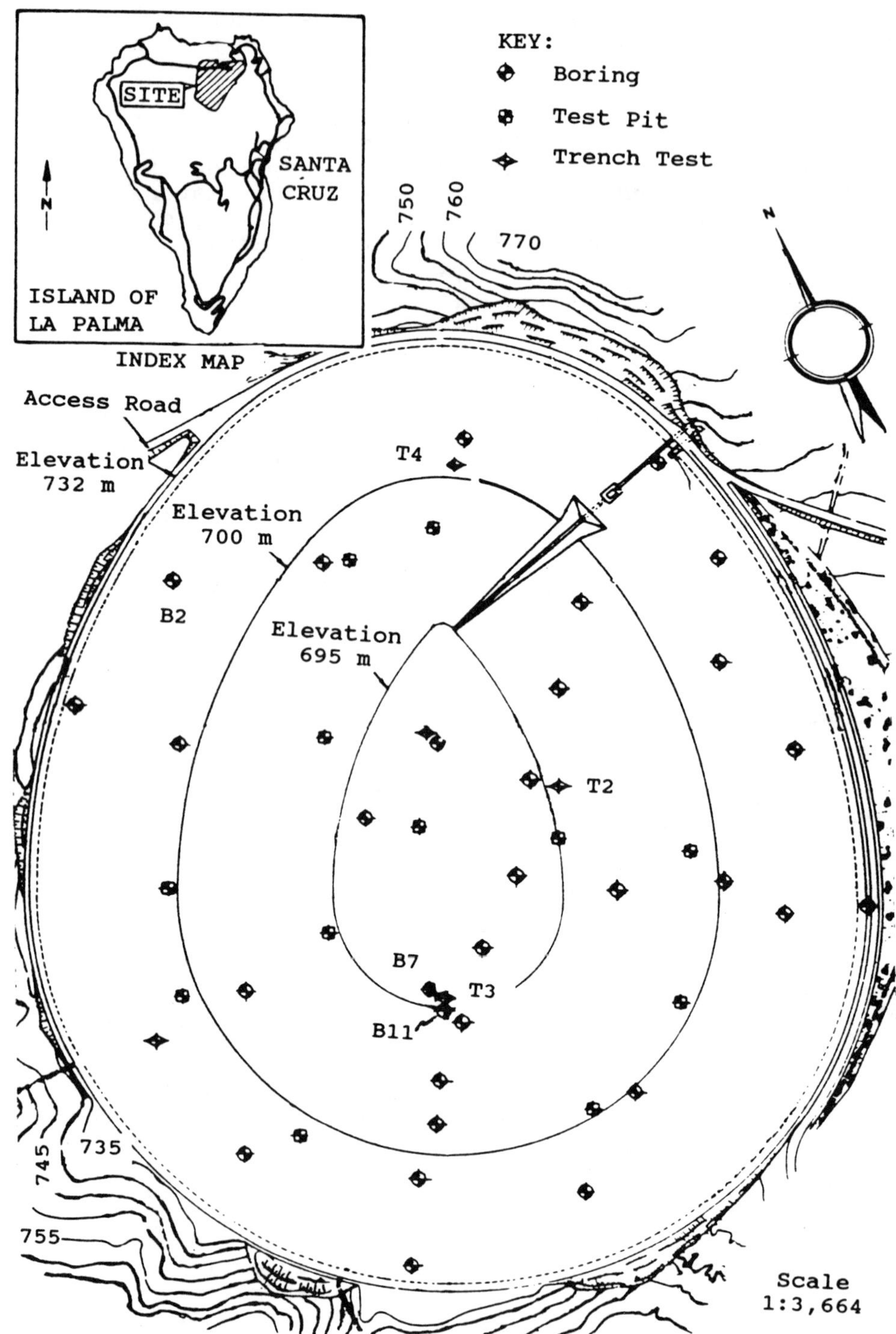

FIG.1--Index Map and Site Investigation Plan

The presence of a ground water surface was not detected in any of the boring or exploratory pits. Furthermore, water districts in the regional area have excavated several infiltration galleries in the vicinity of the site to collect ground water; specifically, two galleries run directly underneath the site. These galleries have cross-sections of 1.8 m x 1.8 m (6 ft x 6 ft), are nearly horizontal, and are located 100 m (330 ft) and 400 m (1300 ft) below the basin. The presence of these galleries is justified by the fractured nature of the basaltic rock and can help explain why no ground water surface was encountered at the site.

Geotechnical Characterization of the Silt Deposit

For purposes of describing the surficial soil stratum, index tests have been performed; grain-size distribution, moisture content, dry density, specific gravity of the solids, Atterberg limits and mineralogical identifications.

The grain-size distribution was determined with a combined sieve and hydrometer analysis. The silt fraction makes up more than 45% of the total dry weight of the soil. The percentage made up by the clay size fractions ranges from 15% to 35%, of the dry weight. The remaining 20% to 40% consists of sand sizes. The grain size distribution of the reddish-brown material does not differ significantly from that of the yellowish brown sediments.

The gravimetric moisture contents of the silt deposit are extremely high; in general, ranging from 55% to 80% with a mean value of 65%. All moisture content determinations are shown in Figure 2 plotted versus the elevation above sea level of the sample in the field. Also shown in this figure are the liquid limits of the same samples. The field moisture content of many specimens is higher than the liquid limit. Figure 2 shows that the moisture contents do not exhibit any trend of decreasing moisture content with depth. Furthermore, there is no significant difference between the reddish-brown and the yellowish-brown sediments.

The results of Atterberg limits determinations show that the silts have very high liquid and plastic limits. The mean values are 58 for the liquid limit and 10 for the plasticity index. While the liquid limit ranged from 50 to more than 80, the plasticity index was always very small, never exceeding 15. These Atterberg limits are characteristic of silts of high plasticity.

The dry densities of the silt deposit are very low, ranging from 800 kg/m^3 (50 pcf) to 1,200 kg/m^3 (75 pcf) with a mean value of 970 kg/m^3 (61 pcf). The results of all determinations performed have been plotted versus the field elevation of the sample in Figure 3. These results illustrate the lack of any trend of variation of the dry density with elevation. Furthermore, the densities of the

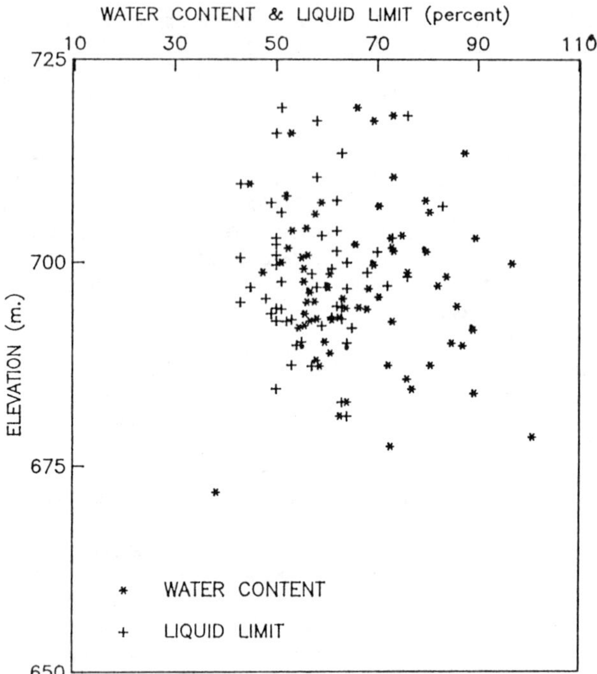

FIG.2--Moisture content and liquid limit of silt specimens.

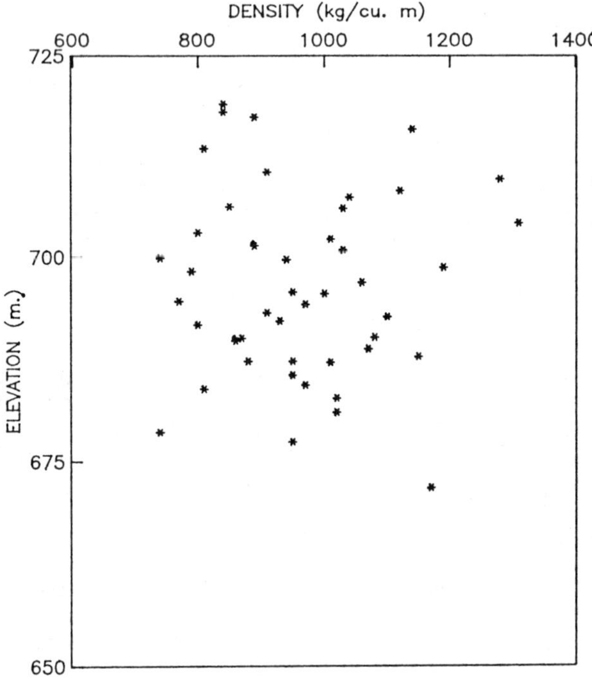

FIG.3--Dry density of silt specimens.

reddish-brown and yellowish-brown sediments do not exhibit any significant differences.

Two specimens of the silt sediments, one of the reddish-brown and the other from the yellowish-brown silts, were prepared and observed in a scanning electron microscope. The tubular particles typical of halloysite were clearly apparent in both specimens. These "SEM" observations are the only mineralogical identification tests attempted on soils from this site.

The <u>specific gravity of the solids</u> ranges from 2.62 to 3.00. The results obtained on specimens with simultaneous determinations of specific gravity, moisture content, and dry density are presented in Table 1. These results have been used to calculate the percentage of saturation of the specimen. The calculated degrees of saturation are also listed in Table 1. The degrees of saturation range from 95% to 105%. The anomaly of degrees of saturation higher than 100% is attributed to the precision of the measurements involved, especially for the specific gravity of solids. Nevertheless, since the results hover around 100%, it is clear that the silt deposits were very nearly saturated in the field.

Table 1--Approximate degrees of saturation

Elevation (m)	Moisture Content (%)	Dry Density (kg/m^3)	Specific Gravity of Solids	Degree of Saturation (%)
713	87.4	810	2.90	98
710	44.8	1,280	2.98	100
708	51.9	1,120	2.97	94
703	89.6	800	2.92	99
702	65.7	1,010	2.99	100
701	73.3	890	2.87	95
700	69.3	940	2.98	95
696	62.5	910	2.67	86
696	63.2	1,000	2.93	96
695	89.1	800	2.62	102
693	84.8	870	2.69	109
692	87.0	860	2.98	105
690	59.6	1,080	3.00	101

LABORATORY MEASUREMENTS

Specimens from samples recovered from the boreholes and exploratory pits have been subjected to laboratory determinations of hydraulic conductivity. The purpose of the test program was several fold: 1) to obtain representative values of the hydraulic conductivity of the silt deposit, 2) to assess the effect that hydraulic gradients might have on the hydraulic conductivity of the silts, and 3) to assess the isotropy/anisotropy of the silt deposit.

Test Procedure

The specimens were tested under constant head conditions in fixed and flexible wall permeameters. In the fixed-wall permeameter, the specimens were subjected to back pressures of 100 kPa (14psi); 200 kPa (29 psi) and 350 kPa (51 psi). The flexible wall permeameters consisted of triaxial cells. In these, the specimen was subjected to cell pressures 50 kPa (7 psi) larger than the tail water pressure applied on the specimen. The test was initiated by testing at 150 kPa (21 psi) tail water pressure. Then the test was repeated for increasing tail water pressures such as 350 kPa (51 psi), 550 kPa (78 psi), 750 kPa (106 psi), and 950 kPa (135 psi).

Tests on Undisturbed Specimens

Eleven undisturbed specimens recovered from the borings were tested in constant head permeameters and two were tested in a triaxial cell. The tail water pressures ranged from 100 kPa to 350 kPa. Four specimens were cut from a block sample recovered from an exploratory pit; three specimens were carved vertically, and the fourth specimen was carved horizontally. The results of all these tests are summarized in Figure 4. In this figure, the hydraulic conductivity is plotted versus the porosity of the specimen.

In general, these results show a consistent trend of increasing hydraulic conductivity for increasing porosity; the most notorious exception being the specimens cut from the block sample. The results obtained with fixed and flexible wall permeameters are not significantly different.

The hydraulic conductivities for the average porosities range from 1 x 10^{-6} cm/sec to 3 x 10^{-5} cm/sec. The results obtained on the specimens cut from block samples show changes over two orders of magnitude. The hydraulic conductivity of the horizontal specimen was found to be in the same order of magnitude as for the vertical specimens, thus, suggesting that the silt deposit is isotropic. This fact could be expected from the origin of the soil, which was weathered in-place rather than being eroded, transported and sedimented.

Hydraulic Gradient Effects

The influence of the overburden stresses and the potential hydraulic gradients on the silt stratum were evaluated by testing specimens in a triaxial cell under increasing tail water pressures. The cell pressure was always kept 50 kPa (7 psi) above the corresponding tail water pressure. The specimen section near the exit side is under the effect of a large effective confining stress caused by the presence of the hydraulic gradient.

The ratios of the hydraulic conductivity for an intermediate tail water pressure to the hydraulic conductivity measured at the lowest tail water pressure are illustrated in Figure 5. A six fold increase of the tail water pressure results in a decrease of the hydraulic conductivity of 30% for most of the specimens tested. Only for one of the specimens did the hydraulic conductivity decrease one order of magnitude.

These results illustrate that the overburden pressure does not significantly alter the hydraulic conductivity of the deposit. This effect is indirectly confirmed by the dry densities and moisture contents of undisturbed specimens illustrated in Figures 2 and 3, since a trend of increasing dry density or decreasing moisture content with depth is not apparent in these figures.

BOREHOLE CONSTANT HEAD INFILTRATION TESTS

To investigate the variability/homogeneity of the silt deposit with depth, several borehole infiltration tests

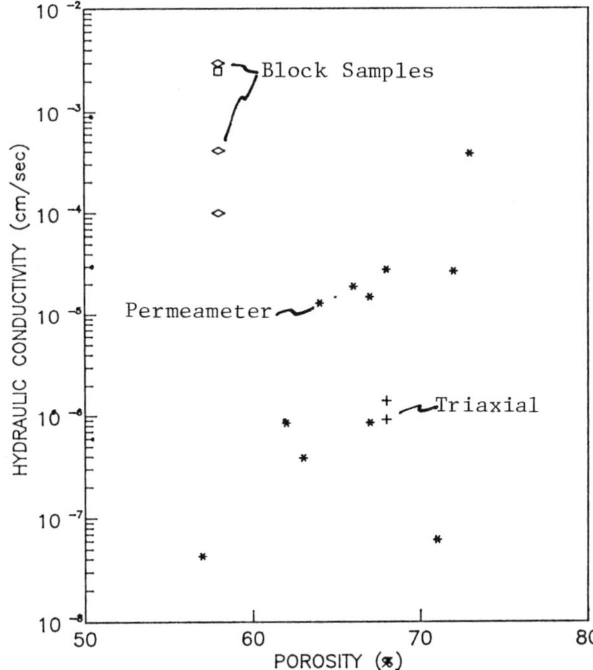

FIG.4--Hydraulic conductivity of undisturbed silt specimens

were performed at the site. The conditions of these tests are listed in Table 2.

Table 2--Conditions of borehole infiltration tests

Boring Designation		Elevation Interval Tested (m)	Conditioning Time (hr)	Total Volume Infiltrated (m^3)	Duration of Test (hr)
B-2		715/709	1.0	0.5	5.91
B-7	(a)	698/683	1.0	0.5	4.32
	(b)	693/683	73.0	0.3	6.79
B-11		702/699	1.0	0.02	6.61

Boring B-2 was drilled in the northern half of the basin and thus coincided within the reddish-brown silts. Borings B-7 and B-11 were drilled in the yellowish-brown silt zone.

Test Procedure

After the 0.1 m (4 in.) diameter boring was drilled, the uncased borehole was filled with water to a level 1 m

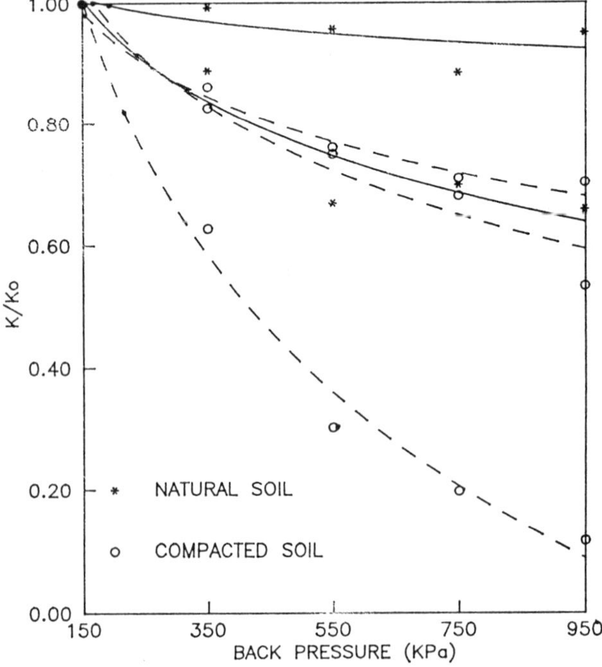

FIG.5--Hydraulic conductivity changes due to gradients

(3.3 ft) above the test level. Then the water level was allowed to drop to the desired test level. This phase was intended to condition the soil by initiating the infiltration before performing the actual permeability test. As soon as the water level reached the test levels, constant head tests were performed.

In this test, the water level inside the borehole is maintained at a constant elevation by regulating the flow rate into the borehole. The raw data collected is the time required for a volume of water to seep out of the borehole. The water levels inside the borings were monitored by electric sounding devices. The specified volume of water was held in an independent tank and the flow towards the borehole was regulated by manually operated valves.

Test Interpretation and Results

The field moisture contents were very high, sometimes were larger than the corresponding liquid limit. This suggests that the soil suction at the site was very low. This is further confirmed by the degrees of saturation presented in Table 1. These considerations indicate that the flow of water out of the boreholes would not be influenced significantly by soil suction.

Several methods to calculate the saturated hydraulic conductivity from the seepage velocity into boreholes are discussed in reference [1]. This study indicated that, for very large ratios of water depth over borehole radius, the best method had been developed, from the original work of Nasberg [2], by Terletskaya [3]. This model, assumes an infinite depth of vadose zone with negligible capillary pressure, and evaluates the flow when the steady state has been reached.

An alternative method has been proposed by Phillips [4] to include the effect of capillary pressures. In this method, the effect of capillary pressure, is included through the "sorptive mumber" that is large when the gravitational flow predominates over the capillary pressure induced flow. For fine grained soils with small effects of the capillary pressure, the "sorptive number" has a typical value of 1.0 m^{-1} [5]. These two methods of analysis were used at the site as summarized in Table 3.

It is worthwhile to notice the negligible difference between the conductivities calculated from the tests in B-7(a) and B-7(b). The main difference between these tests being the length of conditioning time (1 hr vs three days). Based on this fact, it has been assumed that after the conditioning times used the soils surrounding the borings were saturated and that the assumed steady state regime had been reached in all cases. Nevertheless, the results obtained with Phillips approach [4] indicate that small differences in the degree of saturation would have negligible effects on the calculated conductivities.

Table 3--Results from borehole infiltration tests

Boring Designation	Ignoring Capillarity (cm/sec)	Including Capillarity (cm/sec)
B-2	8.09×10^{-5}	6.75×10^{-5}
	7.34×10^{-5}	6.12×10^{-5}
B-7 (a)	2.05×10^{-5}	2.07×10^{-5}
	1.51×10^{-5}	1.52×10^{-5}
	1.46×10^{-5}	1.48×10^{-5}
	1.37×10^{-5}	1.38×10^{-5}
	1.51×10^{-5}	1.52×10^{-5}
B-7 (b)	1.16×10^{-5}	1.11×10^{-5}
	1.19×10^{-5}	1.14×10^{-5}
	1.25×10^{-5}	1.20×10^{-5}
B-11	0.82×10^{-5}	0.57×10^{-5}
	0.83×10^{-5}	0.57×10^{-5}
	0.81×10^{-5}	0.56×10^{-5}
	0.82×10^{-5}	0.57×10^{-5}

TRENCH PERCOLATION TESTS

The large variability observed in the hydraulic conductivity of undisturbed specimens had suggested the need to measure the hydraulic conductivity with field tests to find representative field averages. The infiltration in boreholes indicated that variations of the hydraulic conductivity with depth were not significant. The question still remained about horizontal variations of the silt deposit. To address this aspect several trench percolation tests were performed. A summary of the more relevant conditions of the three trenches tested is presented in Table 4. Trenches T-2 and T-3 were dug into yellowish-brown silts and trench T-4 was dug into the reddish-brown silts. The tests performed on trench T-3 were repeated several times after increasing the length of the trench. Trenches T-2 and T-4 were 1.70 m (5.6 ft) wide and trench T-3 was 1.13 m (3.7 ft) wide.

Test Procedure

Three trenches of rectangular plan shapes and vertical walls were excavated to depths ranging from 1.80 m (6 ft) to 4.27 m (14 ft). Upon completion of the excavation, the trench was filled with water to a depth of 1.0 m (3.3 ft) or 2.0 m (6.6 ft) below the ground surface. The water level was maintained at this level for the conditioning times listed in Table 4.

The test consisted of recording the time needed for a fixed volume of water to seep out of the trench while the water level within the trench was being maintained at a

Table 4--Conditions of the trench percolation tests

Trench Designation		Trench Length (m)	Elevation Interval Tested (m)	Conditioning time (days)	Volume of Seepage (m^3)	Test Duration (hr)
T-2	(a)	4.95	694/692	2	3.2	0.88
	(b)	4.95	693/692	5	3.2	1.97
T-3	(a)	2.75	701/699	10	3.2	2.45
	(b)	4.78	701/699	2	3.2	3.38
	(c)	6.70	701/699	0.6	3.2	0.72
T-4		5.05	705/704	3	3.2	0.81

constant elevation. The duration of the tests was determined by the time necessary for a volume of 3.2 m^3 (4.2 y^3) to seep out of the trench. During the test, the time necessary to consume 0.2 m^3 (0.26 y^3) was recorded.

Test Interpretation

The flow rate per unit length of trench was used in conjunction with the geometry of the trench to calculate the hydraulic conductivity of the surrounding soil from the theory of seepage from canals [6]. Specifically, the approach used the theory proposed by Vedernikov [7] to quantify the seepage from trapezoidal canals into infinitely deep horizontal drainage layers. The analysis considers that the steady state has been established and neglects the effect of capillary pressures in the surrounding soil.

According to this analysis the quantity of seepage per unit length of trench is given by the following formula:

$$q = K (B + A \cdot H) \qquad (1)$$

where:

- q seepage per unit length of trench,
- K hydraulic conductivity,
- B trench width at the water surface,
- H depth of water within the trench, and
- A constant based on the geometry of the trench (B/H).

Values of the constant A were published by Vedernikov [7]. However, the published values considered a maximum slope of the canal walls of 1:1 (45 degrees). Thus, appropriate values of "A" for vertical trench walls were derived in this study. This was accomplished by numerical evaluation of the three improper integrals involved in the calculation of "A" as discussed in [6]. The results derived for nearly vertical trench walls are listed in Table 5.

Table 5--Values of parameter "A" for vertical trench walls

Width/Height (B/H)	Parameter A
0.322	2.394
0.338	2.423
0.363	2.462
0.399	2.512
0.455	2.565
0.523	2.636
0.612	2.720
0.727	2.819
0.877	2.937
1.077	3.079
1.349	3.257
1.735	3.487
2.310	3.804

These values of the parameter "A" were used with the quantities of seepage measured to calculate the hydraulic conductivities. In all these calculations, it has been assumed that the seepage from the trench is two dimensional, and thus, the ends effect has been neglected. However, the tests on trench T-3 have been performed for several trench lengths in an attempt to quantify this effect.

Test Results

The calculated hydraulic conductivities from the six trench tests are presented in Table 6. In this table, it is clear that there is no pattern of decreasing hydraulic conductivity with time. This fact seems to indicate that the steady state regime had been achieved in all the tests, despite that the conditioning times ranged from a few hours to ten days. Based on this fact, it has been assumed that, after conditioning, the soils were saturated and the steady state regime had been stablished.

The trench percolation tests yielded surprisingly close values of hydraulic conductivity, but in stark contrast to the measurements on undisturbed specimens and in borehole infiltration tests.

The trench in the reddish-brown silt (T-4) yielded hydraulic conductivities twice as large as those of the yellowish-brown silt. A very similar pattern was present in the results from borehole infiltration tests shown in Table 3. In this table, the results from the borehole drilled in the reddish-brown silt (B-2) yielded hydraulic conductivities three times larger than the results from boreholes drilled in the yellowish-brown silts.

The tests in trench T-3 and borehole B-11 were performed in very close vicinity and the test elevation interval and the heights of water were kept the same in

Table 6--Hydraulic conductivities (cm/sec) from trench percolation tests

T-2 (a)	T-2 (b)	T-3 (a)	T-3 (b)	T-3 (c)	T-4
2.25E-3	1.77E-3	1.98E-3	1.01E-3	3.02E-3	5.54E-3
2.29E-3	1.81E-3	2.15E-3	3.32E-3	3.32E-3	4.98E-3
2.79E-3	1.89E-3	2.97E-3	1.01E-3	2.80E-3	5.17E-3
	1.75E-3	2.27E-3	0.83E-3	3.22E-3	5.32E-3
	1.82E-3	2.29E-3	0.82E-3	3.12E-3	5.03E-3
2.59E-3	1.79E-3	2.13E-3	0.83E-3	3.15E-3	5.11E-3
2.62E-3	1.77E-3	2.18E-3	0.84E-3	3.08E-3	5.03E-3
2.87E-3	1.80E-3	2.20E-3	0.82E-3	3.15E-3	5.11E-3
3.04E-3	1.80E-3	2.19E-3	0.83E-3	2.82E-3	5.14E-3
2.99E-3	1.56E-3	2.26E-3	0.86E-3	2.74E-3	5.48E-3
2.75E-3	1.72E-3	2.16E-3	0.84E-3	3.04E-3	4.90E-3
2.75E-3	1.64E-3	2.11E-3	0.90E-3	3.12E-3	4.75E-3
2.74E-3	1.77E-3	2.31E-3	1.13E-3	3.72E-3	4.90E-3
	1.77E-3	2.19E-3	1.14E-3	3.32E-3	5.00E-3
	1.76E-3	2.27E-3	1.97E-3	2.93E-3	5.09E-3

both tests. However, the hydraulic conductivities calculated from the trench percolation test is 250 times larger than the value figured from the borehole infiltration test. Although some part of this difference can be attributed to the two dimensional interpretation of the trench percolation test, it is believed that this would be a small percentage of the difference.

The tests performed on T-3 were implemented for three different lengths of the trench. It was expected that, for the longer trenches, the hydraulic conductivities would progressively decrease. This trend is clear for tests T-3 (a) and T-3 (b). However, for test T-3 (c) the hydraulic conductivity increases. These results show that neglecting the ends effect does not introduce large errors in the resulting hydraulic conductivities. In fact, the horizontal variability of the silt deposit seems to have a much more influential effect.

DISCUSSION AND CONCLUSIONS

The moisture content and dry density of undisturbed specimens, revealed a soil deposit with large variations in properties. Nevertheless, this variation appears to be random without any apparent horizontal layering or otherwise. Although no ground water surface was encountered, the soils had very large moisture contents, which in general were higher than the liquid limit. This indicates that the soil deposit had to be very nearly at field capacity. Fact that was confirmed by the degrees of saturation calculated for specimens on which specific gravity of the solids had been determined.

Laboratory determinations of hydraulic conductivity on undisturbed specimens showed a very large range of hydraulic conductivities. This fact is consistent with the variability observed in moisture content and dry density described above.

The results of laboratory tests for different hydraulic gradients and overburden pressures showed that both effects combined have only a very limited effect, less than one order of magnitude, on the hydraulic conductivity of undisturbed and compacted specimens. This suggests that increasing overburden pressures will not have the effect of inducing soil zones with significantly different hydraulic conductivities.

The hydraulic conductivities calculated from borehole infiltration tests yielded very consistent values with differences smaller than one order of magnitude. In fact, the higher conductivities were obtained from a test on a reddish-brown silt, while the results on the yellowish-brown silt were within a narrower range of variation around 1.0×10^{-5} cm/sec.

The result from trench percolation tests were very repeatable within ranges similar to those shown by the borehole infiltration test results. The trench test also indicated that the reddish-brown silt was somewhat more permeable than the yellowish-brown silt. However, in stark contrast to the borehole test, the hydraulic conductivities calculated from trench tests are two orders of magnitude larger than the results from the boreholes.

Potential explanations for these differences based on anisotropy of the soil deposit or a scale effect due to the "in-situ" soil suction do not have any merit because both would have to have affected the borehole and trench tests in a very similar fashion. In conclusion the only plausible explanation of the differences in apparent hydraulic conductivity has to be found in the variability of the soil deposit as indicated by the results of the index laboratory tests. Since the trenches affect much larger soil areas and volumes than the borehole, it is more probable that a highly permeable zone will affect seepage from the trench than from a small soils borings. Thus resulting in the higher hydraulic conductivities calculated from the trench tests.

REFERENCES

[1] Stephens, D.B., and Neuman, S.P., "Vadose Zone Permeability Tests: Summary," Journal of the Hydraulics Division, ASCE, Vol. 108, No HY5, 1982, pp 623-639.

[2] Nasberg, V.M., "The Problem of Flow in an Unsaturated Soil for Injection Under Pressure," Ixvestja Akademia Nauk, SSSR Odt Tekh Nauk, No 9, 1951, pp 1415-1429, as cited in [1].

[3] Terletskaya, N.M., "Determination of Permeability in Dry Soils," Hydroelectric Waterworks, No 2, 1954, Moscow, USSR, as cited in [4].

[4] Phillip, J.R., "Approximate Analysis of the Borehole Permeameter in Unsaturated Soil," Water Resources Research, Vol. 21, No 7, 1985, pp 1025-1033.

[5] Daniel, D.E., "In Situ Hydraulic Conductivity Tests for Compacted Clay," Journal of Geotechnical Engineering, Vol. 115, No 9, 1989, pp 1205-1226.

[6] Harr, M.E., Groundwater and Seepage, McGraw-Hill Book Co., New York, 1962, pp 231-246.

[7] Vedernikov, V.V., "Seepage from Channels," Versickerungen ans Kanalen, Wasserkraft and Wasserwirtschaft, No 5, 11, 12, and 13, 1934, as Ocited in [6].

Alan J. Lutenegger[1] and Don J. DeGroot[1]

MEASUREMENT OF HYDRAULIC CONDUCTIVITY IN CLAY USING PUSH-IN PIEZOMETERS

REFERENCE: Lutenegger, A. J. and DeGroot, D. J., "Measurement of Hydraulic Conductivity in Clay Using Push-In Piezometers," Current Practices in Ground Water and Vadose Zone Investigations, ASTM STP 1118, David M. Nielsen and Martin N. Sara, Eds., American Society for Testing and Materials, Philadelphia, 1992.

ABSTRACT: The hydraulic performance of push-in standpipe piezometers was evaluated at a site consisting of medium stiff and soft varved clay in western Massachusetts. The results of falling head hydraulic conductivity tests performed in the piezometers were compared with results of tests performed in prebored and sand-packed piezometers and laboratory flexible wall tests performed on undisturbed samples obtained at the site. A comparison of the field and laboratory tests conducted throughout the upper portion of the profile indicates that while tests performed in the prebored piezometers generally gave hydraulic conductivity results higher than laboratory tests, results from tests performed in push-in piezometers were closer to laboratory tests.

KEYWORDS: hydraulic conductivity, ground-water investigations, piezometers, monitoring wells.

Monitoring wells are commonly used in ground-water investigations to determine water levels or pore water pressures at a point in the subsurface, obtain samples for water quality analyses and determine strata hydraulic conductivity by performing in situ conductivity tests. In clay soils, conventional monitoring wells for hydrogeologic investigations are often constructed using a slotted screen or porous element installed in a drilled borehole and then sand packed using clean sand. An isolation seal and grouting or backfilling around the pipe/borehole annulus completes the construction.

This type of installation may be undesirable or even prohibitive in some cases for several reasons;

1. Disposal of drill cuttings, especially if contaminated soils are involved,

[1]Associate and Assistant Professor of Civil Engineering, respectively, at the University of Massachusetts, Amherst, MA 01003.

2. Difficulty of construction, for example artesian conditions or unstable soils,

3. Possibility of inadvertant cross contamination of aquifers by drilling an open hole.

Additionally, the accuracy of the hydraulic performance of a well installed in a predrilled borehole may often be in question, a result of suffering from a number of unknowns including:

1. Proper positioning and alignment of the screen,

2. Proper construction of the sand pack, and

3. Positive performance of the isolation seal.

To avoid some of these potential problems, an alternative construction is to use a pushed or driven piezometer which can be installed without predrilling. In addition to being faster and easier to install than a predrilled piezometer, a push-in piezometer reduces the possibility of a hydraulic short circuit between the isolation seal and borehole wall or standpipe which can go otherwise undetected in a predrilled piezometer. Push-in piezometers may also provide a considerable cost benefit to an overall groundwater investigation because of the relative ease of installation and reduction in time.

BACKGROUND

Push-in piezometers have been used in geotechnical engineering projects to monitor in situ pore water pressures for at least the past 25 years [1] and are still an integral part of many instrumentation programs in soft clays [2]. The technology of these instruments is not new, but there are obvious questions concerning the reliability of hydraulic conductivity measurements obtained from tests performed using a piezometer which has been installed in such a way as to displace the soil during installation.

Tavenas et al. [3] presented the results of a field investigation of in situ permeability measurements in soft sensitive clays in Eastern Canada and Sweden. They found that pushed piezometers were subject to clogging and caused extensive remolding of the clays which in turn reduced the void ratio and hydraulic conductivity. In comparison to tests obtained using a self-boring instrument, push-in piezometers gave results which were reduced by about a factor of between 2 and 17. The largest differences appeared to be in the clays which were the most sensitive. Laboratory values of the hydraulic conductivity obtained during consolidation tests on undisturbed samples were in between the push-in and self-boring tests.

Tremblay and Eriksson [4] found similar results between laboratory tests and field tests using push-in piezometers at five sites in Sweden. In these tests, there appeared to be very little difference in hydraulic conductivity obtained using push-in probes ranging in diameters from 25 mm to 35 mm. Similar comparative results between push-in piezometers and laboratory tests have been reported by Handfelt et al. [2] and by Leroueil et al. [5] with mixed results.

Jezequel and Mieussens [6] compared the results of self-boring permeameter tests in which the instrument was both jacked into the ground and installed in a self-boring manner. At a clay site in which the average hydraulic conductivity from all field tests was 8×10^{-7} cm/sec., the ratio between self-bored and direct penetration varied from 2 to 13, suggesting remolding or a disturbance effect from installation.

FIELD INVESTIGATIONS

Test Site

In order to evaluate the hydraulic performance of push-in piezometers in cohesive soils, tests were performed at a site consisting predominantly of medium stiff to soft varved clay. The site is located in the Connecticut River Valley in Amherst, Massachusetts on the University of Massachusetts campus as shown in Fig. 1. In addition to the work described in this paper, the test site has been used to perform other field studies related to geotechnical and hydrogeologic behavior of these soils.

The geologic origin of the soils at the site are related to lacustrine deposition of Glacial Lake Hitchcock during late Pleistocene - early Holocene time. The soils generally consist of varved clays with alternating silt layers with occasional find sand partings within the silt layers. The thickness of individual layers typically ranges from 2 to 8 mm. As a result of desiccation, the upper portion of the profile at the site has developed considerable stiffness from overconsolidation with the soils grading into a more normally consolidated state with increasing depth.

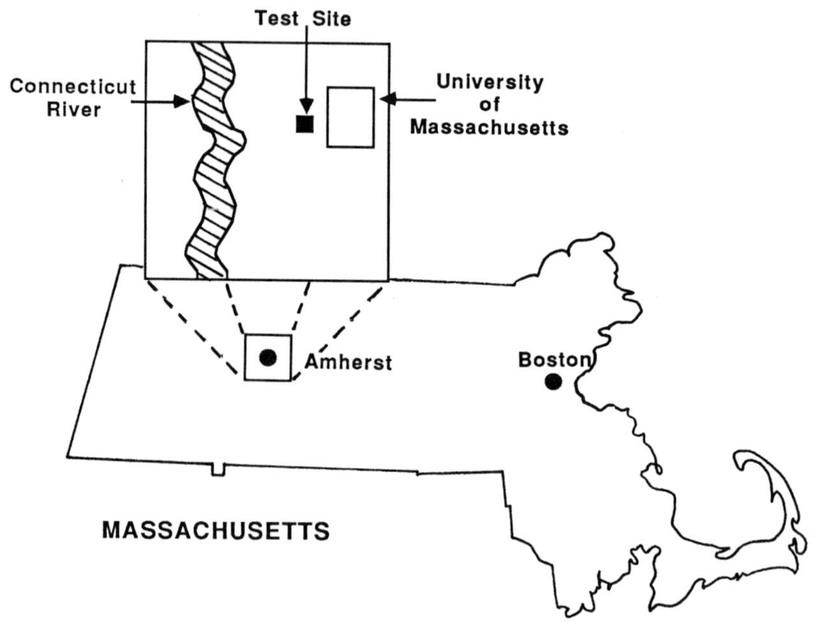

Fig. 1 Location of Test Site

A summary of the site stratigraphy in the upper 10 m and bulk geotechnical characteristics of the soils is shown in Fig. 2. Beneath the surficial fill, the bulk composition is shown to be relatively uniform but as noted the stress history indicates stiff to medium stiff materials to a depth of about 5 m. At the time of field investigations, the water table fluctuated from between 1.2 and 2.5 m below the surface, the fluctuations being related to seasonal variations in local precipitation. Typical variations in the water table are shown in Fig. 3 which presents water level measurements obtained from predrilled piezometers located at a depth of 3 m over a one year period. Pneumatic piezometers installed at the site indicate approximately hydrostatic pore pressure conditions at the site.

Piezometer Installation

Predrilled Piezometers: Piezometers were installed in predrilled holes in order to provide a typical construction which would likely duplicate a common technology currently being used in ground-water investigations. Although most production monitoring wells would be installed in boreholes drilled by a drill rig using continuous flight or hollow stem augers, predrilled holes for this investigation were drilled by hand using a 76 mm diameter open bucket hand auger, except for the piezometer at a depth of 9.1 m. This borehole was drilled using wash boring techniques, flushing with clean water prior to installing the piezometer.

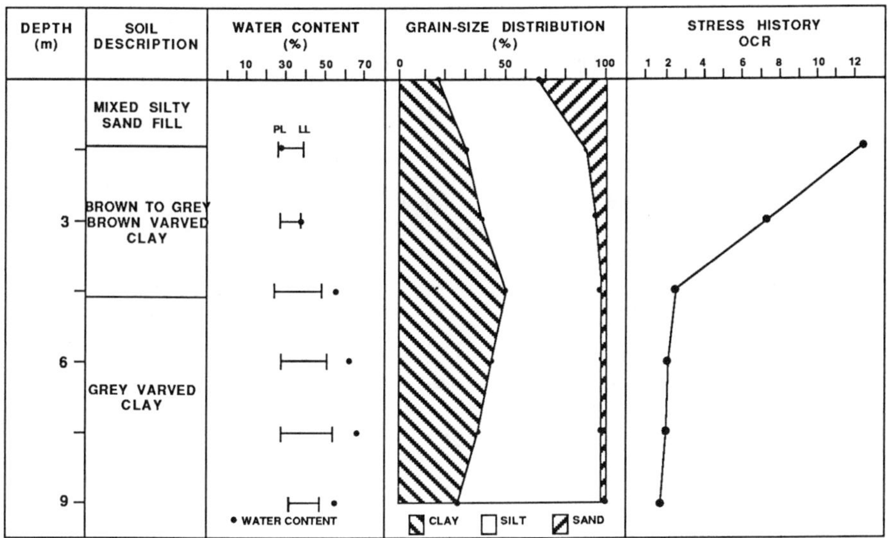

Fig. 2 Soil Properties at Test Site

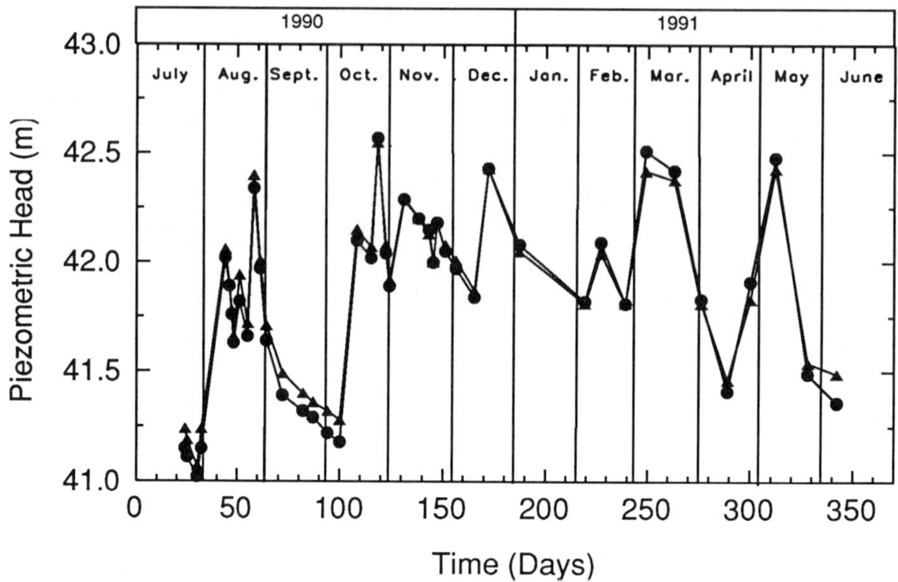

Fig. 3 Groundwater Fluctuation at Test Site

At each test depth, piezometers were constructed of a 0.3 m section of slotted PVC pipe. At depths of 3.0 and 6.1 m, replicate piezometers were constructed of a porous plastic filter element to provide some redundancy of measurements and a comparison of hydraulic response. A schematic of the piezometer dimensions and construction is shown in Fig. 4. Sandpacks were constructed using a uniform clean Ottawa sand with a median grain size of 0.6 mm. A bentonite isolation seal with a minimum length of 1.0 m was placed above the sand pack in each piezometer using bentonite pellets.

Push-In Piezometers: Push-in piezometers installed at the site were of two configurations. The majority of tests were performed using a commercial piezometer (Geonor M-206) fitted with a sintered bronze porous element with an effective opening size of 50 μm flush mounted onto the body. A 20° conical tip at the front of the piezometer displaces the soil during installation. Four Geonor piezometers were installed for this investigation. Dimensions of the piezometers are shown in Fig. 4.

A second style of push-in piezometer was also obtained commercially (Soilinst) and was constructed with a conical tip of larger diameter than the piezometer body. The body of the piezometer was not constructed of continuous porous material but instead had a series of holes connected to an internal porous body. A schematic is shown in Fig. 4.

Push-in piezometers were installed using the hydraulic feed from a drill rig. Holes were predrilled in the upper unsaturated zone prior to installation to remove surficial fill. A penetration rate of about 2 cm/sec. was used and extension pipes were filled with water and capped during installation. A total time of about 10 min. was needed to install each piezometer.

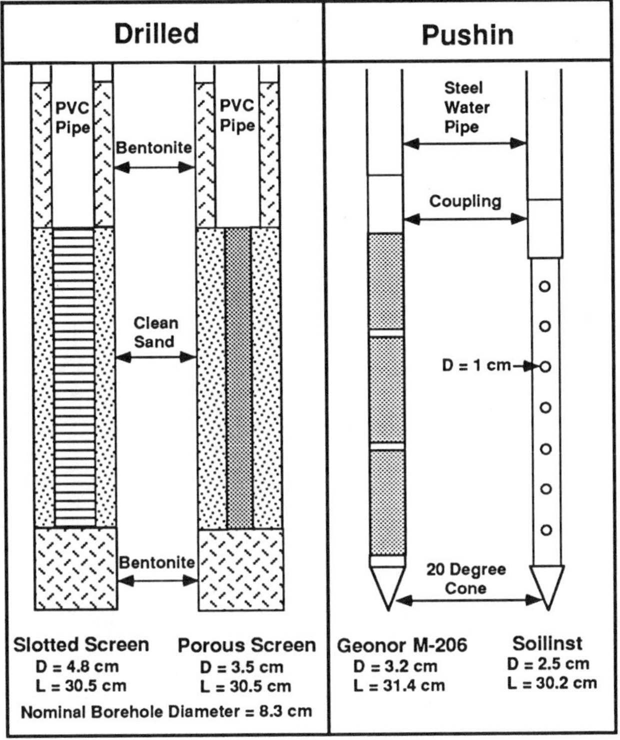

Fig. 4 Schematic of Piezometer Construction

Hydraulic Conductivity Tests

Falling head hydraulic conductivity tests were performed in each of the piezometers installed by adding a "slug" of water to the standpipe. Changes in the total head were obtained at various times using an electrical water level indicator in order to determine the average hydraulic response of the piezometer. Tests were normally conducted for 1 to 2 days, depending on the initial excess head applied and the hydraulic response. Test results were plotted in the form of the change in head as a function of time as shown in Fig. 5. The hydraulic conductivity, K, can be determined from:

$$K = -2.3 \frac{a\ m}{F} \qquad [1]$$

where: a = cross sectional area of standpipe
m = slope of head loss curve
F = piezometer shape factor

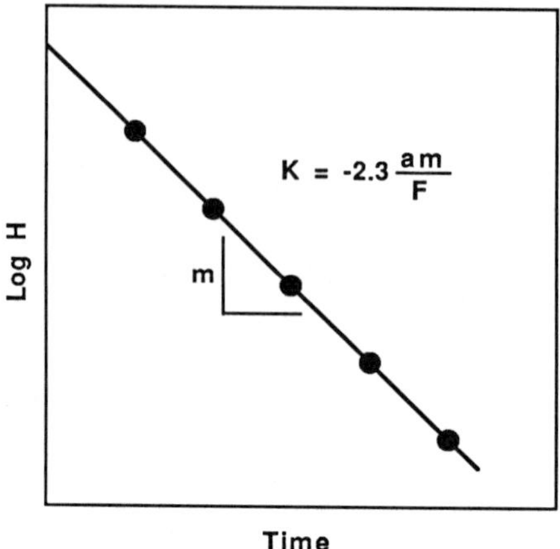

Fig. 5 Data Interpretation of Falling Head Test

Chapius [7] has presented a detailed summary of the available shape factors for use in determining hydraulic conductivity in piezometers. Considering the scatter in reported shape factors for various piezometer geometries and the fact that the hydraulic flow is predominantly in the horizontal direction for L/D's used, for this investigation the shape factor was obtained from:

$$F = 2\pi L/\ln\ [L/D + [1 + L^2/D^2]^{0.5}]\ -2.75D \qquad [2]$$

In order to calculate the shape factor for the predrilled piezometers, it was assumed that the hand augering process produced a borehole slightly larger (83 mm) than the actual auger head (76 mm) to account for auger head drift during rotation.

Multiple tests were performed in each piezometer in order to account for installation and ensure reproducible test procedures to obtain reliable values of hydraulic conductivity. Results of hydraulic conductivity tests performed in two of the piezometers are shown in Fig. 6. Repeatability of the tests is indicated to be very good and was independent of initial excess head and test sequence. Based on estimates of the in situ lateral stress, initial excess heads were kept low enough to preclude hydraulic fracture of the soil.

Laboratory constant head hydraulic conductivity tests were performed on undisturbed samples obtained using a 76 mm diameter fixed piston sampler. In order to determine the horizontal flow characteristics, samples were trimmed on a 90° orientation to allow water flow parallel to the varves, simulating field flow conditions. Flexible wall tests were performed in triaxial chambers using effective confining stresses equal to the vertical in situ effective stress at each sample depth. Back pressure was applied to ensure sample saturation.

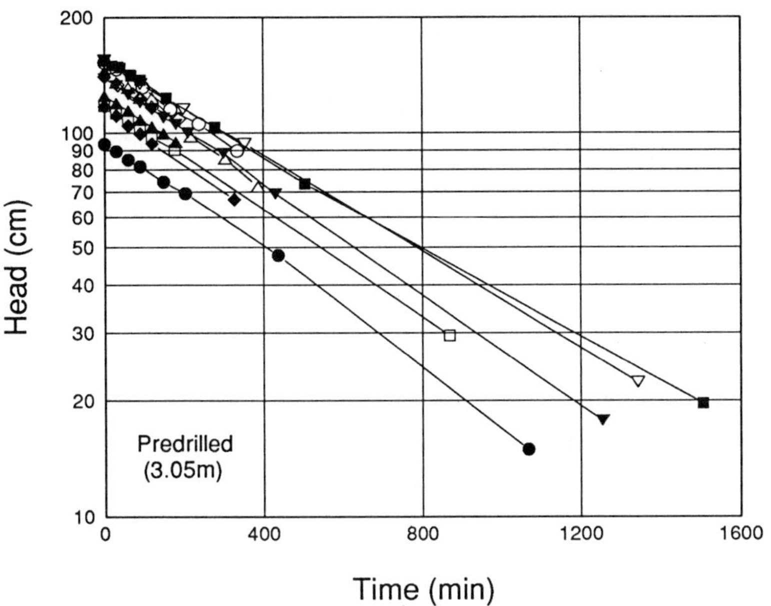

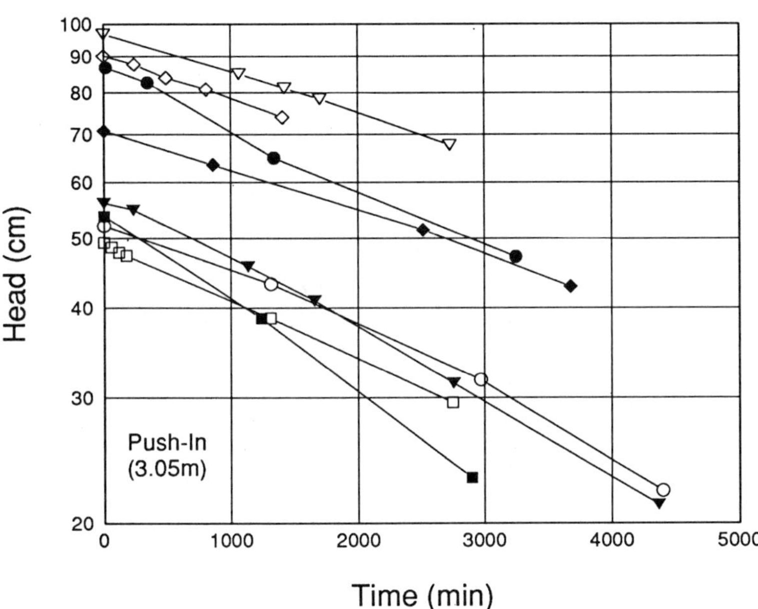

(Note: different symbols indicate different tests)

Fig. 6 Typical Response of Predrilled and Push-In Piezometers

Summary of Test Results

Figure 7 presents a summary of the field falling head tests performed in both the predrilled and Geonor push-in piezometers and the laboratory constant head tests. The results show a clear trend throughout the profile with the predrilled piezometer tests giving the highest values of hydraulic conductivity and the push-in piezometer tests giving the lowest values. The difference between these tests is about one order of magnitude. With the exception of the test at 1.5 m, laboratory horizontal flow test results fall in between the results of piezometer tests, but tend to be much closer to push-in results.
The ranges shown in Fig. 7 for each piezometer type are the maximum and minimum values obtained during the testing period. Only one laboratory test was performed on each undisturbed sample at each depth. Replicate porous plastic predrilled piezometers at depths of 3.0 and 6.1 m gave results within the range of slotted predrilled piezometers. The Soilinst push-in piezometers gave a very slow response because of the small amount of exposed filter, therefore the results were not used in the present study.

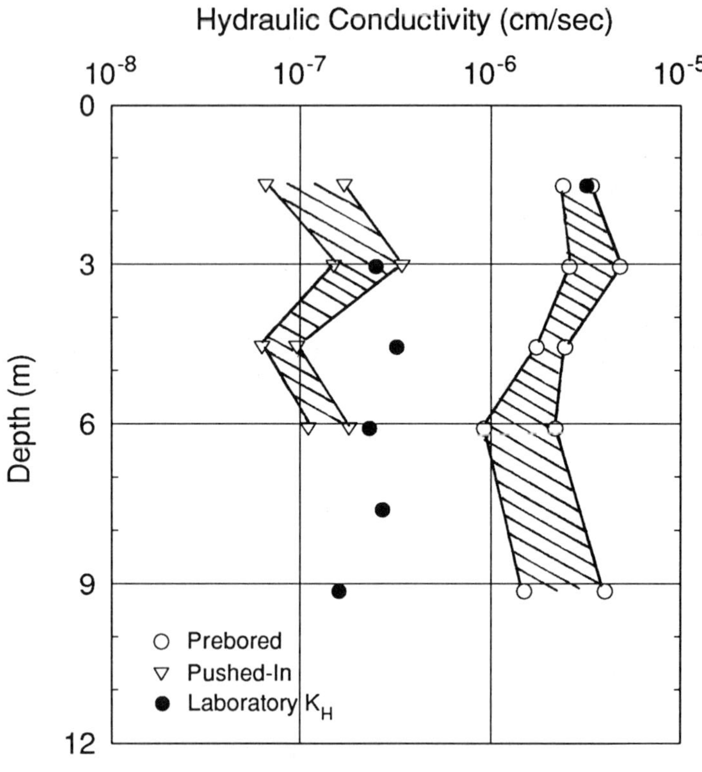

Fig. 7 Summary of Hydraulic Conductivity Tests

DISCUSSION OF RESULTS

The results presented in Fig. 7 indicate different hydraulic properties obtained from the predrilled and push-in piezometers. They also show that the results of field predrilled piezometers give hydraulic conductivity values which are greater than laboratory measured values. These trends should be considered in terms of their accuracy at predicting the "true" value of hydraulic conductivity.

A number of investigations, recently summarized by Olson and Daniel [8], have shown that the ratio of field to laboratory hydraulic conductivity is typically greater than 1 and may range from 2 to 1000. The most often quoted reason for this difference is the larger scale represented by field tests. That is, the laboratory values, which are obtained from small samples, only reflect the microscale matrix hydraulic conductivity of the soil, while the field tests, which presumably involve a much larger volume soil during the flow process, represent a larger scale hydraulic conductivity. Therefore, any inherent macrofabric features, such as joints, fissures, lenses of different material, etc. would influence the measured hydraulic response, which would normally result in a higher hydraulic conductivity from a field test. Using this logic, even larger scale tests, such as well pumping tests, would result in hydraulic conductivity values which are higher yet, and would reflect a "formation" hydraulic conductivity.

This may help explain the differences measured between prebored piezometers and laboratory tests, assuming that the prebored piezometers are constructed with no inherent problems, e.g. hydraulic short circuit, and the assumed geometry is in fact accurate. Since extreme care was taken to provide for a uniform sand pack and a good isolation seal, it is believed that the construction of the prebored piezometers is very good. However, it should be noted that even under ideal conditions, problems with predrilled piezometers can occur. For example, two attempts were made to install a predrilled piezometer at a depth of 4.5 m, however, drilling difficulties and borehole squeezing prevented a proper installation.

There are three potential problems which may contribute to inaccurate hydraulic response of push-in piezometers. These include:

1. Clogging of the filter element during pushing,

2. Consolidation of soil as a result of increases in effective stress from installation,

3. Soil disturbance (smearing or remolding) during installation.

It is difficult to separate these effects however the importance of each can be qualitatively assessed in order to make some logical analysis of the test results.

Clogging of the porous filter element can occur if clay particles are forced into the element during penetration. Since the element is flush-mounted, one might expect this to occur, except than the predominant stress acting along the element during installation is shear and not normal stress. By having the standpipe filled with fluid and capped, the tendency for ingestion of soil into the filter is counteracted. As further evidence against this problem, results of piezocone tests, in which a porous filter mounted behind the cone tip is intimately in contact with the soil during penetration rarely show any significant effects of soil clogging on either the measured dynamic pore pressure measured during penetration or dissipation response. Similarly, there is an obvious lack of reported cases of wick drain clogging during penetration in soft clays.

Consolidation of clay soils as a result of direct penetration under undrained conditions would occur <u>after</u> penetration has stopped. Reduction of the porosity (voids ratio) and hence hydraulic conductivity of the soil results from any increase in effective stress which can be attributed to the installation. In overconsolidated clays, direct push installation results in an increase in effective stress surrounding the probe, however this is rarely in excess of the yield stress, which means that any reduction in porosity is small, since compression is in the reload range. In normally consolidated clays, the installation creates large positive pore water pressures which tends to reduce the effective stress. During reconsolidation after installation, the effective stress generally returns to a level equal to or only slightly above the preinstallation level. Again, this would tend to produce only a small change in hydraulic conductivity if any.

Soil disturbance (remolding) results from destructurization of the natural fabric of the soil produced by the direct push penetration. In massive (non macro-structured) clays, this would generally consist of remolding the soil at constant porosity, which would typically produce a reduction in hydraulic conductivity by a factor of about 2 [9, 10]. In soils which have macro-structure related to either depositional environment or post-depositional changes, the influence will be greater, and may result in a reduction of the hydraulic conductivity by an order of magnitude or more depending on the nature of change in macro-structure. The same would be true of sensitive soils which undergo destructurization as a result of penetration, which may help explain the data presented by Tavenas et al. [3].

The soils at the site tested have macro-structure in the form of layering as a result of lacustrine deposition. No other features, such as fissures or joints, have ever been encountered during the past two years of extensive test drilling at the site. During penetration of the push-in piezometers, it is likely that the layering was destroyed so that some smearing of the flow channels occurred adjacent to the piezometers. If a finite skin of mixed soil is present next to the piezometer, the hydraulic conductivity measured by the push-in piezometers would be reduced, which appears to be the case. It is of some interest to note that hydraulic conductivity values obtained using the push-in piezometers compare very well with laboratory values obtained from <u>vertical</u> flow (i.e., across the varves) conductivity tests.

Indirect evidence of the influence of direct penetration on the change in horizontal hydraulic conductivity may be obtained by examining the pertinent literature regarding the installation of sand drains or prefabricated wick drains to enhance the consolidation rate in soft clays.

Casagrande and Poulos [11] reported on a case where sand drains had been installed in a varved clay in New Jersey where two 14 in. diameter sand drains had been installed by washing a pipe down and by driving a closed-end pipe. The hydraulic conductivity measure in the driven hole was 10 times lower than that measured in the washed hole. The difference may be the result of distortion or disturbance of the layering in the varved clay, but also may relate to soil remolding.

Results of laboratory consolidation tests presented by Hansbo [12] and Akagi [13] both show that below the preconsolidation stress, the difference in coefficient consolidation between an undisturbed and a completely remolded clay varies by a ratio of between 4 and 25. Once the preconsolidation stress is passed, i.e., in the normally consolidated range of loading, there is essentially no difference between undisturbed and remolded behavior. Samples disturbed by installation of a displacement drain consistently showed a decrease in the coefficient of consolidation by a factor of 3 in comparison to the undisturbed soil. Again, once the preconsolidation stress is passed, the soil looses it's initial fabric and the results give identical values.

In terms of the hydraulic response of a push-in piezometer, the issue is much more complicated than one original assumes with the original soil fabric, stress history, sensitivity and initial hydraulic conductivity all affecting the measured response. The response is no doubt a combined response of a zone of disturbed and possibly remolded soil immediately adjacent to the piezometer surrounded by a zone of undisturbed soil. Therefore, a more rigorous analytical procedure which takes into account initial soil conditions is needed in order to make more rational interpretation of test results.

CONCLUSIONS

Results of hydraulic conductivity tests performed in conventional predrilled piezometers, push-in piezometers and on undisturbed specimens in laboratory flexible wall cells have been compared for a site consisting of medium stiff to soft varved clay.

The test results indicate that:

1. Push-in piezometers gave the lowest values of hydraulic conductivity while the predrilled piezometers gave the highest values. Laboratory test results fell in between the two types of field tests, but generally closer to the push-in.

2. The hydraulic response of the push-in piezometers is probably reduced by a zone of disturbed soil adjacent to the filter element, which in this case accentuates the difference between push-in and predrilled tests because of the varved nature of the soil.

3. Push-in piezometers were much easier to install than the predrilled, with only about 10 minutes needed for each piezometer. Therefore, this may be a distinct advantage for this kind of technology in certain cases.

REFERENCES

[1] Bjerrum, L., Kenney, T.C. and Kjaernsli, B., Measuring Instruments for Strutted Excavating, Jour. of the Soil Mechanics and Foundations Division, ASCE, 1965, Vol. 91, pp. 111-141.

[2] Handfelt, L.D., Koutsoftas, D.C., and Foott, R., Instrumentation for Test Fill in Hong Kong, Jour. of Geotechnical Engineering, ASCE, Vol. 113, pp. 127-146, 1987.

[3] Tavenas, F., Tremblay, M., Larouche, G., and Lerouel, S., 1986. In Situ Measurement of Permeability in Soft Clays, Use of In Situ Tests in Geotechnical Engineering, ASCE, pp. 1034-1048.

[4] Tremblay, M. and Eriksson, L., Use of Piezometers for In Situ Measurement of Permeability, Proc. 9th European Conference on Soil Mechanics and Foundation Engineering, 1987, Vol. 1, pp. 23-28.

[5] Lerouel, S., Diene, M., Tavenas, F., Kabbaj, M., and LaRochelle, P., 1988. Direct Determination of Permeability of Clay Under Embankment, Jour. of Geotechnical Engineering, ASCE, Vol. 114, pp. 645-657.

[6] Jezequel, J.F. and Micssens, C., 1975. In Situ Measurement of Coefficients of Permeability and Consolidation in Fine-Grained Soils, In Situ Measurement of Soil Properties, ASCE, Vol. 1, pp. 208-224.

[7] Chapius, R.P., Shape Factors for Permeability Tests in Boreholes and Piezometers, Groundwater, Vol. 27, No. 5, 1989, pp. 647-654.

[8] Olson, R.E. and Daniel, D.E., Measurement of the Hydraulic Conductivity of Fine-Grained Soils, ASTM STP 746, 1981, pp. 18-64.

[9] Mitchell, J.K., The Fabric of National Clays and its Relation to Engineering Properties, Proceedings of the Transportation Research Board, Vol. 35, 1956, pp. 693-713.

[10] Ting, N-H., Onoue, A., Germaine, J.T., Whitman, R.V. and Ladd, C.C., Effects of Disturbance on Soil Consolidation with Vertical Drains, MIT Report No. R90-11, 1990, 406 pp.

[11] Casagrande, L. and Poulos, S., On the Effectiveness of Sand Drains, Canadian Geotechnical Journal, Vol. 6, 1969, pp. 287-326.

[12] Hansbo, S., Consolidation of Clay with Special Reference to Influence of Vertical Sand Drains, Proc. Swedish Geotechnical Institute, No. 18, 1960.

[13] Akagi, T., Effects of Mandrel-Driven Sand Drains on Soft Clay, Proc. 10th International Conf. on Soil Mechanics and Foundation Engineering, 1981, Vol. 3, 581-584.

Statistical Evaluations of Ground-Water Data

Gisella M. Spreizer, Thomas J. Calabrese, and Randall S. Weidner[1]

ASSESSING THE USABILITY OF HISTORICAL WATER-QUALITY DATA FOR CURRENT AND FUTURE APPLICATIONS

REFERENCE: Spreizer, G. M., Calabrese, T. J., and Weidner, R. S., "Assessing the Usability of Historical Water-Quality Data for Current and Future Applications," Current Practices in Ground Water and Vadose Zone Investigations, ASTM STP 1118, David M. Nielsen and Martin N. Sara, Eds., American Society for Testing and Materials, Philadelphia, 1992.

ABSTRACT: Large amounts of laboratory data have been generated as part of environmental investigations, and particularly ground-water investigations conducted during the 1970s and 1980s. These data present a historical water-quality record at a facility by representing, in many cases, the baseline ground-water quality, and changes in water quality due to degradational processes and remedial activities.

Ground-water chemistry data collected during the 1970s and 1980s were often generated using laboratory quality assurance/quality control (QA/QC) procedures considered inadequate by today's standards. In addition, the documentation required to support the quality of the laboratory data may be unavailable. The absence of the supporting QA/QC documentation or the use of different QA/QC procedures than those used today should not be the basis for removal of the data from the facility's ground-water quality database. In a large number of cases, the historical data are "good data." Historical water-quality data are useful and should

[1] Ms. Spreizer and Mr. Calabrese are Senior Scientists with Geraghty & Miller, Inc; Ms. Spreizer is located at the Andover, MA office and Mr. Calabrese is located at the City of Industry, CA office. Mr. Weidner is an Engineering Geologist with Southern California Edison.

be available for use in the design of ground-water monitoring programs, for the selection and design of remedial measures, and to monitor the effectiveness of past and anticipated future remedial activities.

This paper presents a series of techniques that, when used together, can demonstrate the usability of a historical water-quality database where laboratory QA/QC data are unavailable. This paper makes an important assumption that there is a difference between data usability and data validation. The assessment of the usability of a database is performed as a two-part exercise. The first step is to conduct a qualitative pattern recognition procedure and the second is to use more stringent statistical procedures to identify if the data are presenting a consistent description of the ground-water system. This data-usability technique has been successfully applied at a site in California. The results of this case study are also present in this paper.

KEYWORDS: Data usability, data validation, statistics, pattern recognition, historical water-quality, regression analysis, graphical representations, California, pentachlorophenol, remedial programs.

Environmental assessment programs have been conducted since the early 1970s and some of the investigations have continued to the present. During this time period, there has been an evolution of the environmental assessment process. The current review of the results of investigatory programs considered comprehensive in the 1970s and 1980s often reveals inadequacies in these programs when compared to today's standards. Much of the historical data collected during previous investigations are useful and should be, when possible, incorporated into contemporary investigations. However, these historical data often lack supporting quality assurance/quality control (QA/QC) data which would typically be needed to demonstrate the usability of these data [1]. QA/QC data are especially sparse or totally unavailable for historical ground-water chemistry data. In many cases, these data represent baseline ground-water quality data and changes in water quality due to natural degradation processes and, in limited cases, instituted remedial activities. It is often necessary to demonstrate that these historical water-quality data are usable, to document the usability of the data, and to substantiate their usability.

A series of techniques have been developed to demonstrate the usability of historical water-quality data where the laboratory QA/QC information is limited or unavailable. These techniques can also be used to support data validation activities being performed for current investigatory programs. The usability assessment is performed as a two-part exercise. The first part is a screening of the database using qualitative pattern recognition procedures. The second part of the assessment is the use of formal statistical procedures to identify if the data present a consistent description of the ground-water system and primarily relies on regression analysis techniques.

A case study is presented demonstrating the data-usability technique. This technique has been successfully applied to the pentachlorophenol ground-water quality database for a site in California.

BACKGROUND

For the purposes of this data usability assessment a distinction has been made between data usability and data validity. For the context of this paper and application, data validity implies that a specific analytical result can be considered accurate and precise because the field and laboratory methods used to generate that value are known. Data which are considered valid are also considered usable, if the data satisfy the objectives of the data collection [2-4]. However, usable data are not necessarily valid. Data usability implies that a database, which can include historical data, is indicative of an overall condition; no comment regarding the individual data values can be made. For this application, the condition is the spatial and temporal pattern demonstrated by the database. It is a depiction of concentration trends over time and also the areal distribution of the compounds detected in the ground water during previous sampling rounds. Hence, for a water-quality database the usability of the database is considered and not the validity of any single data point.

The objectives of the data usability assessment of the historical water-quality data are to:

- o evaluate the quality of the historical water-quality data by performing a statistical analysis of the trends represented in the data set;
- o describe the relative changes in the concentrations over time;
- o evaluate the effectiveness of the past and existing remedial activities; and

- describe the internal, spatial consistency of the water-quality data in the context of the conceptual hydrostratigraphic model of the area under investigation.

There are limitations of how the water-quality data can be used. No isolated concentration value can be considered as accurate and precise since the validity of the individual data points cannot be assessed. The data should only be used to gauge the effectiveness of the remedial activities currently underway at the site rather than to design new remedial actions. If additional remedial actions are warranted at the site, recently-collected data, with documented QA/QC procedures would be used as part of the design process for expanding the remedial action. The previously-collected data can be used as supplemental data for the recently-collected data by providing a preliminary assessment as to the necessity of additional remedial actions.

METHODOLOGY

As stated earlier, the first part of the data-usability assessment is a screening of the database, where the data included in the assessment are screened to identify which wells would be considered in the evaluation and to identify suspect or outlier data points. The data usability assessment of the subset database then proceeds in two steps: 1) a qualitative assessment utilizing graphical methods to identify both temporal and spatial patterns; and 2) a quantitative statistical evaluation. Each of these steps are described in further detail in the following sections.

Database Screening

It is necessary to screen the database so the data usability assessment can focus on the wells and parameters necessary to meet the objectives of the program. Because the data usability assessment is based on the consistency of the database over time and since no single data point within the database will be considered valid, the size of the database for each well is critical. Both the qualitative and quantitative evaluations would be expected to yield more confident results as the database size increases. It is recommended that the database consists of at least 20 datapoints. However, the application provides the best results when the database contains at least 30 datapoints.

As the initial step in the database screening, the concentrations are plotted as a function of time for these wells, and all obviously suspect or outlier data points are noted. The following are suggested conditions to identify outlier data points:

Condition 1: There is an apparent reporting error likely related to a unit transcription error. Specifically, the data point is off by a factor of 1000 (too high or too low) reflecting a reporting unit error in terms of parts per million (ppm) or parts per billion (ppb).

Condition 2: There is an apparent error by the reporting laboratory in the dilution or concentration factors used in the laboratory analysis and reporting, based on a comparison to analytical results reported for samples collected in the same time period for the same well.

Condition 3: Replicate or duplicate results differ by orders of magnitude and only one of the replicate or duplicate results appear consistent with previous and subsequent analytical results.

Condition 4: The sample identification has some ambiguity in its description.

Condition 5: The sample result is an isolated and substantially high or low reported concentration, and is inconsistent with the existing database.

Condition 6: There are known changes in sampling procedures and responsibilities, or analytical methodologies.

These are only suggested conditions to be used for the screening of the database. Additional conditions can be developed for each particular application; however, attempts should be made to develop them using objective criteria. For those instances where no objective criteria can be developed and applied, the data point should be retained in the database and considered part of the continuing data usability assessment.

Qualitative Pattern Recognition

The qualitative assessment involves the identification of both temporal and spatial patterns. For the temporal analysis, the data for each ground-water monitoring well are plotted as a function of time. The patterns and trends exhibited by each graph are reviewed and the similarity of the patterns among the monitoring wells are compared. For the spatial evaluation, the graphs of wells screened in similar hydrostratigraphic zones were compared.

The review of the internal consistency of these graphs considered the following:

- o Does the database consists primarily of detected, or not detected (ND) and less than the detection limit (LT) results?

- o If the database consists primarily of analytical results at reportable concentrations, are the concentrations decreasing and approaching an apparently asymptotic concentration?

- o Are there limited time periods within the database where the database is obviously inconsistent?

- o What pattern are the data following?

The patterns exhibited for wells screened in similar hydrostratigraphic settings are also compared to each other. The overall trends of the data (increasing or decreasing) for similar wells are considered. Particular attention is focused on those wells where there are limited time periods of inconsistency and whether these time periods appeared similar in any of the wells.

Quantitative Assessment

A quantitative and formal statistical evaluation focusing on the use of regression (trend) techniques is then performed for those wells where the temporal and spatial graphical evaluation present a consistent pattern [5]. The models considered for the quantitative assessment are the general exponential model and a form of the exponential model where an asymptotic concentration is approached. Depending on the patterns identified in the qualitative assessment other models such as linear and higher-order models may also be appropriate.

The equations for the exponential models are presented below:

$$Y = C_o e^{-kx} \quad \text{(General Model)}$$

or

$$Y = C_A + (C_o - C_A) e^{-kx} \quad \text{(Asymptotic Model)}$$

where C_o is the starting concentration, C_A is the concentration that is expected to be approached by the ground-water system as time infinitely increases (i.e. the asymptotic concentration), k is the exponential decay constant (the negative sign reflects that concentrations are decreasing), x is time (presented in days since the initiation of monitoring), and Y is the concentration measured at any time x. For those wells where an increasing concentration trend is present, the sign on the exponential decay constant (**k**) would be positive. Other exponential models are presented below [7].

Model	Asymptotic
$C_t = C_a + (C_o - C_a)t^{-k}$	C_a
$C_t = a/(1 + be^{-kt})$	a
$C_t = C_a + (C_o - C_a)\,k^t$	C_a
$C_t = ka^{b^t}$	k

The exponential models are fit to the database using standard nonlinear regression-analysis techniques [5-6]. The independent variable in this analysis is **x** (time) with the sample collection dates converted to the number of days since the first sample was collected at the specified well. The dependent variable is **Y** (concentration) and is the concentration reported in parts per billion (ppb) for each sampling date. The values for C_o, k and C_A are determined from the nonlinear regression.

Nonlinear regression model fitting techniques are iterative procedures where initial estimates for C_o, k, and C_A (if the asymptotic model is being used) are calculated using graphical methods [5-6]. These initial estimates, and the time and concentration data, are input into the model to obtain the best estimates for C_o, k and C_A. Convergence criteria were used to determine the best estimates. The criteria were selected on mathematical considerations for determining the fit of nonlinear models, and also considering the accuracy and precision of analytical data.

For the regression models used, the fit and confidence of the model should be assessed. A number of factors were considered in the selection of the most appropriate regression model. For each model application, the following should be calculated at a minimum:

- o The coefficient of determination (R^2) which is the total variation of the concentration values about its mean that is explained by the fitted model.

- o The sum of the squares (variability) explained by the model (this is similar to the factor above).

- o The standard error associated with each predicted coefficient.

- o The match between the predicted concentrations from the model and the actual detected concentrations over the entire range of concentrations detected.

- o The distribution and correlation of the model residuals.

Based on the fit and appropriateness of the models to the data, determinations are made regarding the usability of the data. Particularly, recommendations for the types of inferences (qualitative, semi-quantitative or quantitative) that would be warranted based on the database are made.

CASE STUDY

The methodology presented above to demonstrate the data usability for historical water-quality data was successfully applied for pentachlorophenol data collected for a wood-treating facility in California. This facility had been in operation for over 55 years and had an on-going ground-water monitoring program in effect since the early 1970s. The data-usability assessment focused on the pentachlorophenol data for this facility since it was included in the ground-water monitoring program for the facility for a relatively long period of time, it was characteristic of the materials used at the facility, and because the analytical method used for the laboratory analysis was relatively constant over time.

A remedial investigation/feasibility study is being conducted for this facility and to include the historical pentachlorophenol data in the investigation would greatly enhance the available database. Remedial activities were initiated at the site in late 1976 when a slurry wall was installed. The ground-water monitoring data before the installation of the slurry wall and the initiation remedial activities in 1976 represent a continuous record of ground-water conditions from the pre-remediation, remediation, potential upgrades to the remediation, and, eventually, post-remediation periods of the facility operation. The major hydrostratigraphic units present beneath the facility are the vadose zone sediments, the shallow aquifer, the shallow aquitard, the intermediate aquifer, the intermediate aquitard, and the deeper aquifer. Currently, the shallow aquifer has little or no water as a result of the state-wide drought. Ground-water in the intermediate aquifer flows in a west-southwest direction beneath the facility.

For the assessment, 16 wells were considered. The database for each well contain at least 20 data points, and many contain over 70 pentachlorophenol concentration measurements. The database screening removed at a maximum 11 data points from the database for any one well. The smallest database size for the wells included in the quantitative assessment was 22 datapoints. The criterion most often used for the deletion of data points was Condition 3, poor reproducibility of replicate and duplicate results.

The graphical representations of the temporal changes in concentration reflect generally decreasing concentrations which appear to follow an exponentially decreasing curve. The data present consistent patterns and hence a formal trend analysis was performed on the data, with the exception of the wells where the results were reported primarily as ND. For all of these wells, a generalized exponential model was used in the regression analysis. For seven of the wells, both the generalized exponential model and the asymptotic model were fitted to the data.

Five of the shallow wells included in the assessment are located inside the slurry wall. These wells were being monitored prior to the installation of the slurry wall with the exception of one well. The four pre-slurry wall wells all consistently show decreasing concentrations, and substantial decreases after 1977 (when the slurry wall was completed). These data indicate that the combined remedial actions of the installation of the slurry wall with ground-water extraction and treatment have markedly improved ground-water conditions within the slurry wall boundary. Data from all of the deep aquifer wells reflect substantial decreases in pentachlorophenol concentrations starting in Fall 1977.

A total of 12 of the original 16 wells were evaluated during the quantitative assessment phase of the data-usability program. The databases for the other four wells contained primarily not detected results and were therefore excluded from the quantitative assessment step. These 12 wells include three wells from the deep aquifer and nine from the shallow aquifer.

The pentachlorophenol data for the wells listed were initially fitted to the general exponential model. The asymptotic model was used only if the pentachlorophenol concentrations detected in a particular well were above the detection limit concentration. The historical database included the detection limits for the measurements, especially when they were reported as not detected. In most instances, the generalized form of the exponential model was the most appropriate for the data. The asymptotic model was only determined to be applicable for one well (Figure 1). In this instance, an asymptotic concentration of approximately 1,500 ppb is evident graphically as shown on Figure 1. Historical pentachlorophenol data were available from

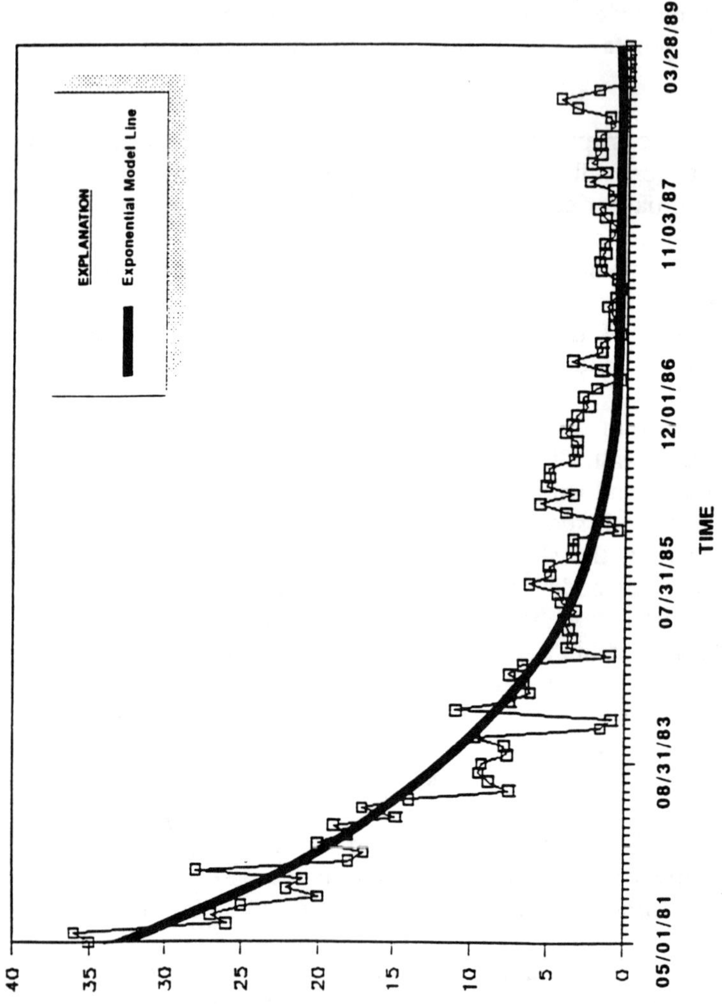
EXPLANATION
Exponential Model Line
PENTACHLOROPHENOL CONCENTRATION (ppb)
(Thousands)
40
35
30
25
20
15
10
5
0
05/01/81
08/31/83
07/31/85
12/01/86
11/03/87
03/28/89
TIME

May 1981 through March 1989. The model determined that the asymptotic concentration is 808 ppb. The fit of the model is good and over 95 percent of the data variability is explained by the model. The concentrations predicted by the model match well with the actually detected concentrations for the entire concentration range (10 ppb to 35,000 ppb) considered in the evaluation.

For the remaining wells, the generalized form of the exponential model was determined to be the most appropriate. However, the fit of the model varied among the wells. The percentage of the variability explained by the model ranges from 18 percent to 87 percent. For the majority of the wells, over 70 percent of the data variability could be explained by the model.

The databases for each of the wells carried into the quantitative phase of the usability assessment were evaluated to identify which are usable and the extent of their usefulness. Based on the overall results of the data usability assessment for the historical pentachlorophenol data, all of the data from the screened database for the 16 wells can be used to qualitatively describe the environmental history and status of cleanup at the facility. The detected concentrations have decreased substantially at all study locations and the decreasing concentrations are apparently related to the operating remedial system at the facility. The database for four wells considered in the qualitative assessment consisted almost entirely of not detected or less than results, and were not included in the quantitative phase of the assessment. The decreasing concentration trends for 12 of the 16 wells were assessed quantitatively. Since these wells represent a wide spatial area near the facility, similar patterns and confidence are expected for the pentachlorophenol data for wells not specifically included in the usability assessment. The quantitative phase of the data usability assessment generally resulted in good fits of the data to the exponential models studied. The results of the usability study for the 12 weeks in the case study are presented in Table 1. Ten of the 12 wells included in the quantitative phase of the assessment had 50 percent or greater of the data variability explained by the model, and the spatial consistency of the data is consistent with the conceptual hydrostratigraphic model. The pentachlorophenol data for two monitoring wells did not fit either of the exponential models with a great deal of confidence. However, the graphical representation of the pentachlorophenol data do indicate decreasing concentrations. Since these data presented a consistent graphical depiction, it is concluded that the data can be used to draw qualitative inferences.

TABLE 1.

SUMMARY OF QUANTITIVE ASSESSMENT OF THE PENTACHLOROPHENOL DATA

		Coefficients			
Well	Model	Ca (ppb)	Co (ppb)	k	Percent of Sum of Squares Explained by Model
Shallow Aquifer					
MW-1	General	--	672	0.00054	18%
MW-2	General	--	58,376	0.0013	64%
MW-3	General	--	10,640	0.0011	86%
MW-4	General	--	126,977	0.0014	76%
MW-5	Asymptotic	808	394,519	0.0015	95%
MW-6	General	—	29,702	0.00066	80%
MW-7	General	—	3,059	0.0017	87%
MW-8	General	--	29	0.0025	50%
MW-9	General	--	7,695	0.0035	84%
Deep Aquifer					
MW-10	General	--	33,370	0.0050	77%
MW-11	General	--	5,046	0.0008	80%
MW-12	General	—	205	0.00049	72%

* For description of model types, see text.

CONCLUSIONS AND RECOMMENDATION

The described statistical procedure to assess the usability of historical water-quality data is an objective methodology to evaluate the usefulness of historical water-quality data. Although procedures for the collection of these historical data may not be documented there is important information contained within the data. This methodology provides a mechanism to assess the reliability of this information. This methodology can be extended to the evaluation of historical data other than water quality.

A potential second application of this methodology is to supplement data validation activities. These activities are generally very time consuming and costly for validation of 100 percent of the data. It is proposed that this data usability methodology be used to allow the validation of less than 100 percent of the data. Individual data points that are considered critical to the investigation (such as for risk and exposure calculations, and for engineering design purposes) must be validated. However, the remaining data could be demonstrated to be usable by evaluating statistically the consistency of the entire database, which would also include the validated data points.

REFERENCES

1. USEPA, 1986, Test Methods for Evaluation Solid Waste. 3rd Edition. November.

2. USEPA, 1987, Data Quality Objectives for Remedial Response Activities and Example Scenario: Volumes I and II. March.

3. USEPA, 1988, Laboratory Data Validation, Functional Guidelines for Evaluating Inorganics Analyses. July.

4. USEPA, 1988, Laboratory Data Validation, Functional Guidelines for Evaluating Organics Analyses. February.

5. Draper, N. and H. Smith, 1981, Applied Regression Analysis, Second Edition, John Wittey & Sons, Inc. New York, New York.

6. SPSS, Inc., SPSS/PC+, 1989 Ver. 2.0 Base Manual and Update for V.30 and V.31, SPSS, Inc., Chicago, Illinois.

7. Spreizer, G.M., Maxim, D., Valkenburg, N., and Hauptmann, M., "How Flat Is Flat? -- Termination of Remedial Ground-Water Pumping," Ground Water and Vadose Zone Monitoring, ASTM STP 1053, D.M. Nielsen and A.I. Johnson, Eds., American Society for Testing and Materials, Philadelphia, 1990, pp. 247-255.

Russell H. Plumb Jr.

AN ALTERNATIVE DATA EVALUATION TECHNIQUE FOR DETECTING DISPOSAL SITE LEAKAGE EVENTS

REFERENCE: Plumb, R. H., Jr., "An Alternative Data Evaluation Technique for Detecting Disposal Site Leakage Events," Current Practices in Ground Water and Vadose Zone Investigations, ASTM STP 1118, David M. Nielsen and Martin N. Sara, Eds., American Society for Testing and Materials, Philadelphia, 1992.

ABSTRACT: A conceptual waste disposal site was compared to a physical model frequently used to demonstrate the properties of entropy. This comparison suggests that dispersion of chemical contaminants due to spontaneous waste disposal site leakage events would be expected to disrupt the ambient chemical homogeneity of an affected aquifer. The resultant changes in aquifer homogeneity can be identified and fingerprinted based on changes in monitoring data variance. Using this alternative data evaluation technique, monitoring data from a RCRA site monitoring program were used to prepare a series of contaminant-specific control charts (monitoring data variance versus time) that identified the occurrence of a leakage event and characterized the contaminants associated with the leakage event.

KEYWORDS: waste site monitoring, monitoring data variance, chemical dispersion, leakage event

One of the more pressing and pervasive environmental problems of the past decade has been the identification and characterization of the impact of waste disposal site operations on the ground-water resource. The magnitude of this problem is illustrated by the fact that there are an estimated 56,000 municipal/sanitary landfills, 225,000 industrial landfills, and 340,000 surface impoundments in this country [1]. Each of these sites represents a potential source of chemical contaminants that may eventually migrate to the ground water and create an adverse impact on human and environmental health. Although more than 1000 different

Dr. Plumb is a senior staff scientist with Lockheed Engineering and Sciences Company, 1050 East Flamingo Road, Suite 126, Las Vegas, NV 89119.

contaminants have been reported in disposal site ground water, the number of contaminants associated with a single site is usually quite limited but variable [2]. However, a common characteristic of every disposal site leakage event is the increased spatial dispersion that occurs with all the chemical contaminants that are involved. As a consequence of this observation, it has been postulated that entropy changes provide a technical basis for understanding the problem of leaking waste disposal sites and developing an effective national site cleanup strategy [3].

Entropy is an extensive property of state that is related to the randomness (homogeneity) of a chemical or energy distribution [4]. The fundamental characteristic of this property is that it is a monotonic function that is maximized for a system at equilibrium. A condition of maximum entropy also coincides with a stable system because of the homogeneous conditions and small energy differentials in the system. When the stability of a state of equilibrium is disrupted by the removal of one or more constraining conditions, spontaneous changes will occur until a new state of thermodynamic equilibrium and maximum entropy is achieved. A consideration of the characteristics of entropy and the relationship between entropy and equilibrium conditions provides a conceptual basis for a data evaluation technique to identify and characterize leakage events at waste disposal sites. The remainder of this paper shall develop this concept.

CONCEPT DEVELOPMENT

A classic physical model that has been used to illustrate the properties of entropy consists of two interconnected glass chambers (Figure 1). An ideal gas introduced into the upper chamber of the model will diffuse throughout the chamber until a uniform partial pressure is attained. At that point, a condition of equilibrium is achieved and the entropy has reached a maximum value. This condition is stable and will persist as long as the boundary conditions remain unchanged. However, if the barrier between the two chambers is removed, the gas will begin to diffuse into the lower chamber (Figure 1b). This process will continue until a new state of equilibrium is attained for which the partial pressure of the gas is the same in both chambers of the model (Figure 1c).

There are three facts that need to be emphasized regarding the example in Figure 1. First, the migration of the gas from the upper chamber to the lower chamber is spontaneous. Second, the Second Law of Thermodynamics states that all spontaneous changes will produce an increase in system entropy. Third, the entropy increase is also accompanied by an increase in chemical homogeneity. Each of these conditions were demonstrated with the ideal gas example but they are also valid for all thermodynamic systems [5].

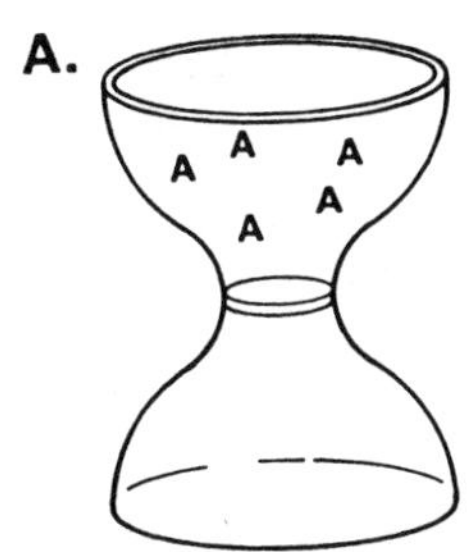

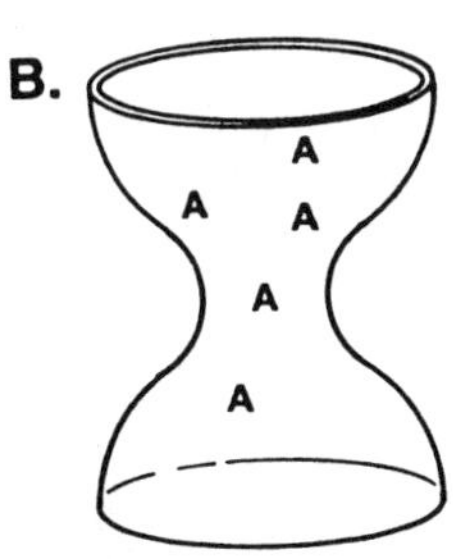

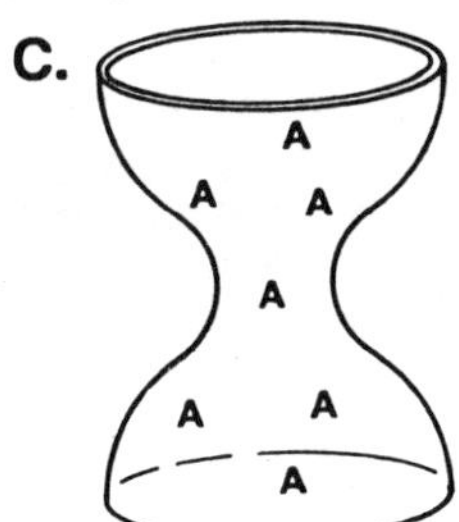

Figure 1. Entropy and homogeneity changes in an ideal gas system.

The classic model used to demonstrate the properties of entropy can be equated with a typical waste disposal site situation. For the sake of discussion, a waste disposal site can be conceptualized as a simple two chamber model (Figure 2). The upper chamber in the ideal gas model is equivalent to a waste disposal site and the second chamber is equivalent to the ground-water aquifer beneath the waste disposal site. As long as the site containment mechanisms remain intact and functional, the site will be in a state of quasi-stability and the waste materials will be contained (Figure 2a). However, when the containment mechanisms fail (barrels and drums leak, liners are punctured or torn, or clay barriers are fractured), the waste products begin to spontaneously migrate towards the aquifer (Figure 2b). Because the process is spontaneous, the migration of waste products will continue until a new state of equilibrium and maximum entropy is attained (Figure 2c).

The similarities between a waste disposal site and the entropy model provides the basis for a data evaluation technique that can be used to identify and characterize leakage events at waste disposal sites. In each instance (Figure 1c and Figure 2c), the removal of the retaining barrier results in increased chemical dispersion until a new state of equilibrium and higher entropy is achieved. Since the entropy of the system is maximized for homogeneous equilibrium conditions, the entropy of each system must be lower for each of the intermediate states (Figure 1b and Figure 2b). Therefore, by inference, each of the intermediate states must be less homogeneous than the final conditions. This line of reasoning leads to the conclusion that changes in ground-water homogeneity can be used to detect disposal site leakage events. A mathematical index that describes the homogeneity of a data set and can be used for this purpose is variance.

A schematic representation of a waste disposal site as shown in Figure 2 is an oversimplification of the hydrologic and geologic factors that can influence contaminant transport. For example, the thickness of the unsaturated zone will determine the distance that contaminants will have to travel before entering ground water. Also, the presence of fractures or zones of high permeability will establish preferred contaminant pathways. Each of these issues are important in assessing and evaluating contaminant behavior but they generally influence the rate of contaminant migration and the location at which contaminants enter the ground water. Despite the site-specific nature of these processes, contaminants leaking from a waste disposal site would still be expected to increase the ambient chemical heterogeneity when they enter ground water in the vicinity of the facility as suggested in Figure 2.

Furthermore, the conventional approach that has been used to monitor waste disposal site ground-water conditions provides the basis for a baseline from which changes in system homogeneity can be measured. Specifically, when a site is not impacting an aquifer, the upgradient and downgradient concentrations of each

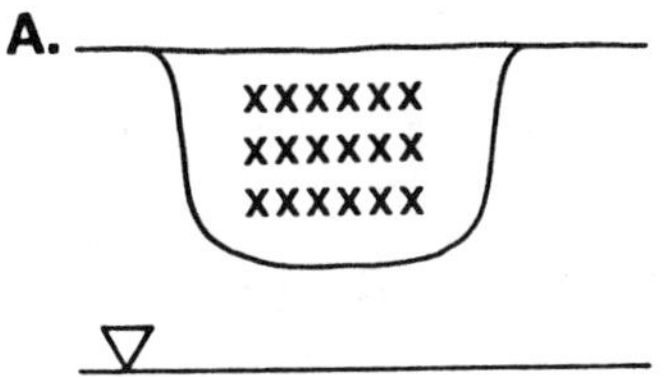

WASTE MATERIALS CONTAINED

GROUNDWATER ENTROPY = S_1
HOMOGENEITY = H_1
VARIANCE = V_1

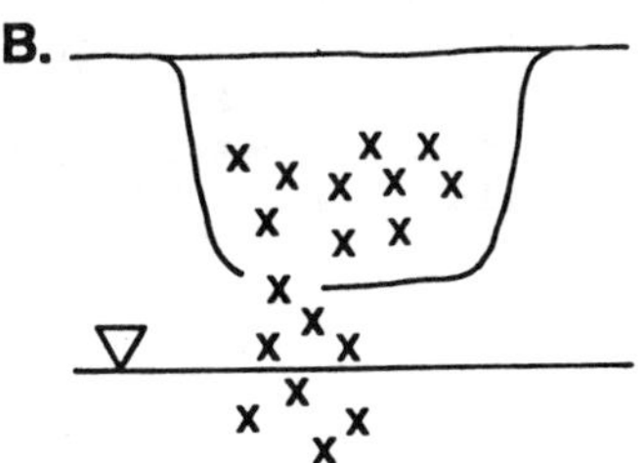

CONTAINMENT FAILS, WASTE MATERIALS REACH GROUNDWATER

GROUNDWATER ENTROPY = S_2
HOMOGENEITY = $H_2 < H_1$
VARIANCE = $V_2 > V_1$

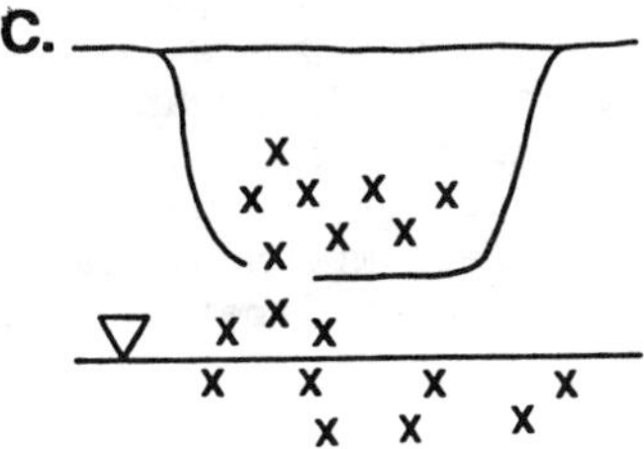

WASTE MIGRATION CONTINUES, NEW EQUILIBRIUM STATE ACHIEVED

GROUNDWATER ENTROPY = $S_3 > S_2$
HOMOGENEITY = $H_3 > H_2$
VARIANCE = $V_3 < V_2$

Figure 2. Entropy and variance changes due to a leakage event at a conceptual waste disposal facility.

constituent being monitored are expected to be equivalent. A set of monitoring data such as this with minimal scatter between sampling locations would be characterized by a minimal variance that ideally approaches a value of zero.

The alternative data evaluation technique that is being suggested as a means to assess disposal site leakage events is summarized in Figure 3. The data are tabulated as a function of sampling location and the sampling event. For each sampling event, the data from each location are used to calculate the inter-well variance. A control chart plot of inter-well data variance versus time for each contaminant can be used to evaluate the impact of the disposal site on local ground water:

1. If a contaminant is not associated with a disposal site leakage event, the inter-well data variance for this contaminant should approach zero. Furthermore, a plot of inter-well variance over time should produce a linear plot with a slope of zero (Figure 3).
2. If a contaminant is migrating from a waste disposal site, the concentration of that contaminant will increase in one or more monitoring wells. As a consequence of the increased inter-well data scatter, the data variance for that constituent will increase above the expected minimum value (approximately zero) associated with non-leaking conditions. As indicated in Figure 3, this change should cause a discontinuity in the variance-time plot.

CONCEPT APPLICATION

Ground-water data generated during a RCRA site monitoring program in Illinois were obtained for use in this paper. This data set consists of approximately 10,000 analytical results that describe the occurrence and distribution of 110 organic priority pollutants [7] in site ground water. The number of individual analyses generally ranged from 75 for each of the pesticide compounds to 125 for each of the volatile compounds. The data were the result of 17 monthly surveys conducted between April, 1985 and December, 1986.

The following approach was used to calculate the monitoring data variance for each organic priority pollutant included in Table 1 and demonstrate the concept:

1. For the first sampling event, the monitoring data for a single constituent (i.e., benzene) at each well in the monitoring network were used to calculate the variance of the data subset.

   For the purpose of these calculations, the following conventions were used. First, all concentrations were expressed in ug/L. Second, results that were below the

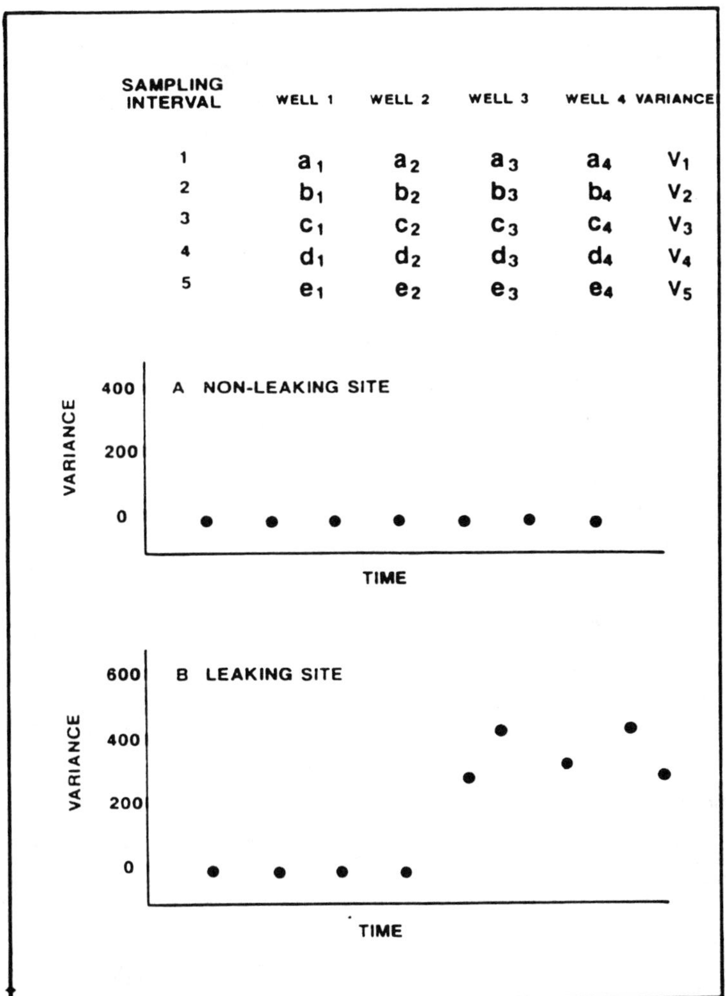

Figure 3. Expected variance patterns as a function of time for (A) a non-leaking site and (B) a leaking site.

TABLE 1 -- Log monitoring data variance as a function of time for each organic priority pollutant at site RCS #28.*

Contaminants	Sampling Date				
	4/85	8/85	12/85	4/86	8/86
Volatile Compounds					
Acrolein	---	X	X	X	X
Acrylonitrile	X	X	X	X	X
Benzene	X	5.4	1.3	4.5	0.6
Carbon tetrachloride	X	X	X	X	X
Chlorobenzene	X	6.0	5.1	5.3	4.6
1,2-Dichloroethane	X	X	X	X	X
1,1,1-Trichloroethane	X	X	X	X	X
1,1-Dichloroethane	X	X	X	X	X
1,1,2-Dichloroethane	---	X	X	X	1.3
1,1,2,2-Tetrachloroethane	X	X	X	X	X
Chloroethane	X	X	X	X	X
2-Chloroethylvinyl ether	X	X	X	1.9	X
Chloroform	X	X	X	X	X
1,1-Dichloroethene	X	X	X	X	X
1,2-trans-Dichloroethene	---	X	X	3.7	X
1,2-Dichloropropane	---	X	X	X	X
1,3-trans-Dichloropropane	---	X	X	X	X
Ethyl benzene	---	6.4	5.9	6.0	5.9
Methylene chloride	X	5.5	6.2	2.9	2.1
Chloromethane	X	X	X	X	X
Bromomethane	X	X	X	X	X
Bromoform	---	X	X	X	X
Bromodichloromethane	---	X	X	X	X
Fluorotrichloromethane	X	X	X	X	X
Dichlorodifluoromethane	---	X	X	X	X
Chlorodibromomethane	---	X	X	X	X
Tetrachloroethene	X	X	X	X	X
Toluene	X	7.6	6.9	6.3	6.3
Trichloroethene	X	X	0.7	2.0	X
Vinyl chloride	---	X	X	X	X
Base/Neutral Compounds					
Acenaphthene	---	1.8	1.8	1.1	X
Benzidine	---	X	X	X	X

(continued)

TABLE 1 -- (Continued)

Contaminants	Sampling Date 4/85	8/85	12/85	4/86	8/86
Base/Neutral Compounds					
1,2,4-Trichlorobenzene	X	X	X	X	X
Hexachlorobenzene	X	X	X	X	X
Hexachloroethane	X	X	X	X	X
bis-(2-Chloroethyl) ether	X	X	X	X	X
2-Chloronaphthalene	---	X	X	X	X
1,2-Dichlorobenzene	X	3.1	2.8	3.0	3.1
1,3-Dichlorobenzene	X	3.2	X	X	X
1,4-Dichlorobenzene	X	X	3.2	3.0	3.2
3,3'-Dichlorobenzidine	---	X	X	X	X
2,4-Dinitrotoluene	X	X	X	X	X
2,6-Dinitrotoluene	---	X	X	X	X
1,2-Diphenylhydrazine	---	X	X	X	X
Fluoranthene	X	1.5	2.1	1.2	2.8
4-Chlorophenyl phenyl ether	---	X	X	X	X
4-Bromophenyl phenyl ether	---	X	X	X	X
bis-(2-Chloroisopropyl) ether		X	X	X	X
bis-(2-Chloroethoxy) ether	---	X	X	X	X
Hexachlorobutadiene	X	X	X	X	X
Hexachlorocyclopentadiene	X	X	X	X	X
Isophorone	---	3.0	2.7	3.5	4.4
Naphthalene	X	5.2	4.5	5.0	4.8
Nitrobenzene	X	X	X	X	X
N-Nitrosodiphenylamine	---	X	X	X	X
N-Nitrosodi-N-propylamine	---	X	X	X	X
bis-(2-ethylhexyl) phthalate		X	3.2	3.4	3.8
Butyl benzyl phthalate	---	X	X	X	X
Di-N-Butyl phthalate	---	X	X	X	X
Di-N-Octyl phthalate	---	X	X	X	X
Diethyl phthalate	---	X	X	2.2	X
Dimethyl phthalate	---	X	X	X	X
Benzo(A)anthracene	X	X	X	X	X
Benzo(A)pyrene	X	X	0.2	X	X
Benzo(A)fluoranthene	X	X	X	X	X
Benzo(K)fluoranthene	---	X	X	X	X
Chrysene	X	X	X	X	X
Acenaphthene	X	1.8	1.7	1.2	1.1
Anthracene	---	X	1.4	0.6	2.7
Benzo(GHI)perylene	---	X	X	X	X

(continued)

TABLE 1 -- (Continued)

Contaminants	Sampling Date				
	4/85	8/85	12/85	4/86	8/86
Base/Neutral Compounds					
Fluorene	---	1.9	2.1	1.4	2.4
Phenanthrene	---	2.4	2.8	2.1	3.1
Dibenzo(A,H)anthracene	X	X	X	X	X
Indeno (1,2,3-CD) pyrene	X	X	X	X	X
Pyrene	---	1.3	1.9	1.0	2.9
Acid Extractable Compounds					
2,4,6-Trichlorophenol	X	X	X	X	X
p-Chloro-m-cresol	X	X	X	X	X
2-Chlorophenol	X	X	X	X	X
2,4-Dichlorophenol	X	X	X	X	X
2,4-Dimethylphenol	X	X	2.4	2.9	3.0
2-Nitrophenol	---	X	1.5	X	X
4-Nitrophenol	X	X	X	X	X
2,4-Dinitrophenol	X	X	X	X	X
4,6-Dinitro-2-methylphenol	X	X	X	X	X
Pentachlorophenol	X	X	1.6	X	X
Phenol	X	6.7	3.1	7.0	6.6
Pesticide Compounds					
Dieldrin	---	X	X	X	X
Chlordane	X	X	X	X	X
4,4'-DDT	---	X	X	X	X
4,4'-DDE	---	X	X	X	X
4,4'-DDD	---	X	X	X	X
Endosulfan I	---	X	X	X	X
Endosulfan II	---	X	X	X	X
Endosulfan sulfate	---	X	X	X	X
Endrin	---	X	X	4.4	X
Endrin aldehyde	---	X	X	X	X
Heptachlor	X	X	X	X	X
Heptachlor epoxide	---	X	X	X	X
alpha-BHC	---	X	X	X	X
beta-BHC	---	X	X	X	X

(continued)

TABLE 1 -- (Continued)

Contaminants	Sampling Date				
	4/85	8/85	12/85	4/86	8/86
Pesticide Compounds					
delta-BHC	---	X	X	X	X
Lindane	---	X	X	X	X
PCB-1242	---	X	X	X	X
PCB-1254	---	X	X	X	X
PCB-1221	---	X	X	X	X
PCB-1232	---	X	X	X	X
PCB-1248	---	X	X	X	X
PCB-1260	---	X	X	X	X
PCB-1016	---	X	X	X	X
Toxaphene	X	X	X	X	X

* For the purpose of these calculations, all analytical results were expressed in ug/L and any reported results below the limit of detection were set equal to zero. A minimum of four analytical results were used in each calculation.

--- Either there were no data or the number of sample results for this parameter was three or less. No variance was calculated for this monitoring period.

X This signifies those data subsets that had a calculated variance of zero (i.e., all the monitoring results were the same) for which the log is not defined. Also, those subsets with a calculated variance less than one and a log of the variance less than zero were included in this category to simplify the data tabulation.

limit of detection were set equal to zero. Since the upgradient and downgradient concentrations at a non-leaking facility should be the same, any convention that is followed consistently will produce a minimal monitoring variance. The selection of zero as a surrogate value simplifies the calculation process and also maximizes the resultant variance when a leakage event occurs. Third, a minimum of four analytical results were used for this calculation. This limitation was selected because the RCRA Subtitle C program requires a minimum of 4 monitoring wells. Therefore, if insufficient data are available to calculate the variance, the adequacy of a monitoring program would be flagged for investigation.

2. The first step was repeated for each constituent that was monitored during the survey.

3. Steps one and two were repeated for each survey.

4. The calculated variance value for each constituent during each survey was tabulated. Because of the observed range of values, this information is presented as the log of the calculated variance in Table 1. Since the log of zero is not defined and the log of variance values less than one were judged insignificant compared to the extreme values in the site monitoring data, these data sets are indicated with an X in Table 1.

Each row of data in Table 1 provides the necessary information to construct a contaminant-specific control chart to assess changes in ground-water conditions with respect to that contaminant. For example, the monitoring data variance for acrolein was uniformly low and unchanging over the period of record (Table 1, Figure 4). This is the type of pattern that would be expected if acrolein was not involved in a leakage event at the facility (Figure 3). A total of 21 volatile compounds, 29 base/neutral compounds, 7 acid extractable compounds, and 23 pesticides listed in Table 1 produced similar data trends over time.

The data for toluene, benzene, and several other priority pollutant compounds produced a different pattern over time (Table 1, Figure 4). During the first sampling period (4/85), the data variance for these compounds was universally low as would be expected for a non-leakage event. However, toluene data variance increased by almost seven orders of magnitude in the next and all subsequent monitoring periods and the benzene data variance increased by more than four orders of magnitude. The observed increase in data variance is consistent with the higher chemical heterogeneity that would be expected if toluene and benzene were migrating from the site (Figure 3). Furthermore, the magnitude of the variance would be indicative of the intensity of the leakage event.

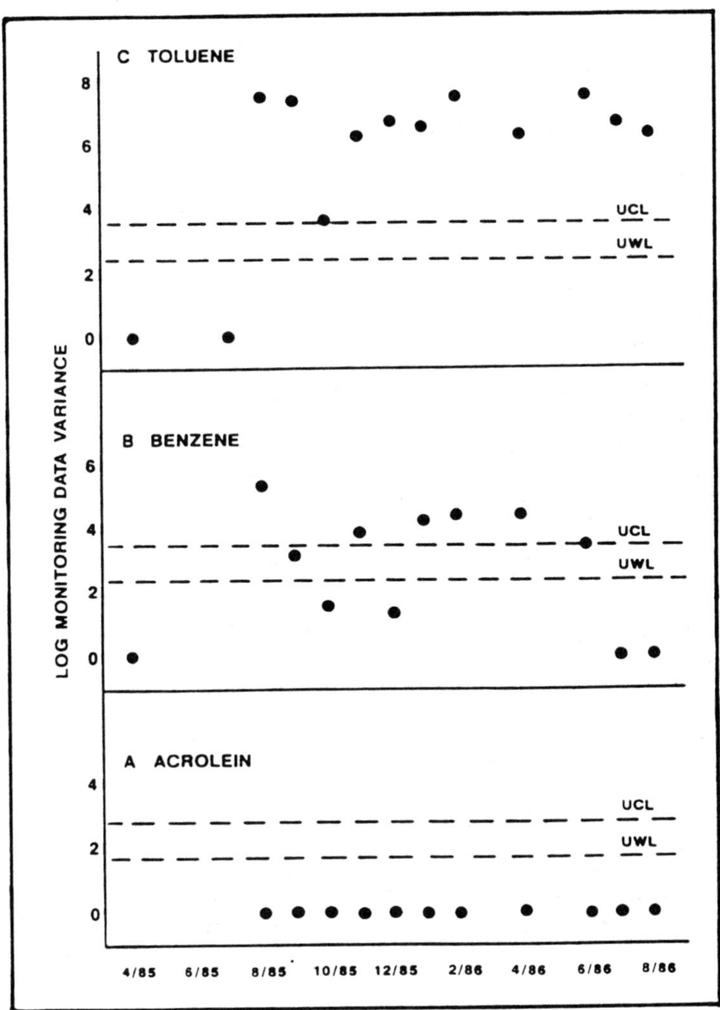

Figure 4. Control charts of monitoring data variance as a function of time for (A) acrolein, (B) benzene, and (C) toluene at site RCS #28.

The summation and presentation of site monitoring data as shown in Table 1 can be used to identify several principal characteristics of a leakage event. First, the homogeneous conditions that prevailed at the site in April, 1985 were disrupted in August of the same year. Thus, the approach identifies the time of occurrence of the leakage event. Second, contaminants associated with the leakage event can be identified as those contaminants that exceed compound-specific control chart limits. For the purpose of this exercise, an overly conservative control chart limit of zero was used for each contaminant included in Table 1. This effort indicates that 9 volatile compounds, 16 base/neutral compounds, 4 acid extractable compounds, and 1 pesticide were associated with the leakage event at the facility. (The development of contaminant specific control chart limits is beyond the scope of this paper. However, it should be possible to use monitoring data to define these limits [6] or to develop upper warning limits and upper control limits based on calculated statistical confidence intervals. If less conservative, non-zero control chart limits are used, the approach would be the same but a smaller number of compounds would be identified as contaminants associated with the leakage event). Third, the approach can be used to identify or prioritize major constituents associated with a leakage event. These compounds must be those with higher variance values. This follows from the fact that the background concentrations and the variance for each of the organic compounds would be expected to be essentially zero when a facility is not leaking. Therefore, the only way to increase monitoring data variance is to increase the maximum observed concentration as a consequence of a leakage event. The major contaminants at the site summarized in Table 1 are chlorobenzene, ethyl benzene, toluene, and phenol because the log variance values for these parameters increased by more than 5 and were detected in each survey after the leakage event occurred. The remaining 26 contaminants identified as constituents of the leakage event had variance values increases that were several orders of magnitude lower than the principal compounds and/or they were not consistently present. Thus, a consideration of the similarities between a waste disposal site leakage event and the fundamental force of entropy leads to an effective method of using site-specific monitoring data to identify the occurrence of a leakage event and fingerprint the resulting contamination.

DISCUSSION

A major environmental issue at waste disposal facilities is the identification of leakage events that result in the migration (dispersion) of contaminants away from the site. Since entropy is a fundamental force that influences dispersion, a physical model that demonstrates the properties/characteristics of entropy was compared with a representation of a waste disposal facility. This approach suggests that a properly functioning waste disposal facility can be considered to be in a state of quasi-equilibrium that does not impact the immediate environment.

However, when the containment mechanisms of the facility fail and the waste components spontaneously migrate away from the site, the result will be to cause an immediate reduction in the entropy and homogeneity of the aquifer into which the contaminants migrate. Based on this analogy, it is proposed that changes in ground-water monitoring data variance, the mathematical inverse of homogeneity, can be used to detect the expected changes in homogeneity that are associated with a disposal site leakage event.

A fundamental purpose of a monitoring program is to identify the occurrence of changes that are a cause for concern. Although variance is considered to be a noisy statistic, there are several arguments that support the use of monitoring data variance as an alternative technique to identify the occurrence of disposal site leakage events. First, it is inconceivable for a leakage event to increase the mean ground-water concentration for a contaminant without also altering the variance. Thus, based on a consideration of the process of dispersion, variance provides a mechanism to monitor conditions that would be directly influenced by a leakage event. Second, the hydrogeologic boundary conditions for a site monitoring program would specify that upgradient and downgradient concentrations occur within a narrow range for each contaminant. The actual concentrations may be as high as several milligram per liter for inorganic constituents or as low as several micrograms per liter for the organic constituents included in Table 1. However, the variance for each contaminant-specific subset should approach zero. As long as the variance for that parameter shows no trend of increasing with time, a leakage event can not be occurring at the site. When the variance increases above an appropriately defined limit, a response can be initiated to distinguish between a potential quality control problem and an actual leakage event. Third, it is recognized that certain factors such as sampling error, analytical error, and temporal variability can cause monitoring data variance to increase above the conceptual limit of zero even though a facility is not leaking. This condition can be addressed through the development of contaminant-specific control chart limits. Thus, common laboratory contaminants such as methylene chloride and bis-(2-ethylhexyl) phthalate would have a higher control chart limits than the other contaminants listed in Table 1

The proposed use of monitoring data variance is a descriptive statistical technique [8,9] that simply organizes and summarizes data in a manner that facilitates identification of trends to be expected after a leakage event occurs. When monitoring data variance is low (i.e., below contaminant-specific control chart limits) and constant over time, the data trends support the inference that a leakage event has not occurred at the facility being monitored or a particular compound was not associated with a leakage event. Under these conditions, more extensive site monitoring would not be considered necessary. However, when one or more control chart limits are exceeded, these trends would support the conclusion that a leakage event has occurred. It would be prudent to institute more extensive site investigation activities when this situation occurs. This method of summarizing and evaluating monitoring data variance is less sophisticated but conceptually similar to the use of geostatistical procedures such as krieging [10]. However, rather than using data variability as a function of distance to estimate optimum sampling locations, the suggested data evaluation technique would utilize changes in variability over time to trigger more extensive monitoring in response to a detected leakage event.

The proposed method of leakage event identification was applied to a set of monitoring data from a waste disposal facility in Illinois. The data were used to (essentially) construct a control chart of monitoring data variance versus time for each constituent being monitored. Collectively, the set of control charts identifies the occurrence of a leakage event and provides a site-specific fingerprint of the contaminants that are involved in the event. This approach suggests that a leakage event occurred during the summer of 1985 at the site and that approximately 30 organic contaminants were associated with the event. (The actual number of contaminants identified with the leakage event will depend on the specific control chart limits that are used). Furthermore, four contaminants can be identified as principal components of the leakage event while the remaining 26 constituents can be characterized as minor or trace components because of the significantly lower variance value increases and/or the intermittent detection of the contaminants in the ground water after the occurrence of the leakage event. This demonstration is not a formal proof of the data evaluation technique but it would be highly unlikely for the monitoring data variance of 30 ground-water contaminants to simultaneously increase, as suggested by the entropy analogy, and persist during monthly surveys for more than a year without a leakage event having actually occurred.

The current regulatory approach to assess the occurrence of a disposal site leakage event is based on a statistical comparison of downgradient data with upgradient data from the site [10]. A critical evaluation of the present RCRA strategy has identified numerous implementation problems related to the implicit program assumptions regarding contaminant behavior and the statistical procedures that have been specified for evaluating ground-water monitoring data [2,6,12,13,14,15,16,17]. It is asserted that the suggested use of monitoring data variance can avoid or alleviate many of the problems associated with the technique currently being used to detect disposal site leakage events.

1. The application of the t-Test as specified in the RCRA program requires the implicit assumption that contaminants can only migrate in the downgradient direction. However, kinetic effects and density differences could alter contaminant behavior, upgradient migration has been simulated in physical laboratory models [18], and high upgradient results have been reported in site monitoring data [2,6,12]. The downgradient migration model would not

correctly assess leakage events in these situations. By comparison, the variance approach is not dependent on a preconceived chemical migration model. As long as the variance remains below appropriately defined limits, contaminants are unlikely to be migrating in any direction from the facility; when high variance values are observed, a change in the ground-water system is indicated. If sampling error and quality control problems can be ruled out as factors responsible for the change, a leakage event is indicated at the facility being monitored. The data can be reviewed to determine the location of the impacted well(s) and the direction of contaminant migration.

2. The use of a t-Test procedure requires replicate analyses for each of the parameters to be analyzed with this technique (Presently, the RCRA program requires 4 replicate analyses for each indicator parameter [11]). However, the water quality parameters, the trace metal analyses, and the pesticide analyses required in the RCRA detection monitoring phase and the Appendix IX analyses that must be submitted during the RCRA assessment monitoring phase are not submitted as replicate analyses [11], in part due to financial considerations, and these data can not be analyzed with the specified statistical procedure. Since the suggested use of data variance can function with single analyses at each sampling location, it is suggested that the variance evaluation technique would permit a more effective utilization of monitoring data to identify and characterize leakage events than the present regulatory approach.

3. Correct utilization of the t-Test requires that the data to be analyzed are normally distributed [14,15], have **zero** temporal and spatial variability [13], and are not auto correlated [14]. Each of these application errors can be avoided by using an assessment of data variance. Furthermore, an implicit assumption in the application of the t-Test is that the variance of the two populations being compared are equivalent. However, based on the monitoring results summarized in Table 1, this condition is not satisfied when a leakage event occurs since the variance increases for each constituent involved in the leakage event.

4. Finally, the current requirement to use the t-Test to evaluate monitoring data does not restrict the quality of the data to be submitted even though this site-specific variability can influence the statistical procedure. By comparison, as long as the monitoring data variance remains below appropriately defined limits, the site can not be leaking and the data must be of acceptable quality. When a change is indicated in the ground-water system being monitored because of the observed variance increase, a structured response can be implemented to insure the data are of acceptable quality and the site is actually leaking:

A. Review the data to assess data quality and insure compliance with established data quality objectives. Correct identified data quality problems and resample the monitoring network;
B. Identify the location(s) contributing to high variance;
C. Respond to the leakage event.

Since the variance technique can simultaneously monitor both data quality and the occurrence of leakage events, the suggested approach would encourage the submission of acceptable quality monitoring data in order to minimize the variability of site monitoring data.

By correcting a number of identifiable problems and enhancing the use of monitoring data, the suggested use of monitoring data variance can improve the effectiveness of the regulatory ground-water monitoring strategy developed for the RCRA program.

SUMMARY

A physical model used to illustrate the basic properties of entropy was compared with a typical waste disposal site situation. The similarities between these two concepts suggests that a spontaneous event such as a leakage event at a waste disposal facility would be expected to cause changes in the homogeneity of an impacted aquifer. Specifically, a leakage event would be expected to increase the heterogeneity of existing ground-water conditions in the vicinity of a waste disposal site. Therefore, it is suggested that changes in monitoring data variance, the mathematical reciprocal of homogeneity, can be used as an alternative data evaluation technique to identify and characterize waste disposal site leakage events.

The entropy-based data evaluation technique was demonstrated with data generated during a RCRA disposal site investigation in Illinois. Results from a series of monthly surveys were presented as control charts (monitoring data variance versus time) for each of the 110 organic contaminants that were monitored. The resultant data summary identified the occurrence of a leakage event (August, 1985), the number of contaminants associated with the leakage event (30), and the four principal components of the leakage event. It is suggested that the entropy-based data evaluation technique is a simple, visual method of summarizing monitoring data. This approach leads to more effective utilization of disposal site monitoring data than the conventional upgradient-downgradient analysis technique because it is not dependent on an assumed chemical migration model and it does not require extensive replicate analyses at each sampling location in the monitoring network.

[1] Office of Technology Assessment, "Superfund Strategy", Report OTA-ITE-252. Office of Technology Assessment, U.S. Congress, Washington D.C. 282p., 1985.

[2] Plumb, R.H., Jr., "Disposal Site Monitoring Data: Observations and Strategy Implications". Proceedings of the Second Canadian/American Conference on Hydrogeology. Hazardous Wastes in Ground Water - A Soluble Dilemma. National Water Well Association/Alberta Research Council. pp. 69-77. 1986.

[3] Hirschhorn, J.S., "Entropy Races Superfund", Eighth National Ground Water Quality Symposium: Anatomy of Superfund. Kansas City, KS. 7p. 23-25 September 1986.

[4] Castellan, G.W., Physical Chemistry, Addison-Wesley Publishing Co., Reading, MA., 717 p., 1964.

[5] Callen, H.B., Thermodynamics, John Wiley and Sons, New York, NY., 376 p., 1970.

[6] Plumb, R.H., Jr. "Monitoring Data Variance - A New Strategy to Identify Subsurface Leakage Events", Presentation at the Conference on Petroleum Hydrocarbons and Organic Chemicals in Ground Water, National Water Well Association/ American Petroleum Institute, Houston, TX, 1987.

[7] Keith, L.H. and Telliard, W.A., "Priority Pollutants. I. A Perspective View", Environmental Science and Technology, Vol 12, No. 4, pp. 416-423, 1979.

[8] Williams, F., Reasoning with Statistics. Simplified Examples in Communication Research, Holt, Rinehart, and Winston, Inc., New York, NY, 182 pp., 1986.

[9] Wallis, W.A. and Roberts, H.V., The Nature of Statistics, The Free Press, New York, NY, 218 pp., 1965.

[10] Clark, I., Practical Geostatistics, Applied Science Publishers Ltd, London, England. 129 pp., 1983.

[11] U.S. Environmental Protection Agency, "Hazardous Waste Management System. Part VII. Standards Applicable to Owners and Operators of Hazardous Waste Treatment, Storage, and Disposal Facilities", Federal Register Vol 98, No. 98, pp. 33154-33258, 1980.

[12] Silver C.A., "Statistical Approaches to Ground-Water Monitoring", Journal of Hazardous Materials, Vol 13, pp. 207-216, 1986.

[13] Starks, T., "RCRA Statistics", Presentation at the U.S. EPA Ground-Water and Subsurface Monitoring Technology Transfer Symposium, Las Vegas, NV., 18-20 November 1986.

[14] Silver, C.A., "Statistical Approaches to Ground-Water Monitoring", Open File Report No. 7 prepared for the Environmental Institute for Waste Management Studies, the University of Alabama, University, AL., 18 pp., 1985.

[15] Montgomery, R.H. and Loftis, J.C., "Applicability of the t-Test for Detecting Trends in Water Quality Variables", Water Resources Bulletin, Vol 23, No 4, pp. 653-662, 1987.

[16] Splitstone, D. "A Statisticians View of Ground Water Monitoring", Proceedings of the National Conference on Hazardous Wastes and Hazardous Materials, Hazardous Materials Control Research Institute, Atlanta, GA., pp. 40-45, 1989.

[17] McNichols, R.J. and Davis, C.B., "Statistical Issues and Problems in Groundwater Detection Monitoring at Hazardous Waste Facilities", Ground Water Monitoring Review, Vol VIII, No. 4, pp. 135-150, 1988.

[18] Schwille, F. Dense Chlorinated Solvents in Porous and Fractured Media. Model Experiments, Lewis Publishers, Inc., Chelsea, MI. 146 p. 1988.

Eric F. Pastor and Douglas R. Frick

CONSIDERATIONS IN SELECTING INDICATOR PARAMETERS FOR THE STATISTICAL EVALUATION OF GROUND-WATER QUALITY

REFERENCE: Pastor, E. F. and Frick, D. R., "Considerations in Selecting Indicator Parameters for the Statistical Evaluation of Ground-Water Quality," Current Practices in Ground Water and Vadose Zone Investigations, ASTM STP 1118, David M. Nielsen and Martin N. Sara, Eds., American Society for Testing and Materials, Philadelphia, 1992.

ABSTRACT: An essential component of a detection ground-water monitoring plan is the establishment of a program for the statistical evaluation of ground-water chemistry data. The objective of such a data evaluation program is to assess whether there is statistically significant evidence of ground-water contamination, based on the indicator parameters for the site being monitored. The occurrence of false conclusions of evidence of contamination (Type I errors) or failures to detect contamination when it actually is present (Type II errors) is minimized by the use of appropriate indicator parameters.

To optimize the effectiveness of a statistical evaluation program, a limited number of indicator parameters should be selected on a site-specific basis, consistent with the following factors:

(1) Concentrations of the parameter in contaminant source and background ground-water samples;

(2) Hydrogeologic conditions of the site being monitored, particularly as they may affect spatial and temporal variability of parameter concentrations in ground water;

(3) Chemical fate and mobility characteristics of the constituents used as indicator parameters;

(4) Precision and limitations of the laboratory analytical procedure specified for the parameter; and

(5) Characteristics of the existing dataset (if any) for the parameter.

Mr. Pastor and Mr. Frick are a project engineer and principal hydrogeologist, respectively, at McCulley, Frick & Gilman, Inc., 5818 Balcones Drive Suite 202, Austin, TX 78731

This paper explains the importance of the above factors and provides examples from a number of case studies to illustrate their use.

KEYWORDS: indicator parameters, statistical evaluation, ground-water monitoring

INTRODUCTION

The evaluation of ground-water chemistry data is the primary means by which judgements are made as to whether or not contamination of ground water has occurred. Legislation and resulting regulations over the past decade have reduced the importance of professional judgement in evaluating ground-water quality to the point where the determination of ground-water contamination has become primarily a statistical exercise. As a consequence of the inherent problems of applying statistical methods to data affected by multiple independent variables, the importance of a well-founded ground-water monitoring program is amplified. The selection of appropriate indicator parameters for statistical evaluation, including the number of parameters to be used as indicators, is a critical ingredient of an effective monitoring program.

The application of statistical methods as a means to detect ground-water contamination was first codified in 1982, when the Environmental Protection Agency (EPA) promulgated standards for ground-water monitoring at facilities regulated under Subtitle C of the Resource Conservation Recovery Act of 1976 (RCRA). Concerns over the limitations and appropriateness of the statistical procedures recommended in these standards led to the amendment of the Subpart F requirements in 1988 [1], which allow a more flexible choice of statistical methods. Most recent journal articles pertaining to the evaluation of ground-water monitoring data have focused on the selection and use of statistical methods, and not on the indicator parameters to be used in the evaluation. Gibbons [2,3,4] has proposed several alternative statistical methods outside of those recommended in the amended Subpart F requirements. Loftis, et al. [5] and Harris, et al. [6] also present several alternative statistical procedures. McNichols and Davis [7] discuss some of the problems and limitations associated with the statistical methods recommended in the Subpart F amendments and provide insights into the realities of applying statistics to ground-water chemistry data, including the importance of selecting appropriate indicator parameters for statistical evaluation.

In most cases, the indicator parameters selected for statistical evaluation are as critical to the success of a monitoring program as the statistical techniques used. The judicious selection of indicator parameters can mitigate potential problems associated with a non-subjective statistical evaluation, and contribute to an accurate assessment of the presence of ground-water contamination.

The purpose of this paper is to describe the factors that should be considered when selecting indicator parameters for statistical evaluation. As described below, these factors include the following:

- site-specific factors, including hydrogeologic and contaminant source considerations;

- parameter-specific factors, including chemical fate and mobility characteristics, laboratory analytical methods and field sampling procedures; and

- dataset-specific factors such as the dataset distribution and the independence of data points.

Literature discussions and examples from several actual ground-water monitoring situations are used to illustrate the parameter selection process. In many cases, such as proposed or newly permitted commercial waste disposal facilities (e.g, hazardous waste or municipal landfills), ground-water chemistry data are limited; hence, dataset-specific factors cannot be evaluated. Similarly, the volumes and chemical characteristics of wastes that will be received at a proposed landfill can only be estimated. As a result, the initial selection of indicator parameters in these situations is often based on parameter-specific factors and site hydrogeologic considerations. However, as a contaminant source becomes better characterized (i.e, wastes are received at the landfill and leachate is sampled, and analyzed) and ground-water chemistry data are obtained, the initially selected list of indicator parameters may require modification. To allow for possible future changes in indicator parameters, it is often advisable to monitor for a number of chemical parameters in addition to those used as indicators for statistical evaluation.

SITE-SPECIFIC FACTORS

Contaminant Source

One of the most important factors to be considered in selecting indicator parameters is the chemical character of the contaminant source. The principal statistical methods used to evaluate ground-water chemistry data are based on the assumption that a statistically significant difference exists between the chemical compositions of contaminated ground water and background (uncontaminated) ground water. This assumption is explicit in inter-well based statistical methods such as analysis of variance (ANOVA), tolerance limit and prediction interval techniques, and is implicit in intra-well based methods such as control charts, which use significant increases in chemical concentrations as evidence of contamination.

The specific chemical compounds associated with a contaminant source, which are often synthetic, are usually the most desirable indicator parameters. Source-specific compounds are nearly always preferable to general water chemistry parameters such as pH, specific conductance, total organic carbon (TOC) and total organic halides (TOX), which traditionally have been used for this purpose. Several studies by Plumb [8,9] involving ground-water chemistry data from hundreds of RCRA and Superfund sites across the nation have shown pH, specific conductance, TOC and TOX to be poor indicators of ground-water contamination by priority pollutants. In most of these cases, volatile organic compounds (VOCs) were the most frequently detected organic compounds (11.7 percent detection frequency to 2.5 percent for acid extractable compounds, 1.1 percent for pesticides and 1.0 percent for base neutral compounds), suggesting their desirability as indicators at sites where organics are of concern.

In most cases involving leaks or spills of compounds such as petroleum hydrocarbons or synthetic organic chemicals, which normally do not occur naturally in near-surface ground-water reservoirs, the selection of indicator parameters is usually very straightforward. Specific contaminant compounds should not, however, be selected without consideration of background conditions at a site. As shown by Slaine and Barker [10], even the traditional gasoline indicators, benzene, toluene and xylenes (BTX), can occur naturally and should not automatically be attributed to a contaminant source.

Selecting indicator parameters at manufacturing facilities can be more problematic, in that most facilities have multiple contaminant sources in relative proximity, including landfills, stockpiles, impoundments, sumps, drains, loading areas, transfer areas, above and below grade piping, storage vessels, and surface drainage collection features. Parameter selection difficulties can be further compounded if the parameters characteristic of a contaminant source are naturally present at significant concentrations in the ground-water zone(s) being monitored, or if the parameter concentrations at the source have varied significantly during the operation of the facility.

The importance of adequately characterizing the contaminant source is illustrated by ground-water monitoring activities at several hazardous waste impoundments in Northern California. Using total dissolved solids (TDS), chloride and sodium as indicator parameters, the consultant evaluating monitoring data from this site concluded that the highly brackish nature of ground-water samples from perimeter monitoring wells indicated leakage from the disposal (evaporative) impoundments, which contained very saline water. These conclusions were used by the regulatory agency as cause to require the facility to initiate an assessment monitoring program. A thorough source characterization relative to background ground water at the site would have indicated that the selected parameters were not suitable as indicators of contamination. Ground waters in the Recent sediments beneath the impoundments are naturally brackish to saline, reflecting their coastal estuarine environment of deposition. VOCs would have been a more appropriate choice of indicator parameters at this site since the impoundment water contained high concentrations of VOCs and semi-volatile compounds, none of which were detected consistently in ground-water samples.

Municipal landfills pose a particularly difficult situation with regard to source characterization, because of the multitude of wastes involved. Typically, a number of general water quality parameters are used at these sites. For example, Texas regulations [11] require that four background ground-water samples be collected at a municipal landfill site prior to deposition of any solid waste at the site. These background samples are analyzed for the parameters listed in Table 1. Following the determination of background concentrations, the list of indicator parameters is reduced to TOC, iron, manganese, pH and chloride. At municipal landfill sites in Wisconsin, the eight chemical parameters listed in Table 2 have most frequently been the focus of monitoring efforts [12]. Of these eight, specific conductance, total alkalinity, total hardness, pH and chemical oxygen demand (COD) have been considered indicator parameters. The ground-water monitoring section of the Subtitle D regulations for municipal landfills [13] lists 62 monitoring parameters (see Table 3) for statistical evaluation. Use of a large number of indicator parameters, as required by the Subtitle D regulations, can increase the probability of false positive errors, particularly when monitoring parameters are applied without consideration of site-specific conditions.

TABLE 1 -- State of Texas background monitoring parameters for municipal solid waste landfills[a]

Heavy Metals		
Arsenic	Copper	Mercury
Barium	Iron	Selenium
Cadmium	Lead	Silver
Chromium	Manganese	Zinc

General Chemistry Parameters	
Calcium	Nitrate
Magnesium	Total Dissolved Solids
Sodium	Phenolphthalein Alkalinity
Carbonate	Total Alkalinity
Bicarbonate	Hardness
Sulfate	pH
Chloride	Specific Conductance
Fluoride	Total Organic Carbon

[a] Reference: Texas Administrative Code 325.74(b)(5)(D)(vii). [11]

TABLE 2 -- Most frequently monitored chemical parameters at landfill sites in Wisconsin[a]

Total Alkalinity	Chemical Oxygen Demand
Hardness	Total Iron
pH	Chloride
Specific Conductance	Sulfate

[a] Reference: Fisher and Potter, 1989. [12]

TABLE 3 -- Subtitle D Phase I monitoring parameters[a]

Inorganic Constituents

(1) Antimony
(2) Arsenic
(3) Barium
(4) Beryllium
(5) Cadmium
(6) Chromium
(7) Cobalt
(8) Copper
(9) Lead
(10) Nickel
(11) Selenium
(12) Silver
(13) Thallium
(14) Vanadium
(15) Zinc

Organic Constituents

(16) Acetone
(17) Acrylonitrile
(18) Benzene
(19) Bromochloromethane
(20) Bromodichloromethane
(21) Bromoform
(22) Carbon disulfide
(23) Carbon tetrachloride
(24) Chlorobenzene
(25) Chloroethane
(26) Chloroform
(27) Dibromochloromethane
(28) 1,2-Dibromo-3-chloropropane
(29) 1,2-Dibromomethane
(30) o-Dichlorobenzene
(31) p-Dichlorobenzene
(32) trans-1,4-Dichloro-2-butene
(33) 1,1-Dichloroethane
(34) 1,2-Dichloroethane
(35) 1,1-Dichloroethene
(36) cis-1,2-Dichloroethene
(37) trans-1,2-Dichloroethene
(38) 1,2-Dichloropropane
(39) cis-1,3-Dichloropropene
(40) trans-1,3-Dichloropropene
(41) Ethylbenzene
(42) 2-Hexanone
(43) Methyl bromide
(44) Methyl chloride
(45) Methylene bromide
(46) Methylene chloride
(47) Methyl ethyl ketone
(48) Methyl iodide
(49) 4-Methyl-2-pentanone
(50) Styrene
(51) 1,1,1,2-Tetrachloroethane
(52) 1,1,2,2-Tetrachloroethane
(53) Tetrachloroethene
(54) Toluene
(55) 1,1,1-Trichloroethane
(56) 1,1,2-Trichloroethane
(57) Trichloroethene
(58) Trichlorofluoromethane
(59) 1,2,3-Trichloropropane
(60) Vinyl acetate
(61) Vinyl chloride
(62) Xylene

[a] Reference: 40 CFR 258. [13]

Clearly, more appropriate indicator parameters for a municipal landfill could be selected on the basis of leachate and background ground-water characteristics. Leachate present in a municipal landfill leachate collection system should be sampled, analyzed and compared to background ground-water values. Compounds showing the greatest contrast between these two sets of data may be the best bets for indicators, provided parameter-specific factors (i.e, fate and mobility, analytical method, etc. as discussed later in this paper) are favorable. In situations where this leachate chemistry information is not available, the use of VOCs as the sole group of indicator parameters may be advisable because: 1) VOCs are generally not present in unimpacted ground water, 2) they are relatively mobile, and 3) they often correlate directly to the ground-water contamination concerns associated with a release (i.e, the toxicity characteristic values listed for many VOCs in 40 CFR 261.24 are lower than those for listed for many other potential leachate components).

Site Hydrogeology

Hydrogeologic conditions, such as the geochemical regime, hydraulic conductivity, primary and secondary porosity, structural features, lithology, and lateral and vertical extent of stratigraphic units at a site, can directly impact the spatial and temporal variability of ground-water chemistry data. The hydraulic relationship between a ground-water zone and recharge/discharge sources such rivers, lakes, springs and wells will also have ground-water chemistry implications. These site hydrogeologic conditions must be carefully considered when selecting indicator parameters.

Spatial variability of ground-water chemistry data is an important concern because of the background (usually upgradient) versus downgradient comparison on which most statistical methods are based. Depending on the degree of spatial variability at a site, it is possible that samples from a "background" monitoring well or wells may not be representative of the range of background concentrations for a particular parameter. Apparent spatial variability can result from differences in the design of wells within a monitoring network. Large screen lengths can mask variable parameter concentrations within a vertical section. Wells screened in different sections of a monitored zone or variations in the depth of the "most permeable" part of the zone can also produce artificial spatial variabilities. An evaluation of the spatial variability of monitoring parameters should begin with an assessment of the monitoring network from which the data were obtained. In the case of a proposed or newly permitted facility with very little ground-water chemistry data, an assessment of the monitoring network can identify areas where spatial variabilities in ground-water chemistry data may be expected.

Graphical techniques, such as box and whisker plots, are often extremely helpful for qualitatively evaluating spatial variability in ground-water chemistry data. In constructing box and whisker plots for the purposes of assessing spatial variability, the chemical parameter data are separated by well, and the key statistics of each dataset such as the median and interquartile range are used to define a "box" for each well. The box plots for the various wells at a site are then compared to assess the spatial variability of the chemical parameter of interest.

The usefulness of box and whisker plots in identifying spatially variable ground-water data is shown in a plot of chloride data from several monitoring wells at a small (less than 1 acre) industrial landfill (Figure 1). These wells are located along three sides of the landfill. All except Well 9 are between the landfill and an adjacent lake. Due to the influence of the constant-level lake, the hydraulic gradient in the vicinity is essentially flat. The box plotted in Figure 1 for Well 8 is clearly different from those plotted for the other wells, suggesting either dramatic spatial variability or evidence of contamination. An evaluation of other general chemistry data (including major anions and cations) showed Well 8 water to be chemically similar to lake water and very different from ground water in the other wells. Spatial variability resulting from the hydraulic influence of the lake made chloride an undesirable indicator parameter. On the basis of their confirmed presence in landfill fluids, VOCs were selected as the appropriate indicator parameters for this landfill.

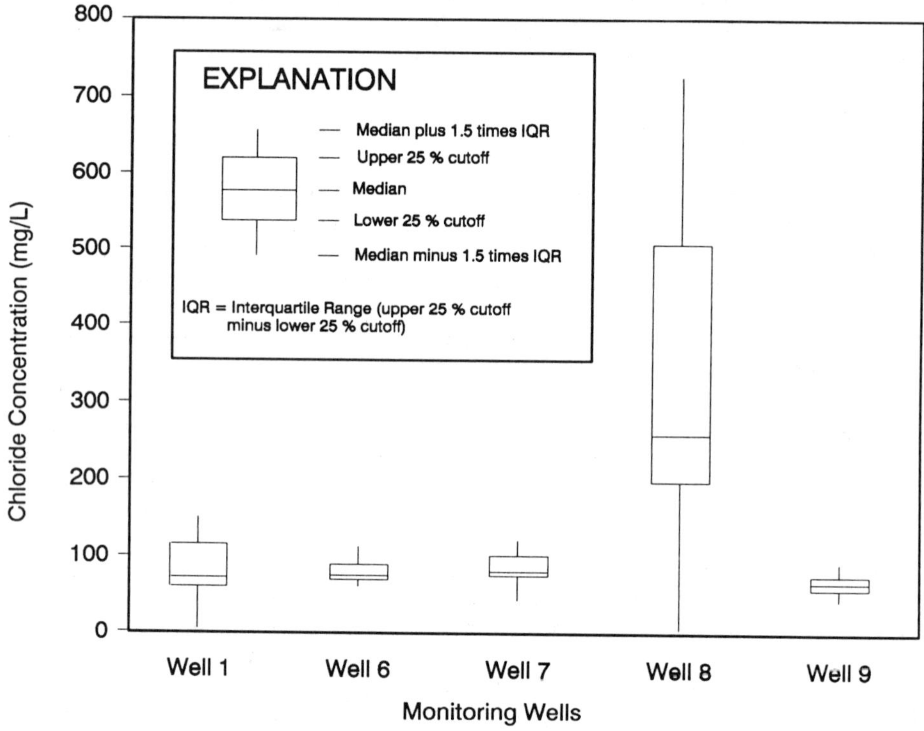

FIGURE 1 -- Box and whisker plot for monitoring wells at an industrial landfill

Natural spatial variabilities are accentuated when background and downgradient monitoring wells are completed in strata composed of several lithologies. As an example, sediments deposited in fluvial-deltaic environments may include, within a small area, channel, overbank (floodbasin), natural levee and crevasse splay deposits, all of which have characteristic lithologies and subsurface geometries. The chemistry of waters contained in these sediments is often quite variable, particularly when comparing sand units and fine-grained strata. Concentrations of major ions and trace elements may be naturally higher in water contained in fine-grained material, reflecting the lack of flushing by ground-water flow and the abundance of source minerals. In fact, the development, purging and sampling of wells in fine-grained strata may have a significant effect on the geochemical equilibrium in the vicinity of the well bore, resulting in significant water chemistry variations as periodic monitoring continues.

The selection process for indicator parameters in complex environments should include consideration of these factors. The implications of well sampling procedures on the desirability of specific indicator parameters are discussed later in this paper.

The potential effects of spatial variability on indicator parameter selection are illustrated by data for a landfill in North Texas. This unlined landfill was constructed within artificial fill material that had been placed over weathered bedrock. The upgradient well at the landfill was completed in the weathered bedrock, because the overlying fill was dry. Downgradient of the landfill, a thin buried soil horizon, which contains considerable organic matter, is present between the fill and weathered bedrock strata. Monitoring wells downgradient of the landfill are completed in the saturated zone, which includes the fill, soil horizon and weathered bedrock strata. Because of concerns regarding naturally elevated TOC concentrations associated with the buried soil horizon in the downgradient wells, TOC was discounted as a possible indicator parameter at this landfill. VOCs, including several compounds identified in a characterization of the landfill contents, were selected as more appropriate indicator parameters. Initial samples collected from the downgradient monitoring wells and analyzed for TOC confirmed the anticipated elevated TOC concentrations.

Potential indicator parameters that show significant spatial variability at a site should not be dismissed routinely, as intra-well statistical techniques, such as control charts can be used. One of the major limitations of using control charts, particularly for initial sampling events, is that a minimum number of data points are required. One practical approach is to use inter-well statistical techniques for the initial sampling events at a well to develop the parameter database and verify that the well is uncontaminated and then switch to intra-well techniques when sufficient data are available.

Use of multiple background wells is another way of allowing the use of a spatially variable indicator parameter. Depending on the statistical method, either the multiple background concentration values or the mean of these measurements can be used in the statistical evaluation. Multiple background wells still have some shortcomings. Use of this approach to resolve spatial variability problems implicitly assumes that the degree of spatial variability is the same for both background and downgradient locations.

Temporal variability in ground-water monitoring parameters can also be an important concern when selecting indicator parameters. Site hydrogeologic conditions can have significant impacts on seasonal variations in ground-water chemistry data. Sites located in areas where ground-water recharge is much greater during certain times of the year (typically associated with rainy seasons, runoff of spring snowmelt in nearby streams or annual irrigation periods) would be expected to exhibit seasonal fluctuations in the chemistry of shallow ground water. Shallow unconfined aquifers are most affected by seasonal recharge events. Montgomery et al. [14] and Barcelona et al. [15] cite several studies where seasonal trends in ground-water chemistry data (nitrate, sulfate and chloride) were attributed to site hydrogeologic conditions. Schwarzenbach et al. [16] attributed seasonal variations in tetrachloroethene (PCE) concentrations in ground-water at one site to water quality variations in infiltrating surface water. At another site, McReynolds [17] associated seasonal trends in trichloroethene (TCE) data with variations in well pumping rates and patterns.

When developing a ground-water monitoring plan at a site where seasonal variations are anticipated and the existing database is limited, it is recommended that indicator parameters be selected subject to ongoing evaluation of seasonal trends as the size of the database is increased. Seasonal patterns in the dataset of potential indicator parameters can be distinguished through the use of time-series concentration graphs or box and whisker plots where the data are grouped by sampling event (as opposed to by monitoring well). Other statistical methods, such as the Kruskal-Wallis and one-way ANOVA tests, can be used to substantiate seasonal patterns observed in the box and whisker plots.

Parameters exhibiting seasonal concentration trends should not automatically be dismissed as potential indicators. If water quality changes by season, it implies that the mean of the distribution is not constant. Therefore, to account for a seasonally shifting mean, the mean and standard deviation can be calculated for each season and used to evaluate data for that season only.

PARAMETER-SPECIFIC FACTORS

The chemical characteristics of a ground-water monitoring parameter must be considered when selecting indicators of ground-water contamination, especially when contaminant sources have not been characterized and/or existing ground-water chemistry data are limited. Consistent with the goals of minimizing the probability of false positive findings and also reducing the laboratory costs associated with ground-water monitoring, the indicator parameter selection process should include an evaluation of which parameters characteristic of a contaminant source are most likely to migrate to downgradient monitoring wells. Such an evaluation requires an adequate assessment of several chemical characteristics for each potential parameter.

Mobility

One of the most important chemical characteristics of an indicator parameter is its mobility in the subsurface. The following discussion relates to transport in saturated flow; however, in many cases, the mobility of a parameter in unsaturated flow conditions as well as in the vapor phase must also be considered. From a mobility standpoint, the most desirable chemical parameters are conservative ones, such as chloride, which are expected to be transported at rates similar to that of ground-water movement. Typically these conservative compounds are not readily sorbed to soil, precipitated or easily degraded. In general, conservative parameters would be expected to be the first constituents of a contaminant source to migrate to downgradient monitoring wells and thus provide the earliest warning of ground-water contamination.

Mobility considerations can often be used to eliminate some parameters from a list of potential indicators. For example, several VOCs (1,1,1-trichloroethane (TCA), benzene, TCE and acetone) and several semi-volatile organic compounds (naphthalene, phenanthrene and pyrene) were detected at similar concentrations in waste samples from an industrial landfill. As reflected by the retardation factors and distribution coefficients summarized in Figure 2, these VOCs are expected to be more mobile than the semi-volatile compounds. On the basis of this mobility difference, the semi-volatile compounds were eliminated from further consideration as potential indicator parameters.

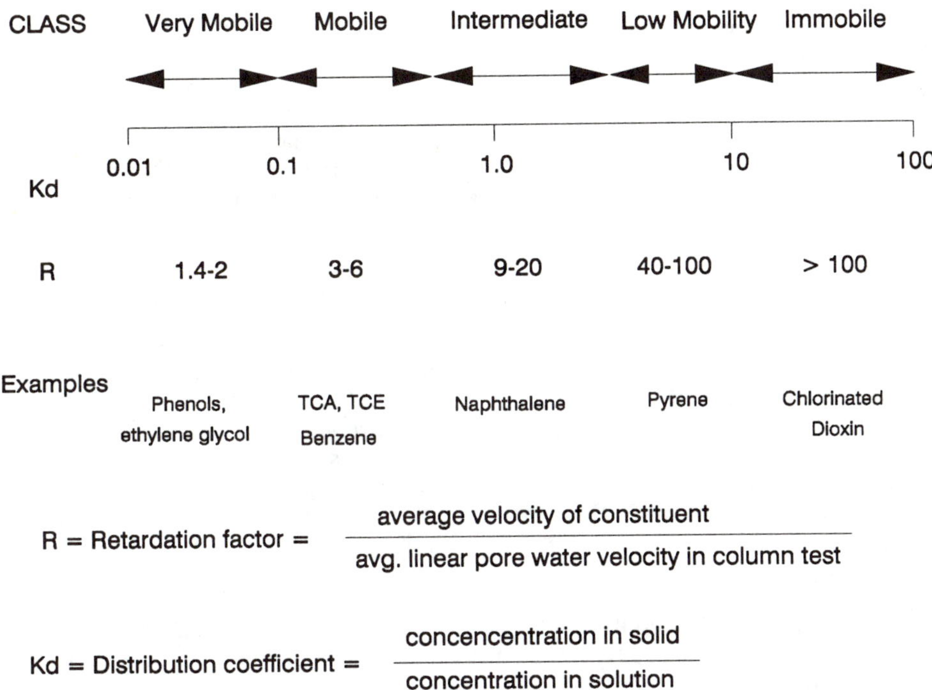

FIGURE 2 -- Mobility classes (adapted from Olsen and Davis [18])

The mobility of an indicator parameter may also be influenced by the subsurface geochemical conditions at a site. The mobility of an inorganic compound is related to its propensity to remain in solution, hence factors such as oxidation-reduction potential, pH, ionic strength, the concentrations of other ions, and the relative abundance of clay minerals and organic material have significant impacts on the mobility of inorganic species. The solubilities of several metals and trace elements are strongly pH-dependent. Selenium, for example, is known to precipitate out of solution under neutral to acidic conditions. Consequently, selenium would not be a good indicator parameter, even if present in high concentrations at the contaminant source, when acidic to neutral ground water is being monitored. Similarly, other metals such as lead and cadmium, are more soluble under acidic conditions and would be appropriate indicators when these conditions are encountered.

Fate

The chemical fate of a potential indicator parameter can also be important. Parameters that are unstable or are readily transformed in the environment are often not desirable indicators. Potential indicators should not, however, be ignored solely on the basis of this criteria. For example, although many VOCs will degrade in the environment, the degradation products are other VOCs. As discussed by Davis and Olsen [19], the degradation of many VOCs to other VOCs under various conditions has been documented by recent research. PCE, for example, can degrade to trichloroethane (TCA) and TCE, both of which can degrade to other VOCs, such as 1,2-dichloroethene (1,2-DCE). Therefore, although a specific VOC may readily degrade to non-detectable concentrations, its use as an indicator may still be appropriate provided its degradation products are also used as indicator parameters.

Laboratory Analytical Methods

Limitations of laboratory analytical methods should be considered when selecting indicator parameters. TOC and TOX are traditional indicator parameters for which analytical methods are of particular concern. The EPA standard analytical method for TOC (Method 9060) is most applicable to organic carbon concentrations above 1 mg/L [20], yet the concentrations of organic priority pollutants at the edge of a contaminant plume would be expected to be two or more orders of magnitude below this value. The standard detection limit for TOX analyses is a more desirable 5 ug/L; however, as noted in the description for EPA Method 9020, this technique has considerable limitations [20].

As indicated by the results of a laboratory performance evaluation reported by Rice, Brinkman and Muller [21], TOC analyses were generally less accurate than analyses of other potential indicator compounds. Of 38 analytes tested at eight different laboratories, TOC had the third lowest percentage of acceptable analytical results (33 % acceptable). Aluminum and calcium had the fewest acceptable analytical results (0 % and 29 %, respectively), suggesting that they also may be undesirable indicator parameters in many situations.

Sample Collection Methods

The effects of sample collection and preservation methods on data variability should also be considered when selecting indicator parameters. For example, possible volatilization of VOCs during well purging, sample collection and sample handling are a concern if these compounds are selected as indicators. Similarly, improper or inadequate well development can result in turbid samples, which may yield unreliable total metals concentration data. Careful adherence to an established sampling plan can help to minimize variability associated with inconsistent sampling procedures. Sampling quality assurance/quality control (QA/QC) procedures such as field blanks, spikes and duplicates can be used to evaluate the effectiveness of sampling protocols.

DATASET-SPECIFIC FACTORS

In most cases, indicator parameters are selected at a time when a site ground-water chemistry database is usually very small. Dataset-specific factors are not important in these instances. However, in situations where a substantial ground-water database exists, the data should be examined carefully during the indicator parameter selection process. The characteristics of a dataset for a specific parameter can indicate how amenable that parameter is to statistical evaluation, which statistical methods may be appropriate, and what transformations, if any, must be made to the dataset. Important characteristics are the distribution of the dataset (e.g., normal, lognormal, etc.) and the independence of the data points within the dataset. The significance of these characteristics as they apply to indicator parameter selected are described below; the characteristics themselves are discussed in detail in numerous references [6, 7, 12, 14, 15 and 22].

DataSet Distribution

Many statistical procedures, especially parametric tests, are based on the assumption that the data are normally distributed. As reported by Fisher and Potter [12] and Montgomery, et al. [14], most ground-water chemistry data are not normally distributed. Despite the frequent violations of the normality assumption, EPA has recommended the use of normality-based statistical tests because only major departures from normality will result in incorrect statistical findings [22].

The distribution of an existing dataset should be checked to identify if a monitoring parameter will be a desirable indicator and if so, what statistical procedures may be appropriate. EPA Guidance [22] presents four methods for checking the normality assumption. The accuracy of some of these methods, particularly the coefficient of variation method, has been questioned, and other methods, such as skewness tests, have been suggested [14]. The best approach for evaluating a dataset is to check its distribution by several methods. A potential indicator parameter need not be dismissed if the existing dataset is non-normally distributed. Dataset transformations can often be performed to obtain a normally distributed dataset for statistical evaluation. If these transformation are not successful, an alternative statistical method that does not require a normally distributed dataset, such as non-parametric procedures, can be used.

Independence of Data Points

Most statistical tests require that the points in a dataset be independent of each other. The independence of data points is a function of the sampling frequency. Methods of testing for serial dependence are described by Harris, et al. [6], Montgomery, et al. [14] and Fisher and Potter [12]. If existing data for a potential indicator parameter show significant autocorrelation, these data may not be useable for statistical evaluation. Potential indicators should not, however, be dismissed on the basis of serial correlation alone, as sampling frequencies can usually be modified for future sampling events to yield independent data.

CONCLUSIONS

Indicator parameters for statistical evaluation should be selected on the basis of all information available. If the contaminant source at a site has been characterized and background ground-water chemistry data exist, these data should be compared. An ideal indicator parameter would be present at significantly different concentrations in contaminant source and background ground-water samples. The spatial and temporal variabilities of potential indicators should be evaluated. An ideal indicator would show little spatial or temporal variability. Dataset characteristics such as the distribution, the independence of data points and the percentage of "non-detects" in the database should be examined to assess which statistical method(s) might be used for potential indicators. Potential indicators characterized by normally distributed independent data points are the easiest to handle using traditional statistical evaluation techniques; however, alternative procedures can be used for parameters that do not possess these characteristics.

When little or no site-specific data are available, indicators must be selected on the basis of parameters-specific factors such as fate and mobility characteristics, laboratory methods and sampling considerations. An ideal indicator parameter would be characterized as stable, mobile, detectable at low concentrations by a reliable laboratory method and not susceptible to potential sampling errors.

In practice there are no ideal indicator parameters. However, our experience has shown VOCs to be the most effective indicators of ground-water contamination at many sites. VOCs are associated with many contaminant sources, are usually not present in background ground waters, and thus are not as susceptible to spatial variability problems. VOCs are more mobile than many other organic and trace metal parameters and, although not particularly stable in many cases, often degrade to other VOCs. The site-specific selection of appropriate indicator parameters such as VOCs can minimize the occurrence of false positive and false negative findings and thus maximize the effectiveness of a ground-water monitoring program.

REFERENCES

[1] United States Environmental Protection Agency, "Statistical Methods for Evaluating Ground-Water Monitoring Data from Hazardous Wastes Facilities, 40 CFR Part 264", Federal Register, Vol. 53, No. 196, October 11, 1988, pp. 39720-39731.

[2] Gibbons, R. D., "Statistical Prediction Intervals for the Evaluation of Ground-Water Quality," Ground Water, Vol. 25, No. 4, July-August 1987, pp. 455-465.

[3] Gibbons, R. L., "Statistical Models for the Analysis of Volatile Organic Compounds in Waste Disposal Sites", Ground Water, Vol. 25, No. 5, September-October 1987, pp. 572-580.

[4] Gibbons, R. D., "A General Statistical Procedure for Ground-Water Detection Monitoring at Waste Disposal Facilities," Ground Water, Vol. 28, No. 2, March-April 1990, pp. 235-243.

[5] Loftis, J. C., Harris, J., and Montgomery, R. H., "Detecting Changes in Ground-Water Quality at Regulated Facilities," Ground-Water Monitoring Review, Vol. 7, No. 1, Winter 1987, pp. 72-76.

[6] Harris, J., Loftis, J. L., and Montgomery, R. H., "Statistical Methods for Characterizing Ground-Water Quality", Ground Water, Vol. 25, No. 2, March-April 1987, pp. 185-193.

[7] McNichols, R. J. and Davis, C. B., "Statistical Issues and Problems in Ground-Water Detection Monitoring at Hazardous Waste Facilities", Ground-Water Monitoring Review, Vol. 8, No. 4, Fall 1988, pp. 135-150.

[8] Plumb, R. H., "Performance Evaluation of RCRA Indicator Parameters," in Proceedings of First Canadian/American Conference on Hydrogeology, Practical Applications of Ground-Water Geochemistry, National Water Well Association, Worthington, Ohio, 1985, pp. 129-137.

[9] Plumb, R. H., "A Practical Alternative to the RCRA Organic Indicator Parameters," in Proceedings of Hazmacon 87, Association of Bay Area Governments, Oakland, California, 1987, pp. 135-150.

[10] Slaine, D. D. and Barker, J. F., "The Detection of Naturally Occurring BTX During a Hydrogeologic Investigation", Ground-Water Monitoring Review, Vol. 10, No. 3, Spring 1990, pp. 88-94.

[11] Texas Administrative Code, 325.74(b)(5)(D)(vii)(IV).

[12] Fisher, S. R. and Potter, K. W., "Methods for Determining Compliance with Ground-Water Quality Regulations at Waste Disposal Facilities," submitted to Wisconsin Department of Natural Resources, Bureau of Water Resources Management and Bureau of Solid and Hazardous Waste Management, 1989.

[13] 40 CFR Part 258

[14] Montgomery, R. H., Loftis, J. C., and Harris, J., "Statistical Characteristics of Ground-Water Quality Variables", Ground Water, Vol. 25, No. 2, March-April, 1987, pp. 176-184.

[15] Barcelona, M. J., Lettenmaier, D. P., and Schock, M. R., "Network Design Factors for Assessing Temporal Variability in Ground-Water Quality," Environmental Monitoring and Assessment, Vol. 12, 1989, pp. 149-179.

[16] Schwarzenbach, R. P., Giger, W., Hoehn, E., and Schneider, J. K., "Behavior of Organic Compounds During Infiltration of River Water to Ground Water: Field Studies," Environmental Science and Technology, Vol. 17, No. 8, 1983, pp. 472-479.

[17] McReynolds, L., "Monitoring Organic Contaminants in Los Angeles San Fernando Valley Ground Water Basin," in Proceedings of the 15th Biennial Conference on Ground Water, California Water Resources Center, San Diego California, 1986, pp. 53-60.

[18] Olsen, R. L. and Davis, A., "Predicting the Fate and Transport of Organic Compounds in Ground Water - Part 1", Hazardous Material Control, Vol. 3, No. 3., May/June 1990, pp. 39-64.

[19] Olsen, R. L. and Davis, A., "Predicting the Fate and Transport of Organic Compounds in Ground Water - Part 2", Hazardous Material Control, Vol. 3, No. 4., July/August 1990, pp. 18-37.

[20] United States Environmental Protection Agency, "Test Methods for Evaluating Solid Waste, Physical/Chemical Methods", SW-846, Office of Solid Waste and Emergency Response, Washington, D. C., 1987.

[21] Rice, G. F., Brinkman, J. E., and Muller, D. L., "Reliability of Chemical Analyses of Water Samples - The Experience of the UMTRA Project", Ground-Water Monitoring Review, Vol. 8, No. 2, Summer 1988, pp. 71-75.

[22] United States Environmental Protection Agency, "Guidance Document on the Statistical Analysis of Ground-Water Monitoring Data at RCRA Facilities", April 1989

Author Index

Subject Index

Q

R

S

T

U

V

W